T0281555

Differential Geometry
of Curves and Surfaces

Differential Geometry of Curves and Surfaces

Third Edition

Thomas F. Banchoff
Stephen Lovett

CRC Press
Taylor & Francis Group
Boca Raton London New York

CRC Press is an imprint of the
Taylor & Francis Group, an **informa** business

A CHAPMAN & HALL BOOK

Third edition published 2023
by CRC Press
6000 Broken Sound Parkway NW, Suite 300, Boca Raton, FL 33487-2742

and by CRC Press
2 Park Square, Milton Park, Abingdon, Oxon, OX14 4RN

© 2023 Stephen Lovett and Thomas F. Banchoff

First edition published by CRC Press 2010
Second edition published by CRC Press 2015

CRC Press is an imprint of Taylor & Francis Group, LLC

ISBN: 9781032281094 (hbk)
ISBN: 9781032047782 (pbk)
ISBN: 9781003295341 (ebk)

DOI: 10.1201/9781003295341

Typeset in Times
by codeMantra

Contents

Preface

What Is Differential Geometry?

Differential geometry studies properties of curves and surfaces, as well as their higher-dimensional generalizations, using tools from calculus and linear algebra. Just as the introduction of calculus expands the descriptive and predictive abilities of nearly every field of scientific study, so the use of calculus in geometry brings about avenues of inquiry that extend far beyond classical geometry.

Before the advent of calculus, much of geometry consisted of proving consequences of Euclid's postulates. Even conics, which came into vogue in the physical sciences after Kepler observed that planets travel around the sun in ellipses, arise as the intersection of a double cone and a plane, two shapes which fit comfortably within the paradigm of Euclidean geometry. One cannot underestimate the impact of geometry on science, philosophy, and civilization as a whole. The geometric proofs in Euclid's *Elements* served as models of mathematical proof for over two thousand years in the Western tradition of a liberal arts education. Geometry also produced an unending flow of applications in surveying, architecture, ballistics, astronomy, and natural philosophy more generally.

The objects of study in Euclidean geometry (points, lines, planes, circles, spheres, cones, and conics) are limited in what they can describe. A boundless variety of curves, surfaces and manifolds arise naturally in areas of inquiry that employ geometry. Various branches of mathematics brought their tools to bear on the expanding horizons of geometry, each with a different bent and set of fruitful results. Techniques from calculus and analysis led to differential geometry, pure set theoretic methods led to topology, and modern algebra contributed the field of algebraic geometry.

The types of questions one typically asks in differential geometry extend far beyond what one can ask in classical geometry and yet the former do not entirely subsume the latter. Differential geometry questions often fall into two categories: local properties, by which one means properties of a curve or surface defined in the neighborhood of a point, or global properties, which refer to properties of the curve or surface taken as a whole. As a comparison to functions of one variable, the derivative of a function f at a point a is a local property since one only needs information f near a whereas the integral of f between a and b is a global property. Some of the most interesting theorems in differential geometry relate local properties to global ones. The culminating theorem in this book, the Gauss-Bonnet Theorem, relates global properties of curves and surfaces to the topology of a surface and leads to fundamental results in non-Euclidean spherical and hyperbolic geometry.

Using This Textbook

This new edition is intended as a textbook for a single semester undergraduate course in the differential geometry of curves and surfaces, with only multivariable calculus and linear algebra as prerequisites. The interactive computer graphics applets that are provided for this book can be used for computer labs, in-class illustrations, exploratory exercises, or simply as intuitive aids for the reader. Each section concludes with a collection of exercises which range from perfunctory to challenging, suitable for daily or weekly problem sets.

However, the self-contained text, the careful introduction of concepts, the many exercises, and the interactive computer graphics also make this text well-suited for self-study. Such a reader should feel free to primarily follow the textbook and use the software as supporting material or primarily follow the presentation in the software package and consult the textbook for definitions, theorems, and proofs. Either way, the authors hope that the dual nature of software applets and classic textbook structure will offer the reader both a rigorous and intuitive introduction to the field of differential geometry.

This book is the first in a pair of books which together are intended to bring the reader through classical differential geometry into the modern formulation of the differential geometry of manifolds. The second book in the pair, by Lovett, is entitled *Differential Geometry of Manifolds*[24]. Neither book directly relies on the other but knowledge of the content of this book is quite beneficial for [24].

Each section ends with a list of Problems, ranging from staightforward applications of formulas to proofs of general results. Problems marked with (*) indicate difficulty, which may be related to technical ability, insight, or length. A few problems require the use of differential equations but we have marked these explicitly with the code (**ODE**). Occasionally a problem involves calculations or integrals that are particularly challenging; we marked these exercises with the code (**CAS**) to indicate that the student should use a computer algebra system to help calculate the desired number or determine an integral.

Computer Applets

An integral part of this book is the access to on-line computer graphics applets that illustrate many concepts and theorems introduced in the text. Though one can explore the computer demos independently of the text, the two are intended as complementary modes of studying the same material: a visual/intuitive approach and an analytical/theoretical approach. The text always motivates various definitions and topics, but the graphical applets allow the reader to explore examples further, and give a visual explanation for definitions or complicated theorems. The ability to change the choice of the parametric curve or the parametrized surface in an applet, or to change other properties allows the reader to explore the concepts far beyond what a static book permits.

Any element in the text (Example, Problem, Definition, Theorem, etc.) that has an associated applet is indicated by the symbol shown in this margin. Each demo comes with some explanation text. The authors intended the applets to be intuitive enough so that after using just one or two (and reading the supporting text) any reader can quickly understand their functionality. At present the applets run on a Java platform but we may add other formats possibly based on standard computer algebra systems. As an interest to an instructor of this course, the Java applets are extensible in that they are designed with considerable flexibility so that the reader can often change whether certain elements are displayed or not. Often, there are additional elements that one can display either by accessing the Controls menu on the Demo window or the Plot/Add Plot menu on any display window. Applets given in a computer algebra system are naturally extensible through the capabilities of the given program.

In previous editions, the applets were delivered via Java embedded into a browser. That is no longer the case. Now all the applets are delivered via a single *Maple* workbook, using embedded components. The user does not need the full *Maple* software, but only needs to download the free MaplePlayer. The reader will find instructions to download the MaplePlayer software and the workbook that accompanies this text at:

> http://cs.wheaton.edu/~slovett/diffgeo

Organization of Topics

Chapters 1 through 4 cover alternately the local and global theory of plane and space curves. In the local theory, we introduce the fundamental notions of curvature and torsion, construct various associated objects (e.g., the evolute, osculating circle, osculating sphere), and present the fundamental theorem of plane or space curves, which is an analogue of the fundamental theorem of calculus. The global theory studies how local properties (esp. curvature) relate to global properties such as closedness, concavity, winding numbers, and knottedness. The topics in these chapters are particularly well suited for computer investigation. Students often make discoveries on their own by manipulating curves or surfaces and various associated objects.

Chapter 5 rigorously introduces the notion of a regular surface, the type of surface on which the techniques of differential geometry are well-defined. Here one first sees the tangent plane and the concept of orientability.

Chapter 6 introduces the local theory of surfaces in $\mathbb{R}^3$, focusing on the metric tensor and the Gauss map from which one defines the essential notions of principal, Gaussian, and mean curvatures. In addition, we introduce the study of surfaces that have Gaussian curvature or mean curvature identically 0. One cannot underestimate the importance of this chapter. Even a reader primarily interested in the advanced topic of differentiable manifolds should be comfortable with the local theory of surfaces in $\mathbb{R}^3$ because it provides many visual and tractable examples of what one generalizes in the theory of manifolds. Here again, as in Chapter 8, the use of the software applets is an invaluable aid for developing a good geometric intuition.

Chapter 7 first introduces the reader to the component notation for tensors. It then establishes the famous Theorema Egregium, the celebrated classical result that the Gaussian curvature depends only on the metric tensor. Finally, it outlines a proof for the fundamental theorem of surface theory.

Another title commonly used for Chapter 8 is *Intrinsic Geometry*. Just as Chapter 1 considers the local theory of plane curves, Chapter 8 starts with the local theory of curves on surfaces. Of particular importance in this chapter are geodesics and geodesic coordinates. The highlight of this chapter is the famous Gauss-Bonnet Theorem, both in its local and global forms, without which no elementary course in differential geometry is complete. The chapter finishes with applications to spherical and hyperbolic geometry.

The book concludes in Chapter 9 with a brief discussion on curves and surfaces in Euclidean n-space. This chapter emphasizes in what ways definitions and formulas given for objects in $\mathbb{R}^3$ extend and possibly change when adapted to $\mathbb{R}^n$.

A Comment on Prerequisites

The mathematics or physics student often first encounters differential geometry at the graduate level. Typically, at that point, one is immediately exposed to the formalism of manifolds, thereby skipping the intuitive and visual foundation that informs the deeper theory. The advent of computer graphics has renewed the interest in classical differential geometry but this pedagogical habit remains. The authors wish to provide a book that introduces the undergraduate student to an interesting and visually stimulating subject that is accessible with only the full calculus sequence and linear algebra as prerequisites.

In calculus courses, students usually do not study all the analysis that underlies the theorems. Similarly, in keeping with the stated requirements, this textbook does not always provide all the topological and analysis background for some theorems. The reader who is interested in all the supporting material is encouraged to consult [24].

A few key results in this textbook rely on theorems from the theory of differential equations. We either spell out the calculations or provide a reference to the appropriate theorem. Therefore, experience with differential equations is occasionally helpful though not necessary. A few exercises require some skills with differential equations but these are clearly marked with the prefix (**ODE**).

Notation

As a comment on vector notation, this book consistently uses the following conventions. A vector or vector function in a Euclidean vector space is denoted by $\vec{v}$, $\vec{X}(t)$, or $\vec{X}(u,v)$. Often γ indicates a curve parametrized by $\vec{X}(t)$ while writing $\vec{X}(t) = \vec{X}(u(t), v(t))$ indicates a curve on a surface. The unit tangent and the binormal vectors of a curve in space are written in the standard notation $\vec{T}(t)$ and $\vec{B}(t)$ but the principal normal is written $\vec{P}(t)$, reserving $\vec{N}(t)$ to refer to the unit normal vector to a curve on a surface. For a plane curve, $\vec{U}(t)$ is the vector obtained by rotating $\vec{T}(t)$ by a positive quarter turn. Furthermore, we denote by $\kappa_g(t)$ the curvature of a plane curve since one identifies this curvature as the geodesic curvature in the theory of curves on surfaces.

In this book, we often work with matrices of functions. The functions themselves are denoted, for example, by a_{ij}, and we denote the matrix by (a_{ij}). Furthermore, it is essential to distinguish between a linear transformation between vector spaces $T : V \to W$ and its matrix with respect to given bases in V and W. Following common notation in current linear algebra texts, if $\mathcal{B}$ is a basis in V and $\mathcal{B}'$ is a basis in W, then we denote by $[T]_{\mathcal{B}'}^{\mathcal{B}}$ the matrix of T with respect to these bases. If the bases are understood by context, we simply write $[T]$ for the matrix associated to T.

Occasionally, there arise irreconcilable discrepancies in definitions or notations (e.g., the definition of a critical point for a function $\mathbb{R}^n \to \mathbb{R}^m$, or how one defines θ and ϕ in spherical coordinates). In these instances the authors made a choice that best suits their purposes and indicated commonly used alternatives.

Section Dependency

The topics in this book primarily oscillate back and forth between local theory and global theory, first of plane curves, then space curves, and then surfaces in $\mathbb{R}^3$. In the authors' perspective, the Gauss-Bonnet Theorem (presented in Sections 8.2 and 8.3) serves as the climax theorem of the textbook. Sections depend on each other according to the following chart.

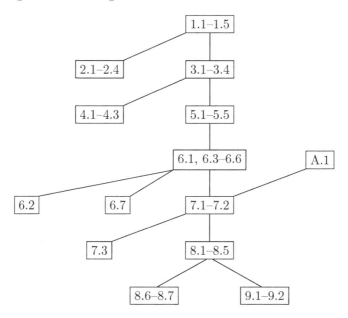

Note that Appendix A.1 on tensor notation is valuable for background on the indices of tensor notation but it is not essential for Chapter 7 and subsequent chapters.

Changes from the Second to the Third Edition

Feedback from faculty who used this textbook has led to the improvements for the third edition. These include:

- adding more exercises, both introductory and advanced;

- improving the explanations in select areas;

- rewriting the computer demos into *Maple*, as described earlier.

Acknowledgments

We would like to thank Dawson Bremner, Jonathan Higgins, and Kevin Tully for their many suggestions for more exercises.

With the transition from the second edition to the third edition, the interactive demos changed from being Java applets embedded in a browser to a self-contained workbook in *Maple*. We are very grateful to Dawson Bremner, Dave Broaddus, Lindsay Fadel, Dawson Miller, and Caitlin Smith for programming all of the demos in the new *Maple* format.

Thomas Banchoff

Our work at Brown University on computer visualizations in differential geometry goes back more than forty-five years, and I acknowledge my collaborator computer scientist Charles Strauss for the first fifteen years of our projects. Since 1982, an impressive collection of students have been involved in the creation and development of the software used to produce the applets for this book. The website for my 65th birthday conference

> http://tombanchoff.com

lists dozens of them, together with descriptions of their contributions. Particular thanks for the applets connected with this book belong to David Eigen, Mark Howison, Greg Baltazar, Michael Schwarz, and Michael Morris. For his work as a student, an assistant, and now as a co-author, I am extremely grateful to Steve Lovett. Special thanks for help in the Brown University mathematics department go to Doreen Pappas, Natalie Johnson, Audrey Aguiar, and Carol Oliveira, and to Larry Larrivee for his invaluable computer assistance. Finally, I thank my wife Kathleen for all her support and encouragement.

Stephen Lovett

I would first like to thank Thomas Banchoff my teacher, mentor, and friend. After one class, he invited me to join his team of students on developing electronic books for differential geometry and multivariable calculus. Despite ultimately specializing in algebra, the exciting projects he led and his inspiring course in differential geometry instilled in me a passion for differential geometry. His ability to introduce differential geometry as a visually stimulating and mathematically interesting topic served as one of my personal motivations for writing this book.

I am grateful to the students and former colleagues at Eastern Nazarene College. In particular I would like to acknowledge the undergraduate students who served as a sounding board for the first few drafts of this manuscript: Luke Cochran, David Constantine, Joseph Cox, Stephen Mapes, and Christopher Young. Special thanks are due to my colleagues Karl Giberson, Lee Hammerstrom, and John Free. In addition, I am indebted to Ellie Waal who helped with editing and index creation.

The continued support from my colleagues at Wheaton College made writing this book a gratifying project. In particular, I must thank Terry Perciante, Chair of the Department of Mathematics and Computer Science, for his enthusiasm and his interest. I am indebted to Dorothy Chapell, Dean of the Natural & Social Sciences, and to Stanton Jones, Provost of the College, for their encouragement and for a grant which freed up my time to finish writing. I am also grateful to Thomas VanDrunen and Darren Craig for helpful comments.

Finally, I cannot adequately express in just a few words how grateful I am to my wife Carla Favreau Lovett and my daughter Anne. While I was absorbed in this project, they provided a loving home, they braved the significant time commitment and encouraged me at every step. They also kindly put up with my occasional geometry comments such as how to see the Gaussian curvature in the reflection of "the Bean" in Chicago.

Second Edition Acknowledgments

In preparation for this second edition, we wish to thank the many people who contributed corrections, improvements, and suggestions. These include but are not limited to Teddy Parker for careful editing, Daniel Flath for many suggestions for improvements, Judith Arms for errata and suggestions, and Robert Ferréol for bringing to our attention an excellent on-line encyclopedia of curves at

http://www.mathcurve.com/ (in French).

We would also like to thank Nate Veldt, Gary Babbatz, Nathan Bliss, Matthew McMillan, and Cole Adams.

Authors

Thomas F. Banchoff is a geometer and a professor at Brown University. Dr. Banchoff was President of the Mathematical Association of America (MAA) from 1999 to 2000. He has published numerous papers in a variety of journals and has been the recipient of many honors, including the MAA's Deborah and Franklin Tepper Haimo Award and Brown's Teaching with Technology Award. He is the author of several books, including *Linear Algebra Through Geometry* with John Wermer and *Beyond the Third Dimension*.

Stephen Lovett is a professor of mathematics at Wheaton College. Dr. Lovett has taught introductory courses on differential geometry for many years. He has given many talks over the past several years on differential and algebraic geometry as well as cryptography. In 2015, he was awarded Wheaton's Senior Scholarship Faculty Award. He is the author of *Abstract Algebra: A First Course, Differential Geometry of Manifolds*, Second Edition, and *Transition to Advanced Mathematics* with Danilo Diedrichs, all published by CRC Press.

CHAPTER 1

Plane Curves: Local Properties

Just as calculus courses introduce real functions of one variable before tackling multivariable calculus, so it is natural to study curves before addressing surfaces and higher-dimensional objects. This chapter presents local properties of plane curves, where by *local property* we mean properties that are defined in a neighborhood of a point on the curve. For the sake of comparison with calculus, the derivative $f'(a)$ of a function f at a point a is a local property of the function since we only need knowledge of $f(x)$ for x in $(a - \varepsilon, a + \varepsilon)$, where ε is any positive real number, to define $f'(a)$. In contrast, the definite integral of a function over an interval is a global property since we need knowledge of the function over the whole interval to calculate the integral. In contrast to this chapter, Chapter 2 introduces global properties of plane curves.

1.1 Parametrizations

Borrowing from a physical understanding of motion in the plane, we can think about plane curves by specifying the coordinates x and y as functions of a time variable t, which give the position of a point traveling along the curve. Thus we need two functions $x(t)$ and $y(t)$. Using vector notation to locate a point on the curve, we often write $\vec{X}(t) = (x(t), y(t))$ for this pair of coordinate functions and call $\vec{X}(t)$ a *vector function* into $\mathbb{R}^2$. From a mathematical standpoint, t does not have to refer to time and is simply called the *parameter* of the vector function.

Example 1.1.1 (Lines) Euclid's first postulate of geometry is that through two distinct points there passes exactly one line. The point slope formula gives a Cartesian equation of a line through two points. In analytic geometry, the approach to describing a line through two points is slightly different. Given two distinct points $\vec{p}_1 = (x_1, y_1)$

DOI: 10.1201/9781003295341-1

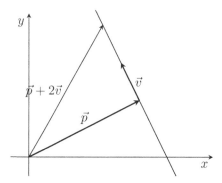

Figure 1.1: A line in the plane.

and $\vec{p}_2 = (x_2, y_2)$, the vector

$$\vec{v} = \vec{p}_2 - \vec{p}_1 = (x_2 - x_1, y_2 - y_1)$$

is called a *direction vector* of the line because all vectors along this line are multiples of $\vec{v}$. Then every point on the line can be written with a position vector $\vec{p}_1 + t\vec{v}$ for some $t \in \mathbb{R}$. Figure 1.1 shows an example with $t = 2$. Therefore, we find that a line can also be described by providing a point and a direction vector.

Using the coordinates of vectors, given a point $\vec{p} = (x_0, y_0)$ and a direction vector $\vec{v} = (v_1, v_2)$, a line through $\vec{p}$ in the direction of $\vec{v}$ is the image of the following vector function:

$$\vec{X}(t) = \vec{p} + t\vec{v} = (x_0 + v_1 t, y_0 + v_2 t) \qquad \text{for } t \in \mathbb{R}.$$

Note that just as the same line can be determined by two different pairs of points, so the same line may be specified by different sets of these equations. For example, using any point $\vec{p}$ on the line will ultimately trace out the same line as t varies through all of $\mathbb{R}$. Similarly, if we replace $\vec{v}$ with any nonzero multiple of itself, the set of points traced out as t varies through $\mathbb{R}$ is the same line in $\mathbb{R}^2$ plane.

 Example 1.1.2 (Circles) The pair of functions

$$\vec{X}(t) = (R\cos t + a, R\sin t + b)$$

traces out a circle of radius $R > 0$ centered at the point (a, b). To see this, note that by the definition of the $\sin t$ and $\cos t$ functions, $(\cos t, \sin t)$ are the coordinates of the point on the unit circle that is also on the ray out of the origin that makes an angle t with the positive x-axis. Thus,

$$\vec{X}_1(t) = (\cos t, \sin t), \qquad \text{with } t \in [0, 2\pi],$$

traces out the unit circle in a counterclockwise manner. Multiplying both coordinate functions by R stretches the circle out by a factor of R away from the origin. Thus, the vector function

$$\vec{X}_2(t) = (R\cos t, R\sin t), \qquad \text{with } t \in [0, 2\pi],$$

has as its image the circle of radius R centered at the origin. Notice also that by writing $\vec{X}_2(t) = (x(t), y(t))$, we deduce that $x(t)^2 + y(t)^2 = R^2$ for all t, which is the algebraic equation of the circle. In order to obtain a vector function that traces out a circle centered at the point (a, b), we must simply translate $\vec{X}_2$ by the vector (a, b). This is vector addition, and so we get

$$\vec{X}(t) = (R\cos t + a, R\sin t + b), \qquad \text{with } t \in [0, 2\pi].$$

Two different vector functions can have the same image in $\mathbb{R}^2$. For example, if $\omega > 0$,

$$\vec{X}(t) = (\cos \omega t, \sin \omega t), \qquad \text{with } t \in [0, 2\pi/\omega],$$

also has the unit circle as its image. Referring to vocabulary in physics, this latter vector function corresponds to a point moving around the unit circle at an angular velocity of ω (lower case Greek "omega").

When trying to establish a suitable mathematical definition of what one usually thinks of as a curve, one does not wish to consider as a curve a set of points that jump around or pieces of segments. We would like to think of a curve as unbroken in some sense. In calculus, one introduces the notion of continuity to describe functions without "jumps" or holes, but one must exercise a little care in carrying over the notion of continuity to vector functions. More generally, we need to define the notion of a limit of a vector function as the parameter t approaches a fixed value. First, however, we remind the reader of the Euclidean distance formula.

Definition 1.1.3 Let $\vec{v}$ be a vector in $\mathbb{R}^2$ with coordinates $\vec{v} = (v_1, v_2)$ in the standard basis. The (Euclidean) length of $\vec{v}$ is given by

$$\|\vec{v}\| = \sqrt{\vec{v} \cdot \vec{v}} = \sqrt{v_1^2 + v_2^2}.$$

If p and q are two points in $\mathbb{R}^n$ with coordinates given by vectors $\vec{v}$ and $\vec{w}$, then the Euclidean distance between p and q is $\|\vec{w} - \vec{v}\|$.

Definition 1.1.4 Let $\vec{X}$ be a vector function from a subset of $\mathbb{R}$ into $\mathbb{R}^n$. We say that the limit of $\vec{X}(t)$ as t approaches a is a vector $\vec{w}$, and we write

$$\lim_{t \to a} \vec{X}(t) = \vec{w},$$

if for all $\varepsilon > 0$ there exists a $\delta > 0$ such that $0 < |t - a| < \delta$ implies $\|\vec{X}(t) - \vec{w}\| < \varepsilon$.

Definition 1.1.5 Let I be an open interval of $\mathbb{R}$, let $a \in I$, and let $\vec{X} : I \to \mathbb{R}^2$ be a vector function. We say that $\vec{X}(t)$ is continuous at a if the limit as t approaches a of $\vec{X}(t)$ exists and

$$\lim_{t \to a} \vec{X}(t) = \vec{X}(a).$$

The above definitions mirror the usual definition of a limit of a real function but must use the length of a vector difference to discuss the proximity between $\vec{X}(t)$ and a fixed vector $\vec{w}$. Though at the outset this definition appears more complicated than the usual definition of a limit of a real function, the following proposition shows that it is not.

Proposition 1.1.6 *Let $\vec{X}$ be a vector function from a subset of $\mathbb{R}$ into $\mathbb{R}^2$ that is defined over an interval containing a, though perhaps not at a itself. Suppose in coordinates we have $\vec{X}(t) = (x(t), y(t))$ wherever $\vec{X}$ is defined. If $\vec{w} = (w_1, w_2)$, then*

$$\lim_{t \to a} \vec{X}(t) = \vec{w} \text{ if and only if } \lim_{t \to a} x(t) = w_1 \text{ and } \lim_{t \to a} y(t) = w_2.$$

Proof: Suppose first that $\lim_{t \to a} \vec{X}(t) = \vec{w}$. Let $\varepsilon > 0$ be arbitrary and let $\delta > 0$ satisfy the definition of the limit of the vector function. Note that $|x(t) - w_1| \leq \|\vec{X}(t) - \vec{w}\|$ and that $|y(t) - w_2| \leq \|\vec{X}(t) - \vec{w}\|$. Hence, $0 < |t - a| < \delta$ implies $|x(t) - w_1| < \varepsilon$ and $|y(t) - w_2| < \varepsilon$. Thus, $\lim_{t \to a} x(t) = w_1$ and $\lim_{t \to a} y(t) = w_2$.

Conversely, suppose that $\lim_{t \to a} x(t) = w_1$ and $\lim_{t \to a} y(t) = w_2$. Let $\varepsilon > 0$ be an arbitrary positive real number. By definition, there exist δ_1 and δ_2 such that $0 < |t - a| < \delta_1$ implies $|x(t) - w_1| < \varepsilon/\sqrt{2}$ and $0 < |t-a| < \delta_2$ implies $|y(t)-w_2| < \varepsilon/\sqrt{2}$. Taking $\delta = \min(\delta_1, \delta_2)$ we see that $0 < |t - a| < \delta$ implies that

$$\|\vec{X}(t) - \vec{w}\| = \sqrt{|x(t) - w_1|^2 + |y(t) - w_2|^2} < \sqrt{\frac{\varepsilon^2}{2} + \frac{\varepsilon^2}{2}} = \varepsilon.$$

This finishes the proof of the proposition.   □

Corollary 1.1.7 *Let I be an open interval of $\mathbb{R}$, let $a \in I$, and consider a vector function $\vec{X} : I \to \mathbb{R}^2$ with $\vec{X}(t) = (x(t), y(t))$. Then $\vec{X}(t)$ is continuous at $t = a$ if and only if $x(t)$ and $y(t)$ are continuous at $t = a$.*

Definition 1.1.5 and Corollary 1.1.7 provide the mathematical framework for what one usually thinks of as a curve in physical intuition. This motivates the following definition.

Definition 1.1.8 Let I be an interval of $\mathbb{R}$. A *parametrized curve* (or *parametric curve*) in the plane is a continuous function $\vec{X} : I \to \mathbb{R}^2$. If we write $\vec{X}(t) = (x(t), y(t))$, then the functions $x : I \to \mathbb{R}$ and $y : I \to \mathbb{R}$ are called the *coordinate functions* or *parametric equations* of the parametrized curve. We call the *locus* of $\vec{X}(t)$ the image of $\vec{X}(t)$ as a subset of $\mathbb{R}^2$.

It is important to note the distinction in this definition between a parametrized curve and its locus. For example, in Example 1.1.1 we show that the parametrization $\vec{X}(t) = t\vec{v} + \vec{p}$ traces out a line L that goes through the point $\vec{p}$ with direction vector $\vec{v}$. However, the line is the *locus* of $\vec{X} : \mathbb{R} \to \mathbb{R}^2$ and not the vector-valued function $\vec{X}$ itself. The vector-valued function is the parametrized curve. Since the function $t \mapsto t^3 + t$ is a bijection from $\mathbb{R}$ to itself, the parametrized curve $\vec{Y}(t) = (t^3 + t)\vec{v} + \vec{p}$ has the same locus, the line L, but is quite different as a function.

The following examples begin to provide a library of parametric curves and illustrate how to construct parametric curves to describe a particular shape or trajectory.

Example 1.1.9 (Graphs of Functions) The graph of a continuous function $f : [a, b] \to \mathbb{R}$ over an interval $[a, b]$ can be viewed as a parametric curve. In order to view the graph of a continuous function as a parametrized curve, we use the coordinate functions $\vec{X}(t) = (t, f(t))$, with $t \in [a, b]$.

Example 1.1.10 (Circles Revisited) The parametrization given for circles in Example 1.1.2 utilizes trigonometric functions. Inspired by Euclid's formula for Pythagorean triples, we can get a parametrization for the unit circle using rational functions:

$$\vec{X}(t) = (x(t), y(t)) = \left(\frac{1 - t^2}{1 + t^2}, \frac{2t}{1 + t^2} \right) \qquad \text{for } t \in \mathbb{R}. \qquad (1.1)$$

It is easy to see that for all $t \in \mathbb{R}$, $x(t)^2 + y(t)^2 = 1$. This means that the locus of $\vec{X}$ is on the unit circle. However, this parametrization does not trace out the entire circle as it misses the point $(-1, 0)$. We leave it as an exercise to determine a geometric interpretation of the parameter t and to show that

$$\lim_{t \to \infty} \vec{X}(t) = \lim_{t \to -\infty} \vec{X}(t) = (-1, 0).$$

This example emphasizes that it is not possible to tell what the parametrization is simply by looking at the locus. A counterclockwise circle is indistinguishable from the circle covered twice, or from the circle traced out in a clockwise fashion, or, as in this example, from a circle covered in a completely different way.

 Example 1.1.11 (Ellipses) Without repeating all the reasoning of the previous exercise, it is not hard to see that

$$\vec{X}(t) = (a\cos t, b\sin t)$$

provides a parametrization for the ellipse centered at the origin with axes along the x- and y-axes, with respective half-axes of length $|a|$ and $|b|$. Note that these coordinate functions do indeed satisfy

$$\frac{x(t)^2}{a^2} + \frac{y(t)^2}{b^2} = 1 \qquad \text{for all } t.$$

 Example 1.1.12 (Lissajous Figures) It is sometimes amusing to see how $\cos t$ and $\sin t$ relate to each other if we change their respective periods. Lissajous figures, which arise in the context of electronics, are curves parametrized by

$$\vec{X}(t) = (\cos mt, \sin nt),$$

where m and n are positive integers. See Figure 1.2 for an example of a Lissajous figure with $m = 5$ and $n = 3$.

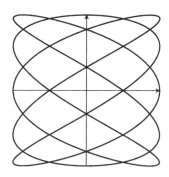

Figure 1.2: A Lissajous figure.

 Example 1.1.13 (Cycloids) We can think of a usual cycloid as the locus traced out by a point of light affixed to a bicycle tire as the bicycle rolls forward. We can establish a parametrization of such a curve as follows.

Assume the wheel of radius a begins with its center at $(0, a)$ so that the part of the wheel touching the x-axis is at the origin. We view the wheel as rolling forward on the positive x-axis. As the wheel rolls, the position of the center of the wheel is $\vec{f}(t) = (at, a)$, where t is the angle measuring how much (many times) the wheel has turned since it started. At the same time, the light—at a distance a from the center of the wheel and first positioned straight down from the center of the wheel—rotates in a clockwise motion around the center of the wheel. See Figure 1.3. The motion of the light with respect to the center of the wheel is

$$\vec{g}(t) = \left(a \cos \left(-t - \frac{\pi}{2} \right), a \sin \left(-t - \frac{\pi}{2} \right) \right) = (-a \sin t, -a \cos t).$$

The locus of the cycloid is the vector function that is the sum of $\vec{f}(t)$ and $\vec{g}(t)$. Thus, a parametrization for the cycloid is

$$\vec{X}(t) = (at - a \sin t, a - a \cos t).$$

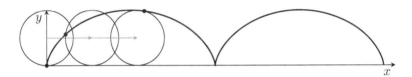

Figure 1.3: Cycloid.

One can point out that reflectors on bicycle wheels are usually not attached directly on the tire but on a spoke of the wheel. We can easily modify the above discussion to obtain the relevant parametric equations for when the point of light is located at a distance b from the center of the rolling wheel. We obtain

$$\vec{X}(t) = (at - b \sin t, a - b \cos t).$$

Whether or not $b < a$, the curve traced out by $\vec{X}(t)$ is called a *trochoid*.

Example 1.1.14 (Heart Curve) The most popular curve around Valentine's Day is the heart shape. Here are some parametric equations that trace out such a curve:

$$\vec{X}(t) = ((1 - \cos^2 t) \sin t, (1 - \cos^3 t) \cos t).$$

We encourage the reader to visit this example in the accompanying software and to explore ways of modifying these equations to create other interesting curves. The Online Encyclopedia of Curves at http://www.mathcurve.com/ credits this curve to Raphaël Laporte who designed it for his "petite amie" (translated "girl friend").

Example 1.1.15 (Polar Functions) When working with polar co-ordinates, we usually consider functions of the radial distance r as a function of the angle θ, written $r = f(\theta)$. The graphs of such functions are as parametrized curves in a natural way. Recall the coordinate transformation

$$\begin{cases} x = r\cos\theta, \\ y = r\sin\theta. \end{cases}$$

Then take θ as the parameter t, and the parametric equations for the graph of $r = f(\theta)$ are

$$\vec{X}(t) = (f(t)\cos t, f(t)\sin t).$$

As an example, the polar function $r = \sin 3\theta$ traces out a curve that resembles a three-leaf flower. (See Figure 1.4.) As a parametric curve, it is given by

$$\vec{X}(t) = (\sin 3t \cos t, \sin 3t \sin t).$$

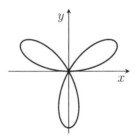

Figure 1.4: Three-leaf flower.

 Example 1.1.16 (Cardioid) Another common polar function is a cardioid, which is the locus of $r = 1 - \cos\theta$. In parametric equations, we have

$$\vec{X}(t) = ((1 - \cos t)\cos t, (1 - \cos t)\sin t).$$

As mentioned in Example 1.1.2, for a parametric curve $\vec{X} : I \to \mathbb{R}^2$, the set of points $\mathcal{C} = \{\vec{X}(t) \mid t \in I\}$ as a subset of $\mathbb{R}^2$ *does not* depend uniquely on the functions $x(t)$ and $y(t)$. In fact, it is important to make a careful distinction between the notion of a parametrized curve as defined above, the image of the parametrized curve as a subset of $\mathbb{R}^2$ (also called the *locus* of the curve), and the notion of a curve, eventually defined as a one-dimensional manifold. (See [24, Chapter 3].)

Definition 1.1.17 Given a parametrized curve $\vec{X} : I \to \mathbb{R}^2$ and any continuous functions g from an interval J onto the interval I, we can produce a new vector function $\vec{\xi} : J \to \mathbb{R}^2$ defined by $\vec{\xi} = \vec{X} \circ g$. The image of $\vec{\xi}$ is again the set $\mathcal{C}$, and $\vec{\xi} = \vec{X} \circ g$ is called a *reparametrization* of $\vec{X}$.

If g is not onto I, then the image of $\vec{\xi}$ may be a proper subset of $\mathcal{C}$. In this case, we usually do not call $\vec{\xi}$ a reparametrization as it does not trace out the same locus of $\vec{X}$.

PROBLEMS

In Problems 1.1.1 through 1.1.4, sketch the curve with the given parametric equations. Indicate with an arrow the direction in which t increases.

1. $\vec{X}(t) = (t^3, t^2)$.

2. $\vec{X}(t) = (t^2, \cos t)$.

3. $\vec{X}(t) = (t \cos t, t \sin t)$.

4. $\vec{X}(t) = (t^2 - 1, t^3 - t)$.

5. Equation (1.1) gave the following parametric equations for a circle:

$$\vec{X} = (x(t), y(t)) = \left(\frac{1 - t^2}{1 + t^2}, \frac{2t}{1 + t^2} \right), \qquad \text{for } t \in \mathbb{R}.$$

   Prove that the parameter t is equal to $\tan \theta$, where θ is the angle shown in the following picture of the unit circle.

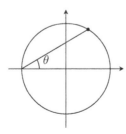

6. Let $\mathcal{C}$ be the circle of center $(0, 2)$ and radius 1. Let $\mathcal{S}$ be the set of points in $\mathbb{R}^2$ that are the same distance from the outside of circle $\mathcal{C}$ as they are from the x-axis. Show the following.

   (a) $\mathcal{S}$ has a parametrization $\vec{X}(t) = \left(\frac{3 \sin t}{1 + \cos t}, 2 - \frac{3 \cos t}{1 + \cos t} \right)$.
      [Hint: Use the parameter t as the angle as shown below.]

   (b) $\mathcal{S}$ is a parabola. Find the Cartesian equation for it.

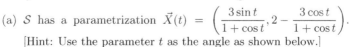

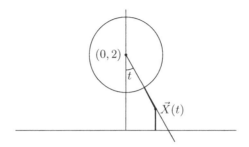

7. An *epicycloid* is defined as the locus of a point on the edge of a circle of radius b as this circle rolls on the outside of a fixed circle of radius a. Supposing that at $t = 0$, the moving point is located at $(a, 0)$. Prove that the following are parametric equations for an epicycloid:

$$\vec{X}(t) = \left((a+b)\cos t - b\cos\left(\frac{a+b}{b}t\right), (a+b)\sin t - b\sin\left(\frac{a+b}{b}t\right) \right).$$

8. A *hypocycloid* is defined as the locus of a point on the edge of a circle of radius b as this circle rolls on the inside of a fixed circle of radius a. Assuming that $a > b$ and that at $t = 0$, the moving point is located at $(a, 0)$, find parametric equations for a hypocycloid.

9. Revisit Example 1.1.13 and find parametric equations for a cycloid resulting from a circle of radius 3 rolling on the line $y = x/2$.

10. Consider the parametrized curve $\vec{X} : \mathbb{R} \to \mathbb{R}^2$ with equations $\vec{X}(t) = \left(\frac{1}{2}(e^t + e^{-t}), \frac{1}{2}(e^t - e^{-t}) \right)$. Prove that the locus is one branch of a unit hyperbola $x^2 - y^2 = 1$. Give a parametrization for the other branch of the hyperbola.

11. Consider the curve $\mathcal{C}$ in $\mathbb{R}^2$ that is the solution to $x^3 + y^3 = 1$.
    (a) Let $\vec{X} : \mathbb{R} - \{-1\} \to \mathbb{R}^2$ be the parametrized curve with $\vec{X}(t) = (1/\sqrt[3]{t^3+1}, t/\sqrt[3]{t^3+1})$. Prove that the locus of $\vec{X}$ is $\mathcal{C}$ except for one point.
    (b) Show that the "missing" point arises as the limit $\lim_{t\to\infty} \vec{X}(t)$ or $\lim_{t\to-\infty} \vec{X}(t)$.
    (c) Let $\vec{Y} : \mathbb{R} \to \mathbb{R}^2$ be the parametrized curve with $\vec{Y}(u) = ((1 + u)/\sqrt[3]{2 + 6u^2}, (1 - u)/\sqrt[3]{2 + 6u^2})$. Prove that the locus of $\vec{Y}$ is on and is in fact all of $\mathcal{C}$.
    (d) Show that (over most of its domain) $\vec{Y}$ is a reparametrization of $\vec{X}$ with $t = (1 - u)/(1 + u)$ and discuss the differences in the domains of $\vec{X}$ and $\vec{Y}$ and why one covers all of $\mathcal{C}$, while another doesn't.

12. Consider the vectors $\vec{a} = (3, 3)$ and $\vec{b} = (-1, 1)$. Explain why the parametric curve $\vec{X}(t) = (\cos t)\vec{a} + (\sin t)\vec{b}$ is an ellipse. Furthermore, give a Cartesian equation of the locus of this parametric curve.

13. Consider the parametric curve $\vec{X}(t) = (t^3 - 5t, 3t^2)$. Graphing it shows that it intersects itself. By solving for t_1 and t_2 the equation $\vec{X}(t_1) = \vec{X}(t_2)$, find the parameters where the curve intersects itself and give the coordinates of the point on the locus where this occurs.

14. Consider the parametrized curve $\vec{X} : [0, 2\pi] \to \mathbb{R}^2$ that gives a deformation of a cardioid $\vec{X}(t) = ((a + \cos t) \cos t, (a + \cos t) \sin t)$, where a is a real number. Show that the curve intersects itself for $|a| < 1$. Describe what happens for $a = 0$.

15. Let Γ be a circle and let O a point on Γ. Set L to be the line tangent to Γ such that its point of tangency is diametrically opposite from O. The *cissoid of Diocles* relative to Γ and O is the set of points P such that $OP = AB$, where A is the intersection of $\overleftrightarrow{OP}$ and Γ and where B is the intersection of $\overleftrightarrow{OP}$ and L. Using the setup in the following diagram, find parametric equations for the cissoid of Diocles.

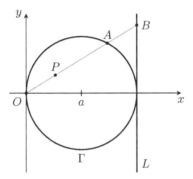

1.2   Position, Velocity, and Acceleration

In physics, one interprets the vector function $\vec{X}(t) = (x(t), y(t))$ as providing the location along a curve at time t in reference to some fixed frame. A *frame* is a (usually orthonormal) basis attached to a fixed origin. The point $O = (0, 0)$ along with the basis $\{\vec{i}, \vec{j}\}$, where

$$\vec{i} = (1, 0) \qquad \text{and} \quad \vec{j} = (0, 1),$$

form the standard reference frame. We call the vector function $\vec{X}(t)$ the *position vector*. When one uses the standard reference frame, it is not uncommon to write

$$\vec{X}(t) = x(t)\vec{i} + y(t)\vec{j}.$$

Directly imitating Newton's approach to finding the slope of a curve at a certain point, it is natural to ask the question, "What is the direction and rate of change of a curve $\vec{X}$ at point t_0?" If we look at two points on the curve, say $\vec{X}(t_0)$ and $\vec{X}(t_1)$, the change in position is given by the vector $\vec{X}(t_1) - \vec{X}(t_0)$, while to specify the rate of change in position, we would need to scale this vector by a factor of $t_1 - t_0$. Hence, the vectorial rate of change between $\vec{X}(t_0)$ and $\vec{X}(t_1)$ is

$$\frac{1}{t_1 - t_0}\left(\vec{X}(t_1) - \vec{X}(t_0)\right).$$

To define an instantaneous rate of change along the curve at the point t_0, we need to calculate (if it exists and if it has meaning) the limit

$$\vec{X}'(t_0) = \lim_{h \to 0} \frac{1}{h}\left(\vec{X}(t_0 + h) - \vec{X}(t_0)\right). \tag{1.2}$$

According to Proposition 1.1.6, if $\vec{X}(t) = (x(t), y(t))$, then the limit in Equation (1.2) exists if and only if $x(t)$ and $y(t)$ are both differentiable at t_0. Furthermore, if this limit exists, then $\vec{X}'(t_0) = (x'(t_0), y'(t_0))$. This leads us to the following definition.

Definition 1.2.1 Let $\vec{X} : I \to \mathbb{R}^2$ be a vector function with coordinates $\vec{X}(t) = (x(t), y(t))$. We say that $\vec{X}$ is differentiable at $t = t_0$ if $x(t)$ and $y(t)$ are both differentiable at t_0. If J is the common domain to $x'(t)$ and $y'(t)$, then we define the derivative of the vector function $\vec{X}$ as the new vector function $\vec{X}' : J \to \mathbb{R}^2$ defined by $\vec{X}'(t) = (x'(t), y'(t))$.

If $x(t)$ and $y(t)$ are both differentiable functions on their domain, the derivative vector function $\vec{X}'(t) = (x'(t), y'(t))$ is called the *velocity vector*. Mathematically, this is just another vector function and traces out another curve when placed in the standard reference frame. However, because it illustrates the direction of motion along the curve, one often visualizes the velocity vector corresponding to $t = t_0$ as based at the point $\vec{X}(t_0)$ on the curve.

Following physics language, we call the second derivative of the vector function $\vec{X}''(t) = (x''(t), y''(t))$ the *acceleration vector* related to $\vec{X}(t)$.

In general, our calculations often require that our vector functions can be differentiated at least once and sometimes more. Consequently, when establishing theorems, we like to succinctly describe the largest class of functions for which a particular result holds.

Definition 1.2.2 Let I be an interval of $\mathbb{R}$, and let $f : I \to \mathbb{R}$ be a function or $\vec{X} : I \to \mathbb{R}^2$ a parametrized curve. We say that f or $\vec{X}$ is of class C^r on I if the rth derivative of $\vec{X}$ exists and is continuous on I. We denote by C^∞ the class of functions or parametrized curves that have derivatives of all orders on I. It is also common to write $f \in C^r(I, \mathbb{R})$ to say that f is a real-valued function of class C^r and to write $\vec{X} \in C^r(I, \mathbb{R}^2)$ to say that $\vec{X}$ is a parametrized curve in the plane of class C^r.

To say that a function is of class C^0 over the interval I means that it is continuous. By basic theorems in calculus, the condition that a function be of a certain class is an increasingly restrictive condition. In other words, as sets of functions defined over the same interval I, the classes are nested according to

$$C^0 \supset C^1 \supset C^2 \supset \cdots \supset C^\infty.$$

Whenever we impose a condition that a function is of class C^r for some r, such a supposition is generically called a "smoothness condition."

Proposition 1.2.3 *Let $\vec{v}(t)$ and $\vec{w}(t)$ be vector functions defined and differentiable over an interval $I \subset \mathbb{R}$. Then the following hold:*

1. *If $\vec{X}(t) = c\vec{v}(t)$, where $c \in \mathbb{R}$, then $\vec{X}'(t) = c\vec{v}'(t)$.*

2. *If $\vec{X}(t) = c(t)\vec{v}(t)$, where $c : I \to \mathbb{R}$ is a real function, then*

$$\vec{X}'(t) = c'(t)\vec{v}(t) + c(t)\vec{v}'(t).$$

3. *If $\vec{X}(t) = \vec{v}(t) + \vec{w}(t)$, then*

$$\vec{X}'(t) = \vec{v}'(t) + \vec{w}'(t).$$

4. *If $\vec{X}(t) = \vec{v}\big(f(t)\big)$ is a vector function and $f : J \to I$ is a real function into I, then*

$$\vec{X}'(t) = f'(t)\vec{v}'\big(f(t)\big).$$

5. *If $f(t) = \vec{v}(t) \cdot \vec{w}(t)$ is the dot product between $\vec{v}(t)$ and $\vec{w}(t)$, then*
$$f'(t) = \vec{v}'(t) \cdot \vec{w}(t) + \vec{v}(t) \cdot \vec{w}'(t).$$

Example 1.2.4 Consider the spiral defined by $\vec{X}(t) = (t \cos t, t \sin t)$ for $t \geq 0$. The velocity and acceleration vectors are

$$\vec{X}'(t) = (\cos t - t \sin t, \sin t + t \cos t),$$
$$\vec{X}''(t) = (-2 \sin t - t \cos t, 2 \cos t - t \sin t).$$

If we wish to calculate the angle between $\vec{X}$ and $\vec{X}'$ as a function of t, we use the dot product as follows. We calculate

$$\|\vec{X}(t)\| = \sqrt{t^2 \sin^2 t + t^2 \cos^2 t} = |t|,$$
$$\|\vec{X}'(t)\| = \sqrt{(\cos t - t \sin t)^2 + (\sin t + t \cos t)^2} = \sqrt{1 + t^2},$$
$$\vec{X}(t) \cdot \vec{X}'(t) = t \cos^2 t - t^2 \cos t \sin t + t \sin^2 t + t^2 \cos t \sin t = t.$$

Thus, the angle $\theta(t)$ between $\vec{X}(t)$ and $\vec{X}'(t)$ is defined for all $t \neq 0$ and is equal to

$$\theta(t) = \cos^{-1}\left(\frac{\vec{X}(t) \cdot \vec{X}'(t)}{\|\vec{X}(t)\| \, \|\vec{X}'(t)\|}\right) = \cos^{-1}\left(\frac{t}{|t|\sqrt{1+t^2}}\right).$$

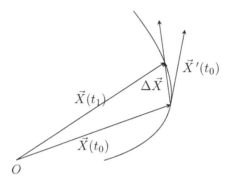

Figure 1.5: Arc length segment.

Let $\mathcal{C}$ be the locus of a vector function $\vec{X}(t)$ for $t \in [a, b]$. Then we approximate the length of the arc $l(\mathcal{C})$ with the Riemann sum

$$l(\mathcal{C}) \approx \sum_{i=1}^{n} \|\vec{X}(t_i) - \vec{X}(t_{i-1})\| \approx \sum_{i=1}^{n} \|\vec{X}'(t_i)\| \Delta t.$$

(See Figure 1.5 as an illustration for the approximation $\|\vec{X}(t_i) - \vec{X}(t_{i-1})\| \approx \|\vec{X}'(t_i)\| \Delta t$.) Taking the limit of this Riemann sum, we obtain the following formula for the arc length of $\mathcal{C}$:

$$l = \int_a^b \sqrt{(x'(t))^2 + (y'(t))^2} \, dt. \tag{1.3}$$

A rigorous justification for the arc length formula, using the Mean Value Theorem, can be found in most calculus textbooks.

In light of Equation (1.3), we define the arc length function $s :$ $[a, b] \rightarrow \mathbb{R}$ related to $\vec{X}(t)$ as the arc length along $\mathcal{C}$ over the interval $[a, t]$. Thus,

$$s(t) = \int_a^t \|\vec{X}'(u)\| \, du = \int_a^t \sqrt{(x'(u))^2 + (y'(u))^2} \, du. \tag{1.4}$$

By the Fundamental Theorem of Calculus, we then have

$$s'(t) = \|\vec{X}'(t)\| = \sqrt{(x'(t))^2 + (y'(t))^2},$$

which, still following the vocabulary from the trajectory of a moving particle, we call the *speed function* of $\vec{X}(t)$.

Example 1.2.5 Consider the parametrized curve defined by $\vec{X}(t) =$ (t^2, t^3). The velocity vector is $\vec{X}'(t) = (2t, 3t^2)$, and thus the arc length function from $t = 0$ is

$$s(t) = \int_0^t \|\vec{X}'(u)\| \, du = \int_0^t \sqrt{4u^2 + 9u^4} \, du = \int_0^t 3|u| \sqrt{\frac{4}{9} + u^2} \, du$$

$$= \text{sign}(t) \left(\left(\frac{4}{9} + t^2 \right)^{3/2} - \frac{8}{27} \right),$$

where $\text{sign}(t) = 1$ if $t > 0$ and -1 if $t < 0$ and $\text{sign}(0) = 0$. The length of $\vec{X}(t)$ between $t = 0$ and $t = 1$ is

$$s(1) - s(0) = \frac{1}{27} (13^{3/2} - 8).$$

The speed function allows us to understand that a parametrized curve $\vec{X} : I \to \mathbb{R}^2$ does not only contain the information that describes the locus of the curve but it also contains information about the speed of travel $s(t)$ along the locus. By looking at the locus, it is impossible to discern the speed function.

Consider a reparametrization of the parametrized curve $\vec{X} \circ f$, where $f : J \to I$ is a differentiable surjective (i.e., onto) function. If $t_0 = f(u_0)$ where f is a differentiable function at u_0, then $\vec{X}(f(u_0)) = \vec{X}(t_0)$ and

$$\left\| \frac{d}{dt} \vec{X}(f(u_0)) \right\| = \|f'(u_0)\vec{X}'(f(u_0))\| = |f'(u_0)| \, \|\vec{X}'(t_0)\|$$

$$= |f'(u_0)| s'(t_0).$$

Consequently, under a reparametrization, the unit vector associated with $\vec{X}'(t)$ only changes by a factor of $+1$ or -1. More precisely, if $\vec{\xi} = \vec{X} \circ f$, at any parameter value t where $f'(t) \neq 0$ and $\vec{X}'(f(t)) \neq \vec{0}$, we have

$$\frac{\vec{\xi}'(t)}{\|\vec{\xi}'(t)\|} = \frac{f'(t)}{|f'(t)|} \frac{\vec{X}'(f(t))}{\|\vec{X}'(f(t))\|}, \tag{1.5}$$

where $f'(t)/|f'(t)|$ is $+1$ or -1. It bears repeating that in order for the right-hand side to be well defined, we need $f'(t) \neq 0$.

Definition 1.2.6 Let $\vec{X} : I \to \mathbb{R}^2$ be a parametrized curve. For a continuously differentiable function f from an interval J onto I, we call $\vec{\xi} = \vec{X} \circ f$ a *regular reparametrization* if for all $t \in J$, $f'(t)$ is well defined and never 0. In addition, a regular reparametrization is called *positively oriented* (resp. *negatively oriented*) if $f'(t) > 0$ (resp. $f'(t) < 0$) for all $t \in J$.

By the Mean Value Theorem, if $f'(t) \neq 0$ for all $t \in J$, then $f : J \to I$ is an injective function. Thus since f is by definition surjective, regular reparametrizations involve a bijective function $f : J \to I$.

Equation (1.5) shows that the unit vector $\vec{X}'(t)/\|\vec{X}'(t)\|$ is invariant under a positively oriented reparametrization and simply changes sign under a negatively oriented reparametrization. Consequently, this unit vector is an important geometric object associated to a curve at a point.

 Definition 1.2.7 Let $\vec{X} : I \to \mathbb{R}^2$ be a plane parametric curve. A point $t_0 \in I$ is called a *critical point* of $\vec{X}(t)$ if $\vec{X}(t)$ is not differentiable at t_0 or if $\vec{X}'(t_0) = \vec{0}$. If t_0 is a critical point, then $\vec{X}(t_0)$ is called a *critical value*. A point $t = t_0$ that is not critical is called a *regular point*. A parametrized curve $\vec{X} : I \to \mathbb{R}^2$ is called *regular* if it is of class C^1 (i.e., continuously differentiable) and $\vec{X}'(t) \neq \vec{0}$ for all $t \in I$. Finally, if t is a regular point of $\vec{X}$, we define the *unit tangent vector* $\vec{T}(t)$ as

$$\vec{T}(t) = \frac{\vec{X}'(t)}{\|\vec{X}'(t)\|}.$$

It is not uncommon to call a property of a curve $\mathcal{C}$ or a property of a point P on a curve $\mathcal{C}$ a *geometric property* if it does not depend on a parametrization of the locus near the point. In particular, a geometric property does not depend on the speed of travel along the curve through the point.

At any point of a curve where $\vec{T}(t)$ is defined, that is to say at any regular point, we can write the velocity vector as

$$\vec{X}'(t) = s'(t)\vec{T}(t). \tag{1.6}$$

This expresses the velocity vector as the product of its magnitude (speed) and direction (unit tangent vector). In most differential geometry texts, authors simplify their formulas by *reparametrizing by arc length*. From the perspective of coordinates, this means picking an "origin" O on the curve $\mathcal{C}$ occurring at some fixed $t = t_0$ and using the arc length along $\mathcal{C}$ between O and any other point P as the parameter with which to locate P on $\mathcal{C}$. In fact, this can always be done.

Proposition 1.2.8 *If $\vec{X}(t)$ is a regular parametrized curve, then there is a regular reparametrization of $\vec{X}$ by arc length. Furthermore, if $\vec{X}$*

is of class C^k, then the arc length reparametrization is also of class C^k.

Proof: Let $s = f(t)$ be the arc length function with $s = 0$ corresponding to some point on the curve. Since $\vec{X}$ is regular over its domain, then $f'(t) = \|\vec{X}'(t)\| > 0$ for all t. Since $f(t)$ is strictly increasing, it has an inverse function $t = h(s)$. By the Inverse Function Theorem, since $f'(t) \neq 0$, then $h(s)$ is differentiable with

$$h'(s) = \frac{1}{f'(h(s))} = \frac{1}{f'(t)}. \tag{1.7}$$

Note that the composite function $\vec{Y}(s) = \vec{X}(h(s))$ satisfies

$$\vec{Y}'(s) = \vec{X}'(h(s))h'(s) = \vec{X}'(t)\frac{1}{f'(t)} = \frac{\vec{X}'(t)}{\|\vec{X}'(t)\|}.$$

If $\vec{X}$ is of class C^k, then $s'(t) = f'(t) = \|\vec{X}'(t)\|$ is of class C^{k-1}. Hence $s(t)$ itself is of class C^k. The Inverse Function Theorem states that the inverse function is also of class C^k. Consequently, the arc length parametrization $\vec{y}(s)$ is of class C^k since it is the composition of $\vec{X}(t)$ of class C^k and $t = h(s)$ of class C^k. $\qquad\square$

The Inverse Function Theorem as used in the above proof (and its more general multivariable counterpart) appear in most introductory texts on analysis. (See [7] for example.) However, the key derivative in Equation (1.7) follows from the chain rule for invertible functions and is a standard differentiation formula.

The habit of reparametrizing by arc length has benefits and drawbacks. The great benefit of this approach is that along a curve parametrized by arc length, the speed function s' is identically 1 and the velocity vector is exactly the tangent vector

$$\vec{X}'(s) = \vec{T}(s).$$

This formulation simplifies proofs and difficult calculations. The main drawback is that, in practice, most curves do not admit a simple formula for their arc length function, let alone a formula that can be written using elementary functions. For example, given an explicit parametrized curve $\vec{X}(t)$, it is often very challenging to find $s(t)$ as defined in Equation (1.4). Furthermore, to reparametrize by arc length, it would be necessary to find the inverse function $t(s)$, representing the original parameter t as a function of s. Even if it is possible to find $s(t)$, determining this inverse function usually cannot be written with elementary functions. Then the parametrization by arc length is $\vec{X}(s) = \vec{X}(t(s))$. Even using a computer algebra systems (CAS), reparametrizing by arc length remains an intractable problem.

The notion of "regular" in many ways mirrors geometric properties of continuously differentiable single-variable functions. In particular, if a parametrized curve $\vec{X}(t)$ is regular at t_0, then locally the curve looks linear. As we will see again in Chapter 3, with the formalism of vector functions it becomes particularly easy to express the equation of the tangent line to a curve at a point in any number of dimensions. If $\vec{X}(t)$ is a curve and t_0 is not a critical point for the curve, then the equation $\vec{L}_{t_0}(t)$ of the tangent line at t_0 is

$$\vec{L}_{t_0}(t) = \vec{X}(t_0) + (t - t_0)\vec{X}'(t_0), \qquad \text{with } t \in \mathbb{R},$$

or alternatively

$$\vec{L}_{t_0}(u) = \vec{X}(t_0) + u\vec{T}(t_0), \qquad \text{with } u \in \mathbb{R}, \tag{1.8}$$

if we do not wish to confuse the parameter of the parametrized curve $\vec{X}(t)$ and the parameter of the tangent line.

On the other hand, if $t_0 \in I$ is a critical point for the curve $\vec{X}(t)$, the curve may or may not have a tangent line at $t = t_0$. If the following one-sided limits exist, we can define two unit tangent vectors at $\vec{X}(t_0)$:

$$\vec{T}(t_0^+) = \lim_{t \to t_0^+} \vec{T}(t) \qquad \text{and} \qquad \vec{T}(t_0^-) = \lim_{t \to t_0^-} \vec{T}(t).$$

Consequently, we can determine the angle the curve makes with itself at the corner t_0 as

$$\alpha_0 = \cos^{-1}\left(\vec{T}(t_0^+) \cdot \vec{T}(t_0^-)\right).$$

To be more precise, the angle from $\vec{T}(t_0^-)$ to $\vec{T}(t_0^+)$ is the exterior angle of the curve at the corner $t = t_0$. One can only define a tangent line to $\vec{X}(t)$ at $t = t_0$ if $\vec{T}(t_0^-) = \vec{T}(t_0^+)$.

Example 1.2.9 Let $\vec{X}(t) = (t^2, t^3)$. We calculate that $\vec{X}'(t) = (2t, 3t^2)$, and therefore $t = 0$ is a critical point because $\vec{X}'(0) = \vec{0}$. However, if $t \neq 0$, the unit tangent vector is

$$\vec{T}(t) = \frac{1}{\sqrt{4t^2 + 9t^4}}(2t, 3t^2) = \frac{t}{|t|\sqrt{4 + 9t^2}}(2, 3t).$$

Then the right-hand and left-hand side unit tangent vectors are

$$\vec{T}(0^-) = (-1, 0) \qquad \text{and} \qquad \vec{T}(0^+) = (1, 0).$$

These calculations indicate that as t approaches 0, $\vec{X}(t)$ lies in the fourth quadrant but approaches $(0, 0)$ in the horizontal direction $(-1, 0)$. From an intuitive standpoint, we could say that $\vec{X}$ stops at $t = 0$, spins around by $180°$, and moves away from the origin in the direction $(1, 0)$, remaining in the first quadrant.

Example 1.2.10 Let $\vec{X}(t) = (t, |\tan t|)$. We calculate that $\vec{X}'(t) = (1, \text{sign}(t) \sec^2 t)$ and deduce that $t = 0$ is a critical point because $\text{sign}(t)$ is not defined at 0, and hence $\vec{X}$ is not differentiable there. However, we also find that

$$\vec{T}(0^-) = \frac{1}{\sqrt{2}}(1, -1) \qquad \text{and} \qquad \vec{T}(0^+) = \frac{1}{\sqrt{2}}(1, 1).$$

Thus,

$$\vec{T}(0^-) \cdot \vec{T}(0^+) = 0,$$

which shows that $\vec{X}$ makes a right angle with itself at $t = 0$. However, one can tell from the explicit values for $\vec{T}(t_0^-)$ and $\vec{T}(t_0^+)$ that the exterior angle of the curve at $t = 0$ is $\frac{\pi}{2}$.

If t_0 is a critical point, it may still happen that

$$\lim_{t \to t_0} \vec{T}(t)$$

exists, namely when $\vec{T}(t_0^+) = \vec{T}(t_0^-)$, which also means $\alpha_0 = 0$. Then through an abuse of language, we can still talk about the unit tangent vector at that point. As an example of this possibility, consider the curve $\vec{X}(t) = (t^3, t^4)$. We can quickly calculate that

$$\lim_{t \to 0} \vec{T}(t) = (1, 0) = \vec{\imath}.$$

When this limit exists, even though t_0 is a critical point, one can use Equation (1.8) as parametric equations for the tangent line at t_0, replacing $\vec{T}(t_0)$ in Equation (1.8) with $\lim_{t \to t_0} \vec{T}(t)$.

Though we will not make use of the following comments until Chapter 2, we finish this chapter on calculus concepts with a quick review of notation for integration.

In geometry, physics, and other applications, we must sometimes integrate a function along a curve. To this end, we use path and line integrals of scalar functions or vector functions, depending on the particular problem. Since we will make use of them, we remind the reader of the notation for such integrals. Let $\mathcal{C}$ be a curve parametrized by $\vec{X}(t)$ over the interval $[a, b]$. Let $f : \mathbb{R}^2 \to \mathbb{R}$ be a real (scalar) function and $\vec{F} : \mathbb{R}^2 \to \mathbb{R}^2$ a vector field in the plane. Then the path integral of f and the line integral of $\vec{F}$ over the curve $\mathcal{C}$ are respectively

$$\int_{\mathcal{C}} f \, ds \stackrel{\text{def}}{=} \int_a^b f(x(t), y(t)) \sqrt{(x'(t))^2 + (y'(t))^2} dt,$$

$$\int_{\mathcal{C}} \vec{F} \cdot d\vec{s} \stackrel{\text{def}}{=} \int_a^b \vec{F}(x(t), y(t)) \cdot \vec{T}(t) \, dt.$$

PROBLEMS

In Problems 1.2.1 through 1.2.4, calculate, the velocity, the acceleration, the speed, and where defined, the unit tangent vector function of the given parametrized curve.

1. $\vec{X}(t) = (t^2 - 1, t^3 - t)$ for $t \in \mathbb{R}$.

2. The circle $\vec{X}(t) = (R \cos \omega t, R \sin \omega t)$.

3. The circle as parametrized in Equation (1.1) in Example 1.1.10.

4. The epicycloids defined in Problem 1.1.7.

5. Using linear algebra, it is possible to prove that the shortest distance between a point (x_0, y_0) and a line with equation $ax + by + c = 0$ is

$$d = \frac{|ax_0 + by_0 + c|}{\sqrt{a^2 + b^2}}.$$

   However, a calculus proof of this result is also possible using parametrized curves.

   (a) Assuming $a \neq 0$, show that $\vec{X}(t) = (-bt - c/a, at)$ with $t \in \mathbb{R}$ is a parametrization of the line with equation $ax + by + c = 0$.

   (b) Find the value t_0 of t that minimizes the function

   $$f(t) = \|\vec{X}(t) - (x_0, y_0)\| = \sqrt{(-bt - c/a - x_0)^2 + (at - y_0)^2},$$

   which gives the distance between a point $\vec{X}(t)$ on the line and the point (x_0, y_0).

   (c) Show that $f(t_0)$ simplifies to the distance formula given above for d.

6. Let $\vec{p}$ be a fixed point, and let $\vec{l}(t) = \vec{a}t + \vec{b}$ be the parametric equations of a line. Prove that the distance between $\vec{p}$ and the line is

$$\sqrt{\|\vec{b} - \vec{p}\|^2 - \frac{(\vec{a} \cdot (\vec{b} - \vec{p}))^2}{\|\vec{a}\|^2}} = \frac{\|\vec{a} \times (\vec{b} - \vec{p})\|}{\|\vec{a}\|},$$

   where for vectors $\vec{v} = (v_1, v_2, 0)$ and $\vec{w} = (w_1, w_2, 0)$ in the plane, we call $\vec{v} \times \vec{w} = (0, 0, v_1 w_2 - v_2 w_1)$, which is the cross product between $\vec{v}$ and $\vec{w}$ when viewed as vectors in $\mathbb{R}^3$. [See the previous problem.]

7. Find the closest point to $(16, 0.5)$ on the curve $\vec{X}(t) = (t, t^2)$.

8. A quadratic Bézier curve with non-collinear control points $\vec{p}_0$, $\vec{p}_1$, and $\vec{p}_2$ is a curve $\vec{X} : [0, 1] \to \mathbb{R}^2$ with end points $\vec{X}(0) = \vec{p}_0$ and $\vec{X}(1) = \vec{p}_2$, whose component functions are quadratic polynomials, and such that the control point $\vec{p}_1$ satisfies $\vec{X}'(0) = \lambda(\vec{p}_1 - \vec{p}_0)$ and $\vec{X}'(1) = \mu(\vec{p}_2 - \vec{p}_1)$, for some positive constants λ and μ.

   (a) Prove that the above conditions imply that $\lambda = \mu = 2$.

   (b) Prove that a parametrization for the quadratic Bézier is $\vec{X}(t) = (1 - t)^2 \vec{p}_0 + 2t(1 - t)\vec{p}_1 + t^2 \vec{p}_0$ with $t \in [0, 1]$.

9. Find the tangent line to $\vec{X}(t) = (\cos(2t), \sin t)$ at $t = \pi/6$. Also give the components of the unit tangent vector there.

10. We say that a plane curve C parametrized by $\vec{X} : I \to \mathbb{R}^2$ intersects itself at a point p if there exist $u \neq t$ in I such that $\vec{X}(t) = \vec{X}(u) = p$. Consider the parametric curve $\vec{X}(t) = (t^2, t^3 - t)$ for $t \in \mathbb{R}$. Find the point(s) of self-intersection. Also determine the angle at which the curve intersects itself at those point(s) by finding the angle between the unit tangent vectors corresponding to the distinct parameters.

11. The parametrized curve $\vec{X}(t) = ((1 + 2\cos t) \cos t, (1 + 2\cos t) \sin t)$ intersects itself at one point. Find this point of intersection and find the angle of self-intersection (i.e., the acute angle between the tangent lines corresponding to the two different parameters t_1 and t_2 of self-intersection).

12. For how many points on the Lissajous curve $\vec{X}(t) = (\cos(3t), \sin(2t))$ does the tangent line go through the point $(3, 0)$?

13. Consider the Lissajous figure $\vec{X}(t) = (\cos(mt), \sin(nt))$. Prove that this curve intersects itself $2mn - (m + n)$ times.

14. What can be said about a parametrized curve $\vec{X}(t)$ that has the property that $\vec{X}''(t)$ is identically 0?

15. Consider the linear spiral given in Example 1.2.4. Suppose that we reparametrize the spiral by $t = f(u) = \tan(u)$ with $0 \leq u < \pi/2$. (a) Show that this reparametrization is regular and positively oriented. (b) Prove that with this reparametrization the angle between the position and the velocity vectors is exactly u.

16. Find the arc length function along the parabola $y = x^2$, using as the origin $s = 0$.

17. Sketch the parametrized curve $\vec{X}(t) = (3t^2, 3t - t^3)$ near the origin and calculate the length of the loop in the curve.

18. Consider the curve $\vec{X}(t) = (t^2, t^3)$ with $t \in \mathbb{R}$. Show that the reparametrization of this curve by arclength is

$$\vec{X}(s) = \left(\frac{1}{9}(27s + 8)^{2/3} - \frac{4}{9}, \operatorname{sign}(s) \left(\frac{1}{9}(27s + 8)^{2/3} - \frac{4}{9} \right)^{3/2} \right).$$

19. (*) Consider the cycloid introduced in Example 1.1.13 given by $\vec{X}(t) = (t - \sin t, 1 - \cos t)$. Prove that the path taken by a point on the edge of a rolling wheel of radius 1 during one rotation has length 8.

20. Calculate the arc length function of the curve $\vec{X}(t) = (t^2, \ln t)$, defined for $t > 0$.

21. Let $\vec{X} : I \to \mathbb{R}^2$ be a regular parametrized curve, and let $\vec{p}$ be a fixed point. Suppose that the closest point on the curve $\vec{X}$ to $\vec{p}$ occurs at $t = t_0$, which is neither of the ends of I. Prove that the line between $\vec{p}$ and the point closest to $\vec{p}$, namely $\vec{X}(t_0)$, is perpendicular to the curve $\vec{X}$ at $t = t_0$.

22. Prove the differentiation formulas in Proposition 1.2.3.

23. Prove that if $\vec{X}(t)$ is a curve that satisfies $\vec{X} \cdot \vec{X}' = 0$ for all values of t, then $\vec{X}$ is a circle. [Hint: Use $\|\vec{X}\|^2 = \vec{X} \cdot \vec{X}$ and apply Proposition 1.2.3 to calculate the derivative of $\|\vec{X}\|^2$.]

24. Consider the ellipse given by $\vec{X}(t) = (a\cos t, b\sin t)$. Find the extremum values of the speed function.

25. Consider the linear spiral of Example 1.2.4. Let $n \geq 0$ be a nonnegative integer. Prove that the length of the nth derivative vector function is given by

$$\|\vec{X}^{(n)}(t)\| = \sqrt{n^2 + t^2}.$$

26. Consider the exponential spiral $\vec{x}(t) = (ae^{bt}\cos t, ae^{bt}\sin t)$ where a and b are constants. Calculate the arc length $s(t)$ function of $\vec{x}(t)$. Reparametrize the spiral by arc length.

27. Consider again the exponential spiral $\vec{x}(t) = (ae^{bt}\cos t, ae^{bt}\sin t)$, with $a > 0$ and $b < 0$.

    (a) Prove that as $t \to +\infty$, $\lim \vec{X}(t) = (0,0)$.
    (b) Show that $\vec{X}'(t) \to (0,0)$ as $t \to +\infty$ and that for any t_0,

$$\lim_{t\to\infty} \int_{t_0}^{t} \|\vec{X}'(u)\|\, du < \infty,$$

    i.e., any part of the exponential spiral that "spirals" in toward the origin has finite arc length.

28. Recall that polar and Cartesian coordinate systems are related as follows:

$$\begin{cases} x = r\cos\theta, \\ y = r\sin\theta, \end{cases} \quad \text{and} \quad \begin{cases} r = \sqrt{x^2 + y^2}, \\ \tan\theta = \frac{y}{x} \end{cases}$$

    Suppose C is a curve in the plane parametrized using polar coordinate functions $r = r(t)$ and $\theta = \theta(t)$ so that one has a parametrization in Cartesian coordinates as

$$\vec{x}(t) = (x(t), y(t)) = (r(t)\cos(\theta(t)), r(t)\sin(\theta(t))).$$

    (a) Express x' and y' in terms of r, θ, r', and θ'.
    (b) Express r' and θ' in terms of x, y, x', and y'.
    (c) Express $\|\vec{x}\|$ and $\|\vec{x}'\|$ in terms of polar coordinate functions.

    (All coordinate functions are viewed as functions of t and so r', for example, is a shorthand for $r'(t)$, the derivative of r with respect to t.)

29. (**ODE**) Suppose a curve C is parametrized by $\vec{x}(t)$ such that $\vec{x}(t)$ and $\vec{x}''(t)$ always make a constant angle with each other. Find the shape of this curve. [Hint: Use polar coordinates and the results of Problem 1.2.28.]

1.3　Curvature

Let $\vec{X} : I \to \mathbb{R}^2$ be a twice-differentiable parametrization of a curve $\mathcal{C}$. As we saw in the previous section, the decomposition of the velocity vector $\vec{X}' = s'\vec{T}$ into magnitude and unit tangent direction separates the geometric invariant (the unit tangent $\vec{T}$) from the dynamical aspect (the speed $s'(t)$) of the parametrization. Taking one more derivative, we obtain the decomposition

$$\vec{X}'' = s''\vec{T} + s'\vec{T}'. \tag{1.9}$$

The first component describes a tangential acceleration, while the second component describes a rate of change of the tangent direction, or in other words, how much the curve is "curving." Since $\vec{T}$ is a unit vector, we always have $\vec{T} \cdot \vec{T} = 1$. Therefore,

$$(\vec{T} \cdot \vec{T})' = 0 \implies 2\vec{T} \cdot \vec{T}' = 0,$$
$$\implies \vec{T} \cdot \vec{T}' = 0.$$

Thus, $\vec{T}'$ is perpendicular to $\vec{T}$.

Just as there are two unit tangent vectors at a regular point of the curve, there are two unit normal vectors as well. Given a particular parametrization, there is no naturally preferred way to define "the" unit normal vector, so we make a choice.

Definition 1.3.1 Let $\vec{X} : I \to \mathbb{R}^2$ be a regular parametrized curve and $\vec{T} = (T_1, T_2)$ the tangent vector at a regular value $\vec{X}(t)$. The *unit normal vector* $\vec{U}$ is

$$\vec{U}(t) = (-T_2(t), T_1(t)).$$

Equivalently, $\vec{U}$ is the vector function obtained by rotating $\vec{T}$ by $\frac{\pi}{2}$. In still other words, if we view the xy-plane in three-dimensional space with x-, y-, and z-axes oriented in the usual way, and if we call $\vec{k} = (0, 0, 1)$ the unit vector along the positive z-axis, then $\vec{U} = \vec{k} \times \vec{T}$.

Since $\vec{T}' \perp \vec{T}$, the vector function $\vec{T}'$ is a multiple of $\vec{U}$ at all t. This leads to the definition of curvature.

Definition 1.3.2 Let $\vec{X} : I \to \mathbb{R}^2$ be a regular twice-differentiable parametric curve. The *curvature* function $\kappa_g(t)$ is the unique real-valued function defined by

$$\vec{T}'(t) = s'(t)\kappa_g(t)\vec{U}(t). \tag{1.10}$$

This definition only gives $\kappa_g(t)$ implicitly, but we can obtain a formula for it as follows. Equation (1.9) becomes

$$\vec{X}'' = s''\vec{T} + (s')^2\kappa_g\vec{U}. \tag{1.11}$$

Viewing the plane as the xy-plane in three-space, we have $\vec{T} \times \vec{U} = \vec{k}$. Thus,

$$
\begin{aligned}
\vec{X}' \times \vec{X}'' &= (s'\vec{T}) \times (s''\vec{T} + (s')^2\kappa_g\vec{U}) \\
&= s's''\vec{T} \times \vec{T} + (s')^3\kappa_g\vec{T} \times \vec{U} \\
&= (s')^3\kappa_g\vec{k}.
\end{aligned}
$$

But $s'(t) = \|\vec{X}'(t)\|$, which leads to

$$\kappa_g(t) = \frac{(\vec{X}'(t) \times \vec{X}''(t)) \cdot \vec{k}}{\|\vec{X}'(t)\|^3} = \frac{x'(t)y''(t) - x''(t)y'(t)}{(x'(t)^2 + y'(t)^2)^{3/2}}. \tag{1.12}$$

As straightforward as Formula (1.12) may seem, we offer a word of wisdom to the student. As with many calculations in differential geometry, it is very useful to simplify algebraic and trigonometric expressions as much as possible. Not doing so can quickly lead to intractable expressions.

The reader might well wonder why in the definition of $\kappa_g(t)$ we include the factor $s'(t)$. It is not hard to confirm (see Problem 1.3.20) that including the $s'(t)$ factor renders $\kappa_g(t)$ independent of any regular reparametrization (except perhaps up to a change of sign).

One should note at this point that if a curve $\vec{X}(s)$ is parametrized by arc length, then by Equations (1.6) and (1.9), the velocity and acceleration have the simple expressions

$$\vec{X}'(s) = \vec{T}(s),$$
$$\vec{X}''(s) = \kappa_g(s)\vec{U}(s).$$

The curvature of a curve at a point has an interpretation that is easy to visualize. We explore this interpretation in the next three examples.

 Example 1.3.3 Consider the vector function that describes a particle moving on a circle of radius R with a nonzero and constant angular velocity $\omega > 0$ given by $\vec{X}(t) = (R\cos\omega t, R\sin\omega t)$. In order to calculate the curvature, we need

$$\vec{X}'(t) = (-R\omega\sin\omega t, R\omega\cos\omega t),$$
$$\vec{X}''(t) = (-R\omega^2\cos\omega t, -R\omega^2\sin\omega t).$$

Thus, using the coordinate form (right-most expression) in Equation (1.12), we get

$$\kappa_g(t) = \frac{R^2\omega^3 \sin^2 t + R^2\omega^3 \cos^2 t}{(R^2\omega^2 \sin^2 t + R^2\omega^2 \cos^2 t)^{3/2}} = \frac{R^2\omega^3}{R^3\omega^3} = \frac{1}{R}.$$

The curvature of the circle is a constant function that is equal to the reciprocal of the radius for all t, regardless of the nonzero angular velocity ω.

The study of trajectories in physics gives particular names for the components of the first and second derivatives of a vector function in the basis $\{\vec{T}, \vec{U}\}$. We already saw that the function $s'(t)$ in Equation (1.6) is called the speed. In Equation (1.11), however, the function $s''(t)$ is called the tangential acceleration, while the quantity $s'(t)^2 \kappa_g(t)$ is called the centripetal acceleration. Example 1.3.3 connects the curvature function to the reciprocal of a radius, so if $\kappa_g(t) \neq 0$, then we define the *radius of curvature* to $\vec{X}(t)$ at t as the function $R(t) = \frac{1}{\kappa_g(t)}$. Using the common notation $v(t) = s'(t)$ for the speed function, one recovers the common formula for centripetal acceleration of

$$(s')^2 \kappa_g = \frac{v^2}{R}.$$

In introductory physics courses, this formula is presented only in the context of circular motion. With differential geometry at our disposal, we see that centripetal acceleration is always equal to v^2/R at all points on a curve where $\kappa_g \neq 0$, where by R one means the radius of curvature.

In contrast to physics where one tends to refer to the radius of curvature, the curvature function is a more geometrically natural quantity to study. Indeed, the radius of curvature of a line segment is undefined even though line segments are such useful geometric objects. On the other hand, a radius of curvature is 0 (and hence the curvature is undefined) at degenerate curves that are points or at critical points.

Example 1.3.4 Let I be a real interval and consider the plane curve given as the graph $y = f(x)$ of a twice-differentiable function $f : I \to \mathbb{R}$. We can parametrize this as $\vec{X}(t) = (t, f(t))$ for $t \in I$. A quick application of (1.12) gives a curvature function of

$$\kappa_g(t) = \frac{f''(t)}{(1 + f'(t)^2)^{3/2}}.$$

Expressing the same function back in the variable x, we have $\kappa_g(x) = f''(x)/(1 + f'(x)^2)^{3/2}$. This calculation underscores that $\kappa_g(x) = 0$

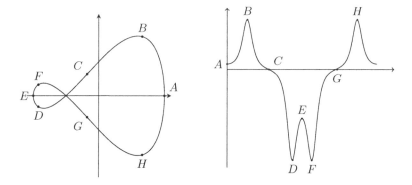

Figure 1.6: Curvature function example.

precisely when $f''(x) = 0$. Recalling from calculus that $\sqrt{1 + f'(x)^2}$ is the speed function $s'(x)$ along the curve, we see from yet another angle that the curvature function of a function graph is $\kappa_g = f''(x)/s'(x)^3$.

Note that the curvature function κ_g may be positive or negative. The previous example along with the typical calculus terminology motivates the following definition.

Definition 1.3.5 Let C be a regular curve of class C^2. An *inflection point* of C is any point where the curvature function κ_g changes sign.

Example 1.3.6 To further develop an intuition for curvature, consider the parametric curve $\vec{X}(t) = (2\cos t, \sin(2t) + \sin t)$ for $t \in [0, 2\pi]$.

Figure 1.6 shows side by side the locus of the curve and the graph of its curvature function, along with a few labeled points to serve as references. First, we see that the curvature function is positive when the curve turns to the left away from the unit tangent vector $\vec{T}$ and negative when the curve turns to the right. Though it is possible for the curvature to become zero but not change signs, at the points labeled C and G, the curvature function changes sign. These are inflection points. Precisely here, the curve changes from curving to the left of $\vec{T}$ to curving to the right of $\vec{T}$, or vice versa. The curve is locally curving the "most" when $\kappa_g(t)$ has a maximum and $\kappa_g(t) > 0$ or when $\kappa_g(t)$ has a minimum and $\kappa_g(t) < 0$. These cases correspond respectively to curving the most to the left and to the right. Besides a situation when the curvature is 0, the curve is locally curving the "least" when $\kappa_g(t)$ has a minimum and $\kappa_g(t) > 0$ or when $\kappa_g(t)$ has a maximum and $\kappa_g(t) < 0$. Figure 1.6 illustrates all five of the different possibilities just described.

Proposition 1.3.7 *A regular parametrized curve* $\vec{X} : I \to \mathbb{R}^2$ *has curvature* $\kappa_g(t) = 0$ *for all* $t \in I$ *if and only if the locus of* $\vec{X}$ *is a line segment.*

Proof: If the locus of $\vec{X}$ traces out a line segment, then (perhaps after a regular reparametrization) we can write $\vec{X}(t) = \vec{a} + \varphi(t)\vec{b}$, where $\varphi(t)$ is a differentiable real function with $\varphi'(t) \neq 0$. Then

$$\vec{X}'(t) = \varphi'(t)\vec{b},$$
$$\vec{X}''(t) = \varphi''(t)\vec{b}.$$

Thus, by Equation (1.12),

$$\kappa_g(t) = \frac{(\varphi'(t)\varphi''(t)\vec{b} \times \vec{b}) \cdot \vec{k}}{(|\varphi'(t)| \, \|\vec{b}\|)^{3/2}} = 0$$

because $\vec{b} \times \vec{b} = \vec{0}$.

Conversely, if $\vec{X}(t) = (x(t), y(t))$ is a curve such that $\kappa_g(t) = 0$, then

$$x'(t)y''(t) - x''(t)y'(t) = 0.$$

We need to find solutions to this differential equation or determine how solutions are related. Since $\vec{X}$ is regular, $\vec{X}' \neq \vec{0}$ for all $t \in I$. Let I_1 be an interval where $y'(t) \neq 0$. Over I_1, we have

$$\frac{x'y'' - x''y'}{(y')^2} = \frac{d}{dt}\left(\frac{x'}{y'}\right) = 0 \Longrightarrow \frac{x'}{y'} = C,$$

where C is a constant. Thus, $x' = Cy'$ for all t. Integrating with respect to t we deduce $x(t) = Cy(t) + D$. Similarly, over an interval I_2 where $x'(t) \neq 0$, we deduce that $y(t) = Ax(t) + B$ for some constants A and B. Since I can be covered with intervals where $x'(t) \neq 0$ or $y'(t) \neq 0$, we deduce that the locus of $\vec{X}$ is a piecewise linear curve. However, since $\vec{X}$ is regular, it has no corners, and hence its locus is a line segment. $\square$

Our formula for curvature given in (1.12) arose from calculating $\vec{X}''$ as the derivative of the expression $\vec{X}' = s'\vec{T}$, which involved finding an expression for $\frac{d}{dt}\vec{T}(t)$. Taking the third derivative of $\vec{X}(t)$ and using the decomposition in (1.11), we get

$$\vec{X}''' = s'''\vec{T} + (3s''s'\kappa_g + (s')^2\kappa_g')\vec{U} + (s')^2\kappa_g\vec{U}'.$$

We now need an expression for the derivative $\vec{U}'(t)$. By definition, for all t, the set $\{\vec{T}, \vec{U}\}$ forms an orthonormal basis and hence

$$\vec{U}' = (\vec{U}' \cdot \vec{T})\vec{T} + (\vec{U}' \cdot \vec{U})\vec{U}.$$

However, since $\vec{U}(t)$ is a unit vector function, we have $\vec{U} \cdot \vec{U}\,' = 0$ for all t. Furthermore, since $\vec{U} \cdot \vec{T} = 0$ for all t, we deduce that

$$\vec{U}\,' \cdot \vec{T} = -\vec{U} \cdot \vec{T}\,' = -s'\kappa_g.$$

Consequently,
$$\vec{U}\,'(t) = -s'(t)\kappa_g(t)\vec{T}(t).$$

Since $\{\vec{T}, \vec{U}\}$ forms an orthonormal basis of $\mathbb{R}^2$ for all t, every higher derivative of $\vec{X}(t)$, if it exists, can be expressed as a linear combination of $\vec{T}$ and $\vec{U}$. The components of $\vec{X}^{(n)}(t)$ in terms of $\vec{T}$ and $\vec{U}$ involve sums of products of derivatives of $s(t)$ and $\kappa_g(t)$. Furthermore, if $\vec{X}$ is parametrized by arc length, then the coefficients of $\vec{X}^{(n)}(s)$ only involve sums of powers of derivatives of $\kappa_g(s)$.

We note that if a curve $\vec{X} : [-a, a] \to \mathbb{R}^2$ is reparametrized by $\vec{X}(-t)$, then the modified curvature function is $-\kappa_g(t)$. Thus, the sign of the curvature depends on what one might call the "orientation" of the curve. Excluding this technicality, curvature has a physical interpretation that one can eyeball on particular curves. If a curve is almost a straight line, then the curvature is close to 0, but if a regular curve bends tightly along a certain section, then the curvature is high (in absolute value). Of particular interest are points where the curvature reaches a local extremum.

Definition 1.3.8 Let C be a regular curve parametrized by $\vec{X} : I \to \mathbb{R}^2$ with curvature function $\kappa_g(t)$. A *vertex* of the curve C is a point $P = \vec{X}(t_0)$ where the curvature function $\kappa_g(t)$ attains an extremum.

By the First Derivative Test, extrema of $\kappa_g(t)$ occur at $t = t_0$, if $\kappa_g'(t)$ changes sign through t_0. This obviously must occur where $\kappa_g'(t_0) = 0$ but the converse is not necessarily true.

PROBLEMS

In Problems 1.3.1 through 1.3.7, calculate the curvature function of the given curve.

1. The curve parametrized by $\vec{X}(t) = (t^m, t^n)$ for $t \geq 0$.

2. The ellipse: $\vec{X}(t) = (a\cos t, b\sin t)$ for $t \in [0, 2\pi]$.

3. The cycloid: $\vec{X}(t) = (t - \sin t, 1 - \cos t)$ for $t \in \mathbb{R}$.

4. The astroid: $\vec{X}(t) = (\cos^3 t, \sin^3 t)$ for $t \in [0, 2\pi]$.

5. The curve parametrized by $\vec{X}(t) = (e^{\cos t}, e^{\sin t})$ for $t \in [0, 2\pi]$.

6. The linear spiral: $\vec{X}(t) = (t\cos t, t\sin t)$ for $t \in \mathbb{R}^{\geq 0}$.

7. The flower curve: $\vec{X}(t) = (\sin(nt)\cos t, \sin(nt)\sin t)$ for $t \in [0, 2\pi]$.

8. In the pictures below, the first row shows three regular plane curves. The second row shows three function graphs of curvature functions.

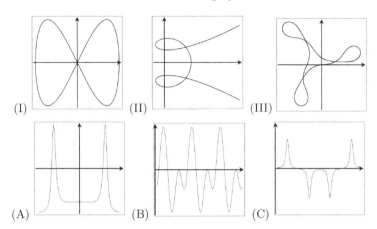

(I)   (II)   (III)

(A)   (B)   (C)

Match the parametrized curve with its corresponding curvature function. Give a sentence of explanation to justify your association.

9. The *cochleoid* is parametrized by $\vec{X}(t) = \left(\dfrac{a \sin t \cos t}{t}, \dfrac{a \sin^2 t}{t} \right)$ for $t \in \mathbb{R}^*$ and $\vec{X}(0) = (a, 0)$, where a is a constant. (a) Calculate the curvature function of the cochleoid for $t \in \mathbb{R}^*$. (b) Determine the values $\vec{X}'(0)$ and $\vec{X}''(0)$ that make $\vec{X}$ into a regular C^2 curve.

10. Consider the curve $\vec{X}(t) = (t^2, t^3 - at)$, where a is a real number. Calculate the curvature function $\kappa_g(t)$ and determine where $\kappa_g(t)$ has extrema (vertices).

11. Find the vertex (there is only one) of the curve $y = e^x$. Prove also that the curvature of the curve $y = e^x$ goes to 0 as $x \to -\infty$ or $x \to +\infty$.

12. Consider the family of curve $\vec{X}_n(t) = (t^n, t^{n+1})$ with $n \in \mathbb{N}^*$. Calculate the curvature function $\kappa_{g,n}(t)$ of $\vec{X}_n(t)$ and show that for all $t \neq 0$ we have $\lim_{n \to \infty} \kappa_{g,n}(t) = 0$. Explain the geometric meaning of this result.

13. Find an equation involving $f(t)$ that describes where the vertices occur for a function graph $\vec{X}(t) = (t, f(t))$. Find an example of a function graph where $\kappa'_g(t) = 0$ but $f'(t) \neq 0$ and vice versa.

14. Prove that the graph of a polar function $r = f(\theta)$ at angle θ has curvature
$$\kappa_g(\theta) = \frac{2f'(\theta)^2 + f(\theta)^2 - f(\theta)f''(\theta)}{(f'(\theta)^2 + f(\theta)^2)^{3/2}}.$$

15. Find the vertices of an ellipse with half-axes $a \neq b$ and calculate the curvature at those points.

16. Calculate the curvature function for all the Lissajous figures: $\vec{x}(t) = (\cos(mt), \sin(nt))$, with $t \in [0, 2\pi]$.

17. Prove by direct calculation that the following formula holds for the ellipse $\vec{X}(t) = (a\cos t, b\sin t)$:

$$\int_0^{2\pi} \kappa_g \, ds = 2\pi.$$

[Hint: Use a substitution involving $\tan\theta = \frac{a}{b}\tan t$.]

18. Calculate the curvature function for the cardioid: $\vec{x}(t) = ((1 - \cos t)\cos t, (1 - \cos t)\sin t)$.

19. Find the vertices of the exponential curve $\vec{X}(t) = (t, e^t)$ for $t \in \mathbb{R}$. Interpret this result in terms of radius of curvature on the curve.

20. Let $\vec{X} : I \to \mathbb{R}^2$ be a parametrized curve and $f : J \to I$ a surjective function so that f makes $\vec{\xi} = \vec{X} \circ f$ a regular reparametrization of $\vec{X}$. Call $t_0 = f(u_0)$ so that $\vec{\xi}(u_0) = \vec{X}(t_0)$. Prove that

$$\kappa_{g,\vec{X}}(t_0) = \begin{cases} \kappa_{g,\vec{\xi}}(u_0) & \text{if } f'(u_0) > 0, \\ -\kappa_{g,\vec{\xi}}(u_0) & \text{if } f'(u_0) < 0. \end{cases}$$

21. Let I be a closed and bounded interval. We define a *parallel curve* to a parametrized curve $\vec{X} : I \to \mathbb{R}^2$ as a curve that can be parametrized by $\vec{\gamma}_\varepsilon(t) = \vec{X}(t) + \varepsilon\vec{U}(t)$, where ε is a real number. Suppose that $\vec{X}$ is a regular curve of class C^2 and assume $\varepsilon \neq 0$. Prove that $\vec{\gamma}_\varepsilon$ is regular if and only if $\frac{1}{\varepsilon} \notin [\kappa_m, \kappa_M]$, where

$$\kappa_m = \min_{t \in I}\{\kappa_g(t)\} \qquad \text{and} \qquad \kappa_M = \max_{t \in I}\{\kappa_g(t)\}.$$

22. Let C be a curve in $\mathbb{R}^2$ defined as the solution to an equation $F(x,y) = 0$. Use implicit differentiation to prove that at any point on this curve, the curvature of C is given by

$$\kappa_g = \frac{1}{(F_x^2 + F_y^2)^{3/2}} \begin{vmatrix} F_{xx} & F_{xy} & F_x \\ F_{yx} & F_{yy} & F_y \\ F_x & F_y & 0 \end{vmatrix}.$$

23. Find the vertices of the curve given by the equation $x^4 + y^4 = 1$.

24. *Pedal Curves.* Let C be a regular curve and A a fixed point in the plane. The *pedal curve* of C with respect to A is the locus of points of intersection of the tangent lines to C and lines through A perpendicular to these tangents. Given the parametrization $\vec{X}(t)$ of a curve C, provide a parametric formula for the pedal curve to C with respect to A. Find an explicit parametric formula for the pedal curve of the unit circle with respect to $(1,0)$.

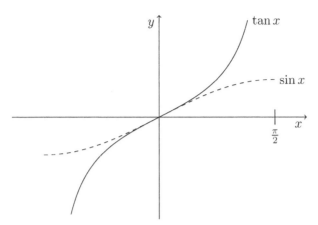

Figure 1.7: Functions $\sin x$ and $\tan x$ have contact order 1 at the origin.

1.4 Osculating Circles, Evolutes, Involutes

Classical differential geometry introduces the notion of order of contact to measure the degree of intersection between two curves or surfaces at a particular intersection point. See Struik in [33, p. 23] for this classical definition. In this text, we provide an alternate but equivalent definition that is more relevant to our approach to always describing curves via a parametrization.

As a motivating example, consider the real functions $f(x) = \sin x$ and $g(x) = \tan x$ near $x_0 = 0$. The graphs of f and g intersect at $(0, 0)$. However, we notice in Figure 1.7 that not only do they intersect, but they share the same tangent line. Even further, consider their sixth-order Taylor polynomial at 0:

$$\sin x \approx x - \frac{1}{6}x^3 + \frac{1}{120}x^5$$
$$\tan x \approx x + \frac{1}{3}x^3 + \frac{2}{15}x^5.$$

In particular, both $\sin x$ and $\tan x$ have the same second-order Taylor polynomial at $x_0 = 0$, but their third-order Taylor polynomials differ. We say that $f(x)$ and $g(x)$ have contact of order 2 at $x_0 = 0$.

Definition 1.4.1 Two functions $f(x)$ and $g(x)$ defined on a neighborhood of x_0 have *contact of order* k if for $i = 0, 1, \ldots, k$, the derivatives $f^{(i)}(x_0)$ and $g^{(i)}(x_0)$ exist and $f^{(i)}(x_0) = g^{(i)}(x_0)$. We say that f and g have contact of strict order k at x_0 if they have contact of order k but do not have contact of order $k + 1$.

Note that a simple monomial function $f(x) = x^n$ with the x-axis at $(0,0)$ has contact of strict order $n-1$. This is because $f^{(i)}(0) = 0$ for $i = 0, 1, \ldots, n-1$ but $f^{(n)}(0) = n!$, which is not equal to 0.

As we attempt to generalize this notion to order of contact between two parametrized curves in the plane, we are immediately met with two obstacles. First, in general the parameter for one parametrized curve has no connection, geometric or otherwise, to the parameter for the second curve. In the definition of order of contact for real-valued functions over the reals, the parameter x in an open interval of $\mathbb{R}$ is used for both functions $f(x)$ and $g(x)$. Secondly, parametrized curves are usually not described as functions with respect to some frame. Consequently, we must make a choice of parameter on both curves simultaneously that has geometric meaning. The typical approach is to use one of the curves parametrized by arc length as a reference.

Definition 1.4.2 Suppose $C_1 : \vec{\alpha}(t)$ and $C_2 : \vec{\beta}(u)$ are two parametrized curves that intersect at a point P, which corresponds to where $t = t_0$ and $u = u_0$. Reparametrize C_1 by arc length and let s_0 be such that $P = \vec{\alpha}(s_0)$. Let $u(s)$ be the function such that the projection of $\vec{\alpha}(s)$ onto C_2 is located at $\vec{\beta}(u(s))$. Then we say that C_1 and C_2 have contact of order n at P if they are of class C^n over an open interval around P and

$$\vec{\alpha}^{(i)}(s_0) = \frac{d^i}{ds^i} \vec{\beta}(u(s)) \Big|_{s_0}$$

for all $0 \leq i \leq n$. Furthermore, C_1 and C_2 have contact of strict order n if they do not also have contact of order $n+1$.

The intuitive picture for order of contact indicates that two intersecting curves with contact of order 1 have the same tangent line at the intersection point. In particular, we leave it as an exercise to prove the fact that a curve and its tangent line have order of contact 1. In contrast, an intersection point between two curves with contact of strict order 0 is said to be a *transversal* intersection.

Note that since the concept of order of contact refers to orders of differentiation, if two curves are not both of class C^n near a point P it does not make sense to discuss contact of order n.

At first glance, this definition of order of contact seems asymmetrical. However, it is possible to prove that using $\vec{\beta}$ as the reference curve and the arc length of C_2 as the reference parameter is equivalent to Definition 1.4.2. (In fact, for a given nonnegative integer n, on the set of parametrized curves in the plane, the relation of contact of order n is an equivalence relation, i.e., is reflexive, symmetric, and transitive. In contrast, the relation of contact of strict order n is reflexive and symmetric but not necessarily transitive.)

Definition 1.4.3 Let C be a curve parametrized by $\vec{X} : I \to \mathbb{R}^2$, and let t_0 be a regular value of the curve. The *osculating circle* to C at the point t_0 is a circle that has contact of order 2 with C at $\vec{X}(t_0)$.

Proposition 1.4.4 *Let $C : \vec{X} : I \to \mathbb{R}^2$ be a parametrized curve and t_0 a regular value. Suppose that $\vec{X}(t)$ is twice differentiable at t_0 and that $\kappa_g(t_0) \neq 0$. Then,*

1. *There exists a unique osculating circle to C at $\vec{X}(t_0)$;*

2. *It is given by the following vector function:*

$$\vec{\gamma}(t) = \vec{X}(t_0) + \frac{1}{\kappa_g(t_0)} \vec{U}(t_0) + \frac{1}{\kappa_g(t_0)} \Big((\sin t)\vec{T}(t_0) - (\cos t)\vec{U}(t_0) \Big).$$

Proof: Without loss of generality, we can assume that $\vec{X}$ is parametrized by arc length s and that we are looking for the osculating circle at $s = 0$. Call $\vec{\gamma}(u)$ the parametrization of the proposed osculating circle, which must have the form

$$\vec{\gamma}(u) = (a + R\cos(u), b + R\sin(u)).$$

For ease of the proof, we will allow R to be any nonzero real number. We assume that $u = u_0$ at the point of contact and that near u_0, the function $u(s)$ gives the projection of $\vec{X}(s)$ onto the curve $\vec{\gamma}$. By Definition 1.4.2, in order for there to exist an osculating circle, the parameters a, b, and R and the function $u(s)$ must satisfy

$$\vec{X}(0) = \vec{\gamma}(u(0)), \qquad \vec{X}\,'(0) = \frac{d}{ds}\vec{\gamma}(u(s))\Big|_0, \qquad \vec{X}\,''(0) = \frac{d^2}{ds^2}\vec{\gamma}(u(s))\Big|_0,$$

which in coordinates is equivalent to

$$a + R\cos(u(0)) = x(0), \qquad (1.13)$$
$$b + R\sin(u(0)) = y(0), \qquad (1.14)$$
$$-Ru'(0)\sin(u(0)) = x'(0), \qquad (1.15)$$
$$Ru'(0)\cos(u(0)) = y'(0), \qquad (1.16)$$
$$-Ru''(0)\sin(u(0)) - Ru'(0)^2\cos(u(0)) = x''(0), \qquad (1.17)$$
$$Ru''(0)\cos(u(0)) - Ru'(0)^2\sin(u(0)) = y''(0). \qquad (1.18)$$

Since $\vec{X}(s)$ is parametrized by arc length, $x'(0)^2 + y'(0)^2 = 1$, from which we conclude that $|u'(0)| = 1/R$. Without loss of generality, we can assume that $u'(0)$ and R have the same sign, so that $u'(0) = 1/R$. Since the curve is parametrized by arc length, $x'(s)^2 + y'(s)^2 = 1$ for all s. Taking a derivative of this equation with respect to s and evaluating at $s = 0$, we deduce that $x'(0)x''(0) + y'(0)y''(0) = 0$. This relation, along with Equations (1.15) through (1.18) leads to $u''(0) = 0$.

We obtain the value of R by noting that after calculation

$$\kappa_g(0) = x'(0)y''(0) - x''(0)y'(0) = R^2(u'(0))^3 = \frac{1}{R}.$$

Using Equations (1.15) and (1.16), we find that

$$(\cos(u(0)), \sin(u(0))) = (y'(0), -x'(0)) = -\vec{U},$$

since $\vec{X}$ is parametrized by arc length. Thus the center of the circle $\vec{\gamma}$ is

$$(a, b) = (x(0), y(0)) - R(\cos(u(0)), \sin(u(0))) = \vec{X}(0) + \frac{1}{\kappa_g(0)}\vec{U}(0).$$

Finally, a quick check that we leave for the reader is to show that given the above values for a, b, and R, Equations (1.13) through (1.18) are redundant given the fact that $\vec{X}(s)$ is parametrized by arc length. These results prove part 1 of the proposition.

Part 2 follows easily from part 1. Since $\vec{X}$ is parametrized by arc length, the unit tangent vector is just $\vec{T}(0) = (x'(0), y'(0))$, giving also $\vec{U}(0) = (-y'(0), x'(0))$. In the solutions for Equations (1.13) and (1.14), we see that the center of the osculating circle is at

$$(a, b) = \left(x(0) - R\cos(\tau(0)), y(0) - R\sin(\tau(0))\right) = \vec{X}(0) + R\vec{U}(0).$$

Furthermore, for any curve parametrized by arc length, $\kappa_g(s) = x'(s)y''(s) - x''(s)y'(s)$. Thus, $1/R = \kappa_g(0)$. □

In light of Proposition 1.4.4 and Example 1.3.3, the curvature function $\kappa_g(t)$ of a plane curve $\vec{X}$ has a nice physical interpretation, namely, the reciprocal of the radius of the osculating circle. A higher order of contact indicates a better geometric approximation, and hence, since there is a unique osculating circle, it is, in a geometric sense, the best approximating circle to a curve at a point. Furthermore, in Problem 1.4.10 of this section, we prove that to a curve $\vec{X}$ at $t = t_0$, there does not necessarily exist a touching circle with order of contact greater than 2. Thus, the curvature is the inverse of the radius of the best approximating circle to a curve at a point.

(From a physics point of view, we obtain an additional confirmation of the above interpretation by considering the units of the curvature function. If we view the coordinate functions $x(t)$ and $y(t)$ with the unit of meters and t in any unit, Equation (1.12) gives the unit of 1/meter for $\kappa_g(t)$.)

Definition 1.4.5 Let $\vec{X}$ be a parametrized curve, and let $t = t_0$ be a regular point that satisfies the conditions in Proposition 1.4.4. The center of the osculating circle at $t = t_0$ is called the *center of curvature*.

Definition 1.4.6 The *evolute* of a curve C is the locus of the centers of curvature.

Proposition 1.4.7 *Let $\vec{X} : I \to \mathbb{R}^2$ be a regular parametric plane curve that is of class C^2, i.e., has a continuous second derivative. Let I' be a subinterval of I over which $\kappa_g(t) \neq 0$. Then over the interval I', the evolute of $\vec{X}$ has the following parametrization:*

$$\vec{E}(t) = \vec{X}(t) + \frac{1}{\kappa_g(t)} \vec{U}(t).$$

Example 1.4.8 Consider the parabola $y = x^2$ with parametric equations $\vec{X}(t) = (t, t^2)$. We calculate the curvature with the following steps:

$$\vec{X}'(t) = (1, 2t),$$
$$\vec{X}''(t) = (0, 2),$$
$$s'(t) = \sqrt{1 + 4t^2},$$
$$\kappa_g(t) = \frac{2}{(1 + 4t^2)^{3/2}}.$$

Thus, the parametric equations of the evolute are

$$\vec{E}(t) = (t, t^2) + \frac{1}{2}(1 + 4t^2)^{3/2} \frac{1}{\sqrt{1 + 4t^2}}(-2t, 1)$$
$$= \left(-4t^3, \frac{1}{2} + 3t^2 \right).$$

The Cartesian equation for the evolute of the parabola is then

$$y = \frac{1}{2} + 3 \left| \frac{x}{4} \right|^{2/3}.$$

Figure 1.8 depicts both the parabola and its evolute.

A closely related curve to the evolute of $\vec{X}(t)$ is the involute, though we leave the exact nature of this relationship to the problems. The involute is defined as follows.

Definition 1.4.9 Let $\vec{X} : I \to \mathbb{R}^2$ be a regular parametrized curve. We call an *involute* to $\vec{X}$ any parametrized curve $\vec{\imath}$ such that for all $t \in I$, $\vec{\imath}(t)$ meets the tangent line to $\vec{X}$ at t at a right angle.

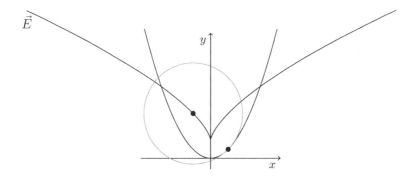

Figure 1.8: Evolute of a parabola.

For all $t \in I$, the point on the involute $\vec{\imath}(t)$ is on the tangent line to $\vec{X}$ at t, so it is natural to write the parametric equations as $\vec{\imath}(t) = \vec{X}(t) + \lambda(t)\vec{T}(t)$. Since we will wish to calculate $\vec{\imath}'(t)$, which involves the derivative $\vec{T}'(t)$, we must assume that the curve $\vec{X}(t)$ is of class C^2, i.e., with a second derivative that exists and is continuous. Definition 1.4.9 requires that the vector $\vec{\imath}'(t)$ be in a perpendicular direction to $\vec{T}(t)$, so we have

$$\vec{T} \cdot \vec{\imath}' = 0 \Longrightarrow \vec{T} \cdot \left(s'\vec{T} + \lambda'\vec{T} + \lambda s' \kappa_g \vec{U} \right)$$
$$\Longrightarrow s' + \lambda' = 0$$
$$\Longrightarrow \lambda(t) = C - s(t),$$

where C is some constant of integration. Therefore, if $\vec{X}$ is a regular parametrized curve of class C^2, then the formula for the involute is

$$\vec{\imath}(t) = \vec{X}(t) + (C - s(t))\vec{T}(t).$$

PROBLEMS

1. Consider the cubic curve $\vec{X}(t) = (t, t^3)$ defined over $\mathbb{R}$.
   (a) Find the osculating circle of $\vec{X}$ at $t = 1$.
   (b) Determine the parametric equations for the evolute of $\vec{X}$.

2. Consider the figure eight curve parametrized by $\vec{X}(t) = (\cos t, \sin 2t)$. Determine parametric equations for the osculating circle to $\vec{X}(t)$ (a) at $t = 0$; and (b) at $t = \pi/4$.

3. Prove the claim that the tangent line to a parametrized curve at a regular point has contact of order 1.

4. Let $\vec{X}$ be a regular parametrized curve and let t_0 be an inflection point, i.e. a point where $\kappa_g(t) = 0$. Prove that the tangent line to $\vec{X}(t)$ at $t = t_0$ has contact of order 2.

5. Prove that the evolute of the ellipse $\vec{X}(t) = (a\cos t, b\sin t)$ has parametric equations

$$\vec{\gamma}(t) = \left(\left(\frac{a^2 - b^2}{a}\right)\cos^3 t, \left(\frac{b^2 - a^2}{b}\right)\sin^3 t\right).$$

6. Consider the catenary given by the parametric equation $\vec{X}(t) = (t, \cosh t)$ for $t \in \mathbb{R}$.
   (a) Prove that the curvature of the catenary is $\kappa_g(t) = 1/\cosh^2 t$.
   (b) Prove that the evolute of the catenary is

$$\vec{E}(t) = (t - \sinh t \cosh t, 2\cosh t).$$

7. Show that the evolute of an exponential spiral parametrized by $\vec{X}(t) = (e^t \cos(\omega t), e^t \sin(\omega t))$ is similar to itself. Describe the similarity as $F(\vec{x}) = A\vec{x} + \vec{b}$, where A is an orthogonal matrix.

8. A general spiral in the plane (centered at the origin) has a parametric equation $\vec{X}(t) = (f(t)\cos(t + \alpha), f(t)\sin(t + \alpha))$, where α is a fixed angle and f is a monotonic function. Exercise 1.3.14 establishes that the curvature function is $(2f'(t)^2 + f(t)^2 - f(t)f''(t))/(f'(t)^2 + f(t)^2)^{3/2}$.
   (a) Set $\alpha = 0$ and obtain the parametric equations for the evolute of $\vec{X}(t)$.
   (b) Show that this evolute is again a spiral only if the ratio $(f(f')^2 - f^2 f'' + (f')^3 + f^2 f')/(f(f')^2 + f^3)$ is constant. (Call it C.)
   (c) Express the above condition as a differential equation in the function $v = f'/f$. (Do not attempt to solve.)
   (d) Show that the constant C is 1 if $f(t) = e^{rt}$ for any real constant r.

9. Show that the evolute of a cardioid is another cardioid rotated, translated and scaled by a factor of $1/3$ from the original cardioid.

10. Prove that if the osculating circle to a regular parametrized curve C at a point P has contact of order 3, then P is a vertex of C. Give an example where this does not occur, thereby proving that at a regular point on a parametrized curve, there does not necessarily exist a circle with contact of order 3.

11. Prove that an osculating circle to a curve C at a point P that is not a vertex lies on either side of the curve. [See Problem 1.4.10.]

12. Let $\vec{X} : I \to \mathbb{R}^2$ be a parametrized curve that does not have any inflection points (i.e., $\kappa_g(t) \neq 0$ for all $t \in I$). Prove that the evolute of the curve has a critical point at t_0 if and only if $\vec{X}(t_0)$ is a vertex of the curve.

13. Consider the circle parametrized by $\vec{X}(t) = (R\cos t, R\sin t)$. Find the equations for the involute of this circle

14. Let $\vec{x}(t) = (t, t^2)$ be the parabola. Define a new curve $\vec{\iota}$ as the involute of $\vec{x}$ such that $\vec{\iota}(0) = \vec{x}(0) = (0,0)$. Calculate parametric equations for $\vec{\iota}(t)$. [Hint: This will involve an integral.]

15. (*) Continuation of the last problem: Calculate parametric equations for the evolute to $\vec{\iota}$.

16. (*) Let Γ be a regular plane curve for which the curvature is never 0. Let $\mathcal{I}$ be an involute of Γ. Prove that the evolute of $\mathcal{I}$ is the original curve Γ.

1.5 Natural Equations

An isometry of the plane is any function $F : \mathbb{R}^2 \to \mathbb{R}^2$ such that for any two points $\vec{p}, \vec{q} \in \mathbb{R}^2$, the distance between them is preserved, i.e.,

$$\|\vec{q} - \vec{p}\| = \|F(\vec{q}) - F(\vec{p})\|.$$

A well-known result about isometries is that F is an isometry if and only if

$$F(\vec{v}) = A\vec{v} + \vec{C},$$

where A is any 2×2 orthogonal matrix and $\vec{C}$ is any constant vector. Isometries of the plane include rotations, translations, reflections, and glide reflections. It is also not hard to show that a composition of isometries is again an isometry.

The condition of orthogonality $A^\top A = I$ implies that $\det(A) = \pm 1$. Consequently, isometries come in two flavors depending on the sign of the determinant of A. If $\det(A) = 1$, we call the isometry a *positive isometry*. These include rotations, translations, and compositions thereof. If $\det(A) = -1$, we call the isometry a *negative isometry*. These include reflections and glide reflections. Positive isometries are also called *rigid motions* because any shape (imagining it to be a solid physical object) can be moved from its original to its image under a positive isometry without bringing the shape out of the plane.

Recall now that for a parametrized curve $\vec{X} : I \to \mathbb{R}^2$ with $\vec{X}(t) = (x(t), y(t))$, the curvature function is given by

$$\kappa_g(t) = \frac{x'(t)y''(t) - x''(t)y'(t)}{\left(x'(t)^2 + y'(t)^2\right)^{3/2}}.$$

Since any rigid motion does not stretch distances, such a transformation should not distort a plane curve. Therefore, as one might expect, the curvature is preserved under rigid motions, a fact that we now prove.

Theorem 1.5.1 *Let $\vec{X} : I \to \mathbb{R}^2$ be a regular plane curve that is of class C^2. Let $F : \mathbb{R}^2 \to \mathbb{R}^2$ be a rigid motion of the plane given by $F(\vec{v}) = A\vec{v} + \vec{C}$, where A is a rotation matrix and $\vec{C}$ is any vector in*

$\mathbb{R}^2$. *The vector function $\vec{\xi} = F \circ \vec{X}$ is a regular parametrized curve that is of class C^2, and the curvature function $\bar{\kappa}_g(t)$ of $\vec{\xi}$ is equal to the curvature function $\kappa_g(t)$ of $\vec{X}$.*

Proof: A rotation matrix in $\mathbb{R}^2$ is of the form

$$A = \begin{pmatrix} a & -b \\ b & a \end{pmatrix},$$

where $a^2 + b^2 = 1$. Let $\vec{C} = (e, f)$. We can write the parametric equation for $\vec{\xi}$ as

$$\vec{\xi}(t) = (ax(t) - by(t) + e, bx(t) + ay(t) + f).$$

Then

$$\vec{\xi}'(t) = (ax'(t) - by'(t), bx'(t) + ay'(t)),$$
$$\vec{\xi}'(t) = (ax''(t) - by''(t), bx''(t) + ay''(t)),$$

and therefore the curvature function of $\vec{\xi}$ is

$$
\begin{aligned}
\bar{\kappa}_g(t) &= \frac{(ax'(t) - by'(t))(bx''(t) + ay''(t)) - (bx'(t) + ay'(t))(ax''(t) - by''(t))}{((ax'(t) - by'(t))^2 + (bx'(t) + ay'(t))^2)^{3/2}} \\
&= \frac{abx'x'' + a^2x'y'' - b^2y'x'' - aby'y'' - abx'x'' + b^2x'y'' - a^2y'x'' + aby'y''}{(a^2(x')^2 - 2abx'y' + b^2(y')^2 + b^2(x')^2 + 2abx'y' + a^2(y')^2)^{3/2}} \\
&= \frac{x'y'' - y'x''}{((x')^2 + (y')^2)^{3/2}} = \kappa_g(t).
\end{aligned}
$$

$\square$

The curvature function is invariant under any positive isometry, i.e., a composition of rotations and translations. Furthermore, in Problem 1.3.20, we saw that the curvature function is invariant under any regular reparametrization, except up to a sign that depends on "the direction of travel" along the curve. Consequently, $|\kappa_g|$ is a geometric invariant that only depends on the shape of the curve at a particular point and not how the curve is parametrized or where the curve sits in $\mathbb{R}^2$.

It is natural to ask whether a converse relation holds, namely, whether the curvature function is sufficient to determine the parametrized curve up to a positive plane isometry. As posed, the question is not well defined since a curve can have different parametrizations. However, we can prove the following fundamental theorem.

Theorem 1.5.2 (Fundamental Theorem of Plane Curves) *If $\kappa_g(s)$ is a piecewise continuous function, there exists a regular curve of class C^2 parametrized by arc length by $\vec{X} : I \to \mathbb{R}^2$ with the curvature function $\kappa_g(s)$. Furthermore, the curve is uniquely determined up to a rigid motion in the plane.*

Proof: If a regular curve $\vec{X}(s)$ is parametrized by arc length, then the curvature formula is $\kappa_g(s) = x'(s)y''(s) - y'(s)x''(s)$. The proof of this theorem consists of exhibiting a parametrization that satisfies this differential equation.

For a regular curve of class C^2 that is parametrized by arc length, we have $\vec{X}' = \vec{T}$, and we can view $\vec{T}$ as a vector function of class C^1 from the interval of definition of $\vec{X}$ into the unit circle. Therefore,

$$\vec{T} = \big(\cos(\theta(s)), \sin(\theta(s))\big) \tag{1.19}$$

for some continuous function $\theta(s)$. However, $\vec{T}'(s) = \kappa_g(s)\vec{U}$, and Equation (1.19) show that

$$\kappa_g(s)\vec{U} = \theta'(s)\big(-\sin(\theta(s)), \cos(\theta(s))\big).$$

Thus, we deduce that $\kappa_g(s) = \theta'(s)$.

The above remarks lead to the following result. Suppose we are given the curvature function $\kappa_g(s)$. Performing two integrations, we see that the only curves $\vec{X}(s)$ with curvature function $\kappa_g(s)$ must be

$$\vec{X}(s) = \left(\int \cos(\theta(s))\, ds + e, \int \sin(\theta(s))\, ds + f\right), \tag{1.20}$$

where

$$\theta(s) = \int \kappa_g(s)\, ds + \theta_0, \tag{1.21}$$

and where θ_0, e, and f are constants of integration. Note that the theorem requires $\kappa_g(s)$ to be piecewise continuous in order to be integrable. Furthermore, the trigonometric addition formulas show that a nonzero constant θ_0 changes $\vec{X}$ by a rotation of θ_0 and the nonzero constants e and f correspond to a translation along the vector $\vec{C} = (e, f)$. $\qquad\qquad\square$

Because of the geometric nature of the curvature function $\kappa_g(s)$ (with respect to arc length) and because there exists a unique plane curve (up to a rigid motion) for a given curvature function, we call $\kappa_g(s)$ the *natural equation* of its corresponding curve. The Fundamental Theorem of Plane Curves is surprising because, a priori, we expect a curve to require two functions (coordinate functions) to define it. However, only one function, $\kappa_g(s)$, is required to uniquely define the shape of a curve.

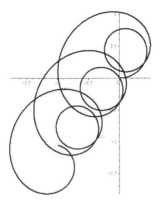

Figure 1.9: Curve with $\kappa_g(s) = 2 + \sin s$.

If we assume that $\vec{X}(s)$ is of class C^{∞} and is equal to its power series in an open interval around $s = 0$, then we can see why Theorem 1.5.2 holds using Taylor series. (This condition is called *real analytic* at $s = 0$.)

Let $\vec{X} : I \to \mathbb{R}^2$ be a real analytic curve parametrized by arc length and assume without loss of generality that $0 \in I$. We can expand the Taylor series of $\vec{X}$ about 0 to get

$$\vec{X}(s) = \vec{X}(0) + s\vec{X}'(0) + \frac{s^2}{2!}\vec{X}''(0) + \frac{s^3}{3!}\vec{X}'''(0) + \cdots \quad .$$

However, since the curve is parametrized by arc length, $\vec{X}'(s) = \vec{T}(s)$, $\vec{X}''(s) = \kappa_g(s)\vec{U}(s)$, $\vec{X}'''(s) = \kappa_g'(s)\vec{U}(s) - (\kappa_g(s))^2\vec{T}(s)$, and so on. The first few terms look like

$$\vec{X}(s) = \vec{X}(0) + \left(s - \frac{1}{6}\kappa_g(0)^2 s^3 + \cdots\right)\vec{T}(0) + \left(\frac{1}{2}s^2 + \frac{1}{6}\kappa_g'(0)s^3 + \cdots\right)\vec{U}(0).$$

Thus, since the normal vector $\vec{U}(s)$ is just a rotation of $\vec{T}(s)$ by $\frac{\pi}{2}$, given a function $\kappa_g(s)$, once one chooses for initial conditions the point $\vec{X}(0)$ and the direction $\vec{T}(0)$, the Taylor series is uniquely determined. The intersection of the intervals of convergence of the two Taylor series in the $\vec{T}(0)$ and the $\vec{U}(0)$ components is a new interval J that contains $s = 0$. It is possible that J could be trivial, $J = \{0\}$, but if it is not, then the Taylor series uniquely defines $\vec{X}(s)$ over J. Choosing a different $\vec{T}(0)$ amounts to a rotation of the curve in the plane, and choosing a different $\vec{X}(0)$ amounts to a translation. Therefore, we see again that making different choices for the initial conditions corresponds to a rigid motion of the curve in the plane.

For even simple functions for $\kappa_g(s)$, it is often difficult to use the approach in the proof of Theorem 1.5.2 to explicitly solve for $\vec{X}(s)$. However, using a CAS with tools for solving differential equations, it is possible to produce a picture of curves that possess a given curvature function $\kappa_g(s)$. One can create a numerical solution using the solution in Equation (1.20) with (1.21). As an equivalent technique, using a CAS, one can solve the system of differential equations

$$\begin{cases} x'(s) & = \cos(\theta(s)), \\ y'(s) & = \sin(\theta(s)), \\ \theta'(s) & = \kappa_g(s), \end{cases}$$

and only plot the solution for the pair $(x(s), y(s))$. A choice of initial conditions determines the position and orientation of the curve in the plane. Figure1.9 gives an interesting example of parametric curves with a given $\kappa_g(s)$.

We encourage the reader particularly in regards to this section, to explore the associated applet provided by the *Maple* workbook. The applet provided allows the reader to explore the relationship between $\kappa_g(s)$ and the curve that it defines.

PROBLEMS

1. Let Γ_1 and Γ_2 be two plane curves that are similar to each other. In other words, there exists a similarity $F : \mathbb{R}^2 \to \mathbb{R}^2$ such that $\Gamma_2 = F(\Gamma_1)$. Recall that a similarity is a function of the form $F(\vec{v}) = rA\vec{v} + \vec{b}$, where A is an orthogonal matrix, $\vec{b}$ is a fixed vector, and r is a positive scalar, called the similarity ratio. Prove that if $P' \in \Gamma_2$ corresponds to $P \in \Gamma_1$ under the similarity, then the curvature of Γ_2 at P' is $\frac{1}{r}$ times the curvature of Γ_1 at P.

2. Suppose a curve has $\kappa_g(s) = A$. Prove that such a curve is a circle.

3. Suppose a curve has $\kappa_g(s) = 2As$. Prove by directly expanding the Taylor series that if such a curve has $\vec{X}(0) = \vec{0}$ and $\vec{X}'(0) = (1, 0)$, then such a curve is

$$\vec{X}(s) = \left(\int_0^s \cos(As^2)\, ds, \int_0^s \sin(As^2)\, ds \right).$$

4. Find parametric equations for a curve satisfying $\kappa_g(s) = 1/(1 + s^2)$ by direct integration, following the method in the proof of Theorem 1.5.2. Show that this curve corresponds to the catenary $y = \cosh x$.

5. Find parametric equations for a curve satisfying $\kappa_g(s) = 1/(s+1)$ by direct integration. Show that the resulting curve is an exponential spiral.

CHAPTER 2

Plane Curves: Global Properties

Most of the properties of curves we have studied so far are called local properties. By definition, a local property of a curve (or surface) is a property that is related to a point on the curve based on information contained just in a neighborhood of that point. By contrast, global properties concern attributes about the curve taken as a whole. In Chapter 1, the arc length of a curve was the only notion introduced that we might consider a global property. Analytically speaking, local properties of a curve at a point involve derivatives of the parametric equations while global properties deal with integration along the curve and topological properties and geometric properties of how the curve lies in $\mathbb{R}^2$. Some of the proofs of global properties rely on theorems from topology. These are supplied concisely in the appendix on topology in [24]. (The interested reader is encouraged to consult some references on basic topology, such as Gemignani [16] or Armstrong [1].)

2.1 Basic Properties

Definition 2.1.1 A parametrized curve C is called *closed* if there exists a parametrization $\vec{X} : [a, b] \to \mathbb{R}^2$ of C such that $\vec{X}(a) = \vec{X}(b)$. A closed curve is of class C^k if, in addition, all the (one-sided) derivatives of $\vec{X}$ at a and at b are equal of order $i = 0, 1, \ldots, k$; in other words, if as one-sided derivatives $\vec{X}'(a) = \vec{X}'(b)$, $\vec{X}''(a) = \vec{X}''(b)$, and so on up to $\vec{X}^{(k)}(a) = \vec{X}^{(k)}(b)$.

We recall that the left-derivative (resp. right-derivative) of a function f at a is the limit

$$\lim_{h \to 0^-} \frac{f(a+h) - f(a)}{h} \qquad \left(\text{resp.} \lim_{h \to 0^+} \frac{f(a+h) - f(a)}{h} \right).$$

The notion of left- and right-sided derivatives of real-valued functions naturally extends to parametrized curves.

DOI: 10.1201/9781003295341-2

The conditions on the derivatives of $\vec{X}$ in the above definition seem awkward but they attempt to establish the fact that the vector function $\vec{X}$ behaves identically at a and at b. Another approach involves using the unit circle $\mathbb{S}^1$ rather than an interval as the domain for the map $\vec{X}$.

Throughout this book, we will regularly use the notation $\mathbb{S}^n$ for the unit sphere in $\mathbb{R}^{n+1}$. To be precise,

$$\mathbb{S}^n \stackrel{\text{def}}{=} \{(x_1, x_2, \ldots, x_{n+1}) \in \mathbb{R}^{n+1} \mid x_1^2 + x_2^2 + \cdots + x_{n+1}^2 = 1\}.$$

However, in order to talk about the continuity of a function $f : \mathbb{S}^n \to \mathbb{R}^m$, we view $\mathbb{S}^n$ as a subset of $\mathbb{R}^{n+1}$. (Using advanced terminology, we say that $\mathbb{S}^n$ has the subspace topology inherited from $\mathbb{R}^{n+1}$.)

For every parametrization of a curve $\vec{X} : [a, b] \to \mathbb{R}^2$, we can use the positive reparametrization $g : [0, 1] \to [a, b]$ with $g(t) = a + (b - a)t$ so that $\vec{\xi} = \vec{X} \circ g : [0, 1] \to \mathbb{R}^2$ has exactly the same locus. In the same way, regardless of the length of a closed plane curve C, we can define it as the image of a continuous function $\vec{X} : \mathbb{S}^1 \to \mathbb{R}^2$.

 Definition 2.1.2 A curve C is called *simple* if there exists a parametrization $\vec{X} : I \to \mathbb{R}^2$ of C such that $\vec{X}$ is an injective (i.e., one-to-one) function. A closed curve C is called simple if there exists a parametrization $\vec{X} : [a, b] \to \mathbb{R}^2$ of C such that $\vec{X}(t_1) = \vec{X}(t_2)$ with $t_1 < t_2$ only when $t_1 = a$ and $t_2 = b$.

Using the language associated to a circle domain, we say that a closed curve C is simple if there exists a bijection $\varphi : \mathbb{S}^1 \to C$ that is continuous and such that φ^{-1} is also continuous.

If a curve is not simple, we intuitively think of the curve as intersecting itself. Using parametrizations, we can give the concept of a self-intersection a precise definition.

Definition 2.1.3 A curve C is said to have a self-intersection at a point $P \in C$ if for every parametrization $\vec{X} : I \to \mathbb{R}^2$ of C, there exist $t_1 \neq t_2$ that are not endpoints of I such that $\vec{X}(t_1) = \vec{X}(t_2) = P$.

Proposition 2.1.4 *Let C be a closed curve with parametrization $\vec{X} : [a, b] \to \mathbb{R}^2$. Then C is bounded as a subset of $\mathbb{R}^2$.*

Proof: Let $\vec{X} : [a, b] \to \mathbb{R}^2$ be a parametrization of C with coordinate functions $x(t)$ and $y(t)$. By the Extreme Value Theorem, since $x(t)$ and $y(t)$ are continuous functions over an interval $[a, b]$, then they respectively attain extrema $x_{\min}$ minimum and $x_{\max}$ maximum of $x(t)$ and $y_{\min}$ minimum and $y_{\max}$ maximum of $y(t)$. The parametrized curve $\vec{X}$ lies entirely in the rectangle $x_{\min} \leq x \leq x_{\max}$ and

$y_{\min} \leq y \leq y_{\max}$. Therefore, the curve lies inside a disc centered around the origin. More precisely, let

$$M_x = \max\{|x_{\min}|, |x_{\max}|\} \qquad \text{and} \qquad M_y = \max\{|y_{\min}|, |y_{\max}|\}.$$

Then for all $t \in [a, b]$,

$$x(t)^2 \leq M_x^2 \qquad \text{and} \qquad y(t)^2 \leq M_y^2.$$

Thus,

$$\|\vec{X}(t)\| \leq \sqrt{M_x^2 + M_y^2},$$

and so the curve C is contained in a disk of finite radius. Hence, C is bounded. $\square$

One of the most fundamental properties of global geometry of plane curves is that a simple, closed plane curve C separates the plane into two open connected components, each with the common boundary of C.

Theorem 2.1.5 (Jordan Curve Theorem) *Suppose that C is a simple closed curve in $\mathbb{R}^2$. Then C separates the plane into precisely two components W_1 and W_2 such that $\mathbb{R}^2 - C = W_1 \cup W_2$ and $W_1 \cap W_2 = \emptyset$. One component is bounded and the other is unbounded.*

This intuitive fact turns out to be rather difficult to prove. Furthermore, a rigorous proof requires a precise definition of component. We refer the reader to Munkres [27] for a detailed discussion. In [24, Section A.4], there is a more readable proof of the weaker statement that only assumes that the curve is regular. The component of $\mathbb{R}^2 - C$ that is bounded is called the *interior* of C and the component of C that is unbounded is called the *exterior*.

We remind the reader of the following theorem from multivariable calculus, which one may view as a global property of plane curves.

Theorem 2.1.6 (Green's Theorem) *Let C be a simple, closed, piecewise regular, positively oriented plane curve with interior region $\mathcal{R}$, and let $\vec{F} = (P(x,y), Q(x,y))$ be a differentiable vector field. Then,*

$$\int_C P\,dx + Q\,dy = \iint_{\mathcal{R}} \left(\frac{\partial Q}{\partial x} - \frac{\partial P}{\partial y}\right) dx\,dy.$$

We remind the reader that a curve is positively oriented if as an object travels along the curve, the interior is to the left. With unit tangent and unit normal vectors, we can say that C is positively oriented if it is parametrized by $\vec{X} : [a, b] \to \mathbb{R}^2$ such that $\vec{U}(t)$ points toward the interior of the curve. We also remind the reader that if $\vec{X} : [a, b] \to \mathbb{R}^2$ is a regular parametrization for C with $\vec{X}(t) = (x(t), y(t))$, then

$$\int_C P\,dx + Q\,dy = \int_C \vec{F} \cdot d\vec{X} = \int_a^b \vec{F}(x(t), y(t)) \cdot \vec{X}'(t)\,dt$$

$$= \int_a^b P(x(t), y(t))x'(t) + Q(x(t), y(t))y'(t)\,dt.$$

Corollary 2.1.7 *Let C be a simple closed regular plane curve with interior region $\mathcal{R}$. Then the area A of $\mathcal{R}$ is*

$$A = \int_C x\,dy = -\int_C y\,dx = \frac{1}{2}\int_C -y\,dx + x\,dy. \qquad (2.1)$$

Proof: In each of these integrals, one simply chooses a vector field $\vec{F} = (P, Q)$ such that

$$\frac{\partial Q}{\partial x} - \frac{\partial P}{\partial y} = 1.$$

For the three integrals, these are, respectively, $\vec{F} = (0, x)$, $\vec{F} = (-y, 0)$, and $\vec{F} = \frac{1}{2}(-y, x)$. Then apply Green's Theorem. □

Example 2.1.8 Consider the ellipse given by $\vec{X}(t) = (a\cos t, b\sin t)$ for $0 \le t \le 2\pi$. The techniques from introductory calculus used to calculate area would lead us to evaluate the integral

$$A = 4\int_0^a b\sqrt{1 - \frac{x^2}{a^2}}\,dx.$$

Though one can calculate this by hand, it is not simple. Much easier would be to use Green's Theorem, which gives

$$A = \int_C x\,dy = \int_0^{2\pi} a\cos t\, b\cos t\,dt$$

$$= ab\int_0^{2\pi} \cos^2 t\,dt = ab\int_0^{2\pi} \frac{1}{2} + \frac{1}{2}\cos(2t)\,dt$$

$$= ab\left[\frac{1}{2}t + \frac{1}{4}\sin(2t)\right]_0^{2\pi} = \pi ab.$$

PROBLEMS

1. Calculate the area of the region enclosed by the cardioid $\vec{X}(t) = ((1 - \cos t)\cos t, (1 - \cos t)\sin t)$.

2. Use Green's Theorem to calculate the area of one loop of $\vec{X}(t) = (\cos t, \sin(2t))$.

3. Consider the function graph of a curve in polar coordinates $r = f(\theta)$ parametrized by $\vec{X}(t) = (f(t)\cos t, f(t)\sin t)$. Suppose that for $0 \le t \le 2\pi$, the curve is closed and encloses a region $\mathcal{R}$. Use Equation (2.1) and Green's Theorem to prove the following area formula in polar coordinates:

$$\text{Area}(\mathcal{R}) = \iint_{\mathcal{R}} r \, dr \, d\theta.$$

4. The *diameter* of a closed curve C is defined as the maximum distance between two points, i.e.,

$$\text{diam}(C) = \max\{d(p_1, p_2) \, | \, p_1, p_2 \in C\}.$$

We call a diameter any chord of C whose length is $\text{diam}(C)$. Let $\vec{X}(t)$ be the parametrization of a closed differentiable curve C. Let $f(t, u) = \|\vec{X}(t) - \vec{X}(u)\|$ be the distance function between two points on the curve.

   (a) Prove that if a chord $[p_1, p_2]$ of C is a diameter of C, then the line (p_1, p_2) is simultaneously perpendicular to the tangent line to C at p_1 and the tangent line to C at p_2.

   (b) Under what other situations is the line (p_1, p_2) simultaneously perpendicular to the tangent line to C at p_1 and the tangent line to C at p_2?

5. (*) Prove that the diameter of the cardioid parametrized by $\vec{X}(t) = ((1 - \cos t)\cos t, (1 - \cos t)\sin t)$ is $3\sqrt{3}/2$. [Hint: Set $F(t, u) = \|\vec{X}(t) - \vec{X}(u)\|^2$ and maximize F. Using trigonometric addition identities, prove that

$$\frac{\partial F}{\partial t} + \frac{\partial F}{\partial u} = 2\sin(t - u)(\cos(2u) - \cos(2t)).$$

Deduce that the maximum distance between $\vec{X}(t)$ and $\vec{X}(u)$ occurs when $u = 2\pi - t$. Use this to find the diameter.]

6. (*) This problem studies the relation between the curvature $\kappa_g(s)$ of a curve C and whether it is closed.

   (a) Prove that if C is closed, then its curvature $\kappa_g(s)$, as a function of arc length, is periodic.

   (b) Assume that $\kappa_g(s)$ is not constant and is periodic with smallest period p. Use the Fundamental Theorem of Plane Curves (Theorem 1.5.2) to prove that if

$$\frac{1}{2\pi} \int_0^p \kappa_g(s) \, ds$$

is an element of $\mathbb{Q} - \mathbb{Z}$, i.e., a rational that is not an integer, then C is a closed curve. [This problem is discussed in [3], though the authors' use of complex numbers is not necessary for this problem.]

2.2 Rotation Index

As a leading example of what we shall term the rotation index of
a curve, consider the circle $C : \vec{X}(t) = (R \cos t, R \sin t)$ with the
defining interval of $I = [0, 2\pi]$. A simple calculation shows that for
all $t \in [0, 2\pi]$, we have a curvature of $\kappa_g(t) = \frac{1}{R}$. Then it is easy to
see that

$$\int_C \kappa_g \, ds = \int_0^{2\pi} \kappa_g(t) s'(t) \, dt = \int_0^{2\pi} \frac{1}{R} \cdot R \, dt = 2\pi.$$

On the other hand, suppose that we use the defining interval
$[0, 2\pi n]$ with the same parametrization. In that case, were we to
evaluate the same integral, we would obtain

$$\frac{1}{2\pi} \int_C \kappa_g \, ds = n.$$

Now consider any regular, closed, plane curve $\vec{X} : I \to \mathbb{R}^2$. To
this curve, we associate the unit tangent vector $\vec{T}(t)$. Placing the
base of this vector at the origin, we see that $\vec{T}$ itself draws out a
curve $\vec{T} : I \to \mathbb{R}^2$. It is called the *tangential indicatrix* of the plane
curve $\vec{X}$.

The tangential indicatrix of a curve lies entirely on the unit circle
in the plane but with possibly a complicated parametrization. Depending on the shape of $\vec{X}$, the tangential indicatrix may change
speed, stop, and double back along a portion of the circle. Nonetheless, we can use the theory of usual plane curves to study the tangential indicatrix. We are in a position to state the main proposition for
this section.

 Proposition 2.2.1 *Let C be a closed, regular, plane curve parametrized by $\vec{X} : I \to \mathbb{R}^2$ of class C^2. Then the quantity*

$$\frac{1}{2\pi} \int_C \kappa_g \, ds$$

is an integer. This integer is called the rotation index *of the curve.*

Proof: Rewrite the above integral as follows:

$$\int_C \kappa_g \, ds = \int_I \kappa_g(t) s'(t) \, dt = \int_I \kappa_g(t) s'(t) \|\vec{U}(t)\| \, dt.$$

The curvature function $\kappa_g(t)$ of plane curves is not always positive,
but since $\vec{X}(t)$ is of class C^2, then $\kappa_g(t)$ is at least continuous, and
the intermediate value theorem applies. Therefore, define two closed
subsets (unions of closed subintervals) of the interval I as follows:

$$I^+ = \{t \in I : \kappa_g(t) \geq 0\} \qquad \text{and} \qquad I^- = \{t \in I : \kappa_g(t) \leq 0\}.$$

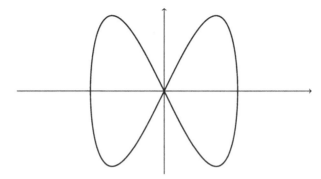

Figure 2.1: The curve $\vec{X}(t) = (\cos t, \sin(2t))$.

Using the definition of curvature from Equation (1.10), we split the above integral into the following two parts:

$$\int_C \kappa_g \, ds = \int_{I^+} \|\vec{T}'(t)\| \, dt - \int_{I^-} \|\vec{T}'(t)\| \, dt.$$

However, the two integrals on the right-hand side give the arc length of the tangential indicatrix traveling in a counterclockwise (resp. clockwise) direction. Since $\vec{X}$ is a closed curve, $\vec{T}$ is as well, and these two integrals represent 2π times the number of times $\vec{T}$ travels around the circle, with a sign indicating the direction. □

Example 2.2.2 Consider $\vec{X}(t) = (\cos t, \sin(2t))$ as the paramtrization for the curve C depicted Figure 2.1. The curvature function is given by

$$\kappa_g(t)s'(t) = \frac{4\sin t \sin(2t) + 2\cos t \cos(2t)}{\sin^2 t + 4\cos^2(2t)} = \frac{6\cos t + 2\cos(3t)}{5 - \cos(2t) + 4\cos(4t)}.$$

Thus, the rotation index n of $\vec{X}$ is

$$n = \frac{1}{2\pi} \int_C \kappa_g(t)s'(t)dt = \frac{1}{2\pi} \int_0^{2\pi} \frac{6\cos t + 2\cos(3t)}{5 - \cos(2t) + 4\cos(4t)} \, dt.$$

Since the integrand is periodic of period 2π, using the substitution $u = t + \frac{\pi}{2}$ and recognizing that we can integrate over any interval of length 2π, we have

$$n = \frac{1}{2\pi} \int_{-\pi}^{\pi} \frac{6\sin u - 2\sin(3u)}{5 + \cos(2u) + 4\cos(4u)} \, du.$$

However, the integrand is now an odd function, so the rotation index of $\vec{X}$ is $n = 0$.

The proof of Proposition 2.2.1 analyzed the integral $\int_C \kappa_g \, ds$ as the signed arc length of the tangential indicatrix $\vec{T}(t)$, which one can view as a map from an interval I to the unit circle $\mathbb{S}^1$. However, the result of Proposition 2.2.1 follows also from a more general result that we wish to explain in detail here. Some of the concepts below come from topology and illustrate the difficulty of analyzing functions from an interval to a circle.

Any path $f : I \to \mathbb{S}^1$ on the unit circle may be described by the angle function $\varphi(t)$, so that

$$f(t) = \big(\cos(\varphi(t)), \sin(\varphi(t)) \big).$$

However, the angle $\varphi(t)$ is defined only up to a multiple of 2π, and hence it need not be a well-defined function, let alone a continuous function. However, since f is continuous, for all $t \in I$ there exists an interval $(t - \varepsilon, t + \varepsilon)$ and a continuous function $\tilde{\varphi}(t)$ such that for all $u \in (t - \varepsilon, t + \varepsilon)$, $\tilde{\varphi}(u)$ differs from $\varphi(u)$ by a multiple of 2π. If $\tilde{\varphi}'(t)$ exists, it is a well-defined function, regardless of any choice made in selecting $\varphi(t)$. Finally, if $I = [a, b]$, we define the *total angle function* related to f as the function

$$\tilde{\varphi}(t) = \int_a^t \tilde{\varphi}'(u) \, du + \varphi(a).$$

By construction, $\tilde{\varphi}(t)$ is continuous, satisfies

$$f(t) = \big(\cos(\tilde{\varphi}(t)), \sin(\tilde{\varphi}(t)) \big),$$

and keeps track of how many times and in what direction the path f travels around $\mathbb{S}^1$. The quantity $\tilde{\varphi}(b) - \tilde{\varphi}(a)$ is called the *total angle swept out* by f.

Definition 2.2.3 Let I be an interval of $\mathbb{R}$. Given a continuous function $f : I \to \mathbb{S}^1$ on the unit circle, the *lifting* of f is the function $\tilde{\varphi}(t)$, with $\varphi(a)$ chosen so that $0 \leq \varphi(a) < 2\pi$. The lifting of f is denoted by $\tilde{f}$ to indicate its unique dependence on f.

For any continuous function between circles $f : \mathbb{S}^1 \to \mathbb{S}^1$, viewing the domain $\mathbb{S}^1$ as an interval $[0, 2\pi]$ with the endpoints identified, we also view f as a continuous function $f : [0, 2\pi] \to \mathbb{S}^1$. Since $f(0) = f(2\pi)$ are points on $\mathbb{S}^1$, then $\tilde{f}(2\pi) - \tilde{f}(0)$ is a multiple of 2π. This leads to the following definition.

Definition 2.2.4 Let $f : \mathbb{S}^1 \to \mathbb{S}^1$ be a continuous map between circles. Let $\tilde{f} : [0, 2\pi] \to \mathbb{R}$ be the lifting of f. Then the *degree* of f is the integer

$$\deg f = \frac{1}{2\pi} \big(\tilde{f}(2\pi) - \tilde{f}(0) \big).$$

Intuitively speaking, the degree of a function $f : \mathbb{S}^1 \to \mathbb{S}^1$ between unit circles is how many times around f winds its domain $\mathbb{S}^1$ onto its target $\mathbb{S}^1$.

Returning to the example of the tangential indicatrix $\vec{T}$ of a regular closed curve C, since C is closed, $\vec{T}$ can be viewed as a function $\mathbb{S}^1 \to \mathbb{S}^1$.

Proposition 2.2.5 *Let C be a regular closed curve parametrized by $\vec{X} : I \to \mathbb{R}^2$. The rotation index of C is equal to the degree of $\vec{T}$.*

Proof: Since $\vec{T}$ is a map $\vec{T} : I \to \mathbb{S}^1$, using Equation (2.2), we can write

$$\vec{T} = \big(\cos(\tilde{\varphi}(t)), \sin(\tilde{\varphi}(t)) \big),$$

so

$$\vec{T}' = \big(-\tilde{\varphi}'(t) \sin(\tilde{\varphi}(t)), \tilde{\varphi}'(t) \cos(\tilde{\varphi}(t)) \big) = \tilde{\varphi}'(t)\vec{U}(t).$$

But $\vec{T}'(t) = \kappa_g(t)s'(t)\vec{U}(t)$, so

$$\kappa_g(t)s'(t) = \tilde{\varphi}'(t). \tag{2.2}$$

Thus, if we call $I = [a, b]$, we have

$$\int_C \kappa_g \, ds = \int_a^b \tilde{\varphi}'(t)\,dt = \tilde{\varphi}(b) - \tilde{\varphi}(a),$$

which concludes the proof.   □

The notion of degree of a continuous function $\mathbb{S}^1 \to \mathbb{S}^1$ can be applied in a wider context than can the rotation index of a curve since the latter requires a curve to be regular, while the former concept, as presented here, only requires $\tilde{\varphi}'(t)$ to be integrable. Another closely related formula to calculate the degree of a curve is the following.

Proposition 2.2.6 *Suppose that a function $f : \mathbb{S}^1 \to \mathbb{S}^1$ is parametrized as $\vec{\gamma} : [a, b] \to \mathbb{R}^2$. If $\vec{\gamma}(t) = (\gamma_1(t), \gamma_2(t))$, then*

$$\deg f = \frac{1}{2\pi} \int_a^b \frac{\gamma_2'(t)}{\gamma_1(t)} \, dt = -\frac{1}{2\pi} \int_a^b \frac{\gamma_1'(t)}{\gamma_2(t)} \, dt.$$

Proof: (Left as an exercise for the reader. See Problem 2.2.7.)   □

In practice, since many functions $f : \mathbb{S}^1 \to \mathbb{S}^1$ as described in the above proposition rely on a parametrization that involves cosine and sine functions, it is very common for the interval $[a, b]$ to be $[0, 2\pi n]$ where n is some positive integer.

Among the many uses of the notion of degree, we can make precise the notion of how often a curve turns around a point.

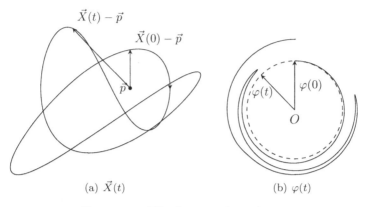

(a) $\vec{X}(t)$ (b) $\varphi(t)$

Figure 2.2: Winding number of -1.

Definition 2.2.7 Let C be a closed, regular, plane curve parametrized by $\vec{X}: I \to \mathbb{R}^2$, and let $\vec{p}$ be a point in the plane. Since C is a closed curve, we may view $\vec{X}$ as a function on $\mathbb{S}^1$. We define the *winding number* $w(p)$ of C around $\vec{p}$ as the degree of the function $f: \mathbb{S}^1 \to \mathbb{S}^1$ that has the parametrization $\vec{\gamma}: I \to \mathbb{R}^2$ with

$$\vec{\gamma}(t) = \frac{\vec{X}(t) - \vec{p}}{\|\vec{X}(t) - \vec{p}\|}.$$

Proposition 2.2.6 gives us a direct method to calculate the winding number of a curve around a point.

Proposition 2.2.8 *Let C be a closed, regular, plane curve parametrized by $\vec{X}: [a, b] \to \mathbb{R}^2$ with $\vec{X}(t) = (x(t), y(t))$, and let $\vec{p}$ be a point in the plane. The winding number of C around $\vec{p}$ is*

$$\frac{1}{2\pi} \int_a^b \frac{(x(t) - p_x)y'(t) - (y(t) - p_y)x'(t)}{\|\vec{X}(t) - \vec{p}\|^2} \, dt.$$

Proof: Applying Proposition 2.2.6 to the calculation of the winding number of C around a point $\vec{p}$, we use the parametrization

$$\vec{\gamma}(t) = (\gamma_1(t), \gamma_2(t)) = \frac{\vec{X}(t) - \vec{p}}{\|\vec{X}(t) - \vec{p}\|} = \left(\frac{x(t) - p_x}{\|\vec{X}(t) - \vec{p}\|}, \frac{y(t) - p_y}{\|\vec{X}(t) - \vec{p}\|} \right)$$

as a parametrization $\gamma: [a, b] \to \mathbb{R}^2$. We calculate

$$\gamma_2'(t) = \frac{y'(t)\|\vec{X}(t) - \vec{p}\|^2 - (y(t) - p_y)\frac{1}{2}(2\vec{X}'(t) \cdot (\vec{X}(t) - \vec{p}))}{\|\vec{X}(t) - \vec{p}\|^3}.$$

After simplification this becomes

$$\gamma_2'(t) = \frac{y'(t)(x(t) - p_x)^2 - (x(t) - p_x)(y(t) - p_y)x'(t)}{\|\vec{X}(t) - \vec{p}\|^3}.$$

Then

$$\frac{\gamma_2'(t)}{\gamma_1(t)} = \frac{(x(t) - p_x)y'(t) - (y(t) - p_y)x'(t)}{\|\vec{X}(t) - \vec{p}\|^2}$$

and the proposition follows from the result of Proposition 2.2.6. □

Example 2.2.9 Consider the ellipse $\vec{X}(t) = (2\cos t, \sin t)$ with $0 \le t \le 2\pi$ and consider the winding number of this ellipse around the point $\vec{p} = (1,0)$. This is a simple configuration and we expect that the winding number should be 1 (or possibly -1 based on the orientation of the parametrization of the ellipse). By Proposition 2.2.8, the winding number is

$$\frac{1}{2\pi} \int_0^{2\pi} \frac{(2\cos t - 1)\cos t - (\sin t)(-2\sin t)}{(2\cos t - 1)^2 + \sin^2 t} \, dt$$

$$= \frac{1}{2\pi} \int_0^{2\pi} \frac{2 - \cos t}{(2\cos t - 1)^2 + \sin^2 t} \, dt.$$

This integral is challenging to evaluate by hand. A computer algebra system evaluates it as exactly 1. This confirms our geometric intuition.

One might expect that the winding number of a curve around a point p depends only on what connected component of the curve's complement p lies in.

In practice, it is often hard to explicitly calculate the winding number of a curve around a point either directly from Definition 2.2.7 or by hand from Proposition 2.2.8. With Proposition 2.2.8 the calculation becomes tractable with a computer algebra system. Interestingly enough, it is usually easy to see what the degree is by plotting out the image $\varphi(t)$. (See Figure 2.2, where in Figure 2.2(b) we have allowed $\varphi(t)$ to come off the circle in order to see its graph more clearly.)

One might expect that the winding number of a curve around a point p depends only on what connected component of the curve's complement p lies in.

Proposition 2.2.10 *Let C be a closed, regular, plane curve parametrized by $\vec{X} : I \to \mathbb{R}^2$, and let p_0 and p_1 be two points in the same connected component of $\mathbb{R}^2 - C$. The winding number of C around p_0 is equal to the winding number of C around p_1.*

Proof: Let $\beta : [0,1] \to \mathbb{R}^2$ be a path such that $\beta(0) = p_0$, $\beta(1) = p_1$, and $\beta(u) \ne \vec{X}(t)$ for any $u \in [0,1]$ and $t \in I$. Define the two-variable function $H : I \times [0,1] \rightarrow \mathbb{S}^1$ by

$$H(t,u) = \frac{\vec{X}(t) - \beta(u)}{\|\vec{X}(t) - \beta(u)\|}.$$

The function H is continuous since $\vec{X}(t) - \beta(u)$ is continuous and $\|\vec{X}(t) - \beta(u)\|$ is also continuous and never 0. For all $u \in [0,1]$, we consider $H(t,u)$ to be a function of t from I to $\mathbb{S}^1$ and define $\tilde{H}(t,u)$ as its lifting, which is a continuous function $\tilde{H} : I \times [0,1] \rightarrow \mathbb{R}$. But then the function of u given by

$$\frac{1}{2\pi}\big(\tilde{H}(2\pi, u) - \tilde{H}(0,u)\big)$$

is continuous and discrete. Thus, it is constant, and hence

$$\frac{1}{2\pi}\big(\tilde{H}(2\pi, 1) - \tilde{H}(0,1)\big) = \frac{1}{2\pi}\big(\tilde{H}(2\pi, 0) - \tilde{H}(0,0)\big),$$

which means that the winding numbers of C around p_0 and around p_1 are equal. □

The winding number of a regular curve around a point offers a strategy to prove the Jordan Curve Theorem (Theorem 2.1.5) in the case when the curve is regular. The strategy is to show that near a point p on the curve C, points on one side of C have a winding number of 0, while on the other side of C points have a winding number of 1 or -1, depending on the orientation of the parametrization given for C. This establishes that a regular, simple, closed curve has at least two connected components. The proof of the Regular Jordan Curve Theorem given in [24, Section A.4] also establishes a fact that is useful in its own right.

Proposition 2.2.11 *Let $\vec{X} : [a,b] \rightarrow \mathbb{R}^2$ be a simple, closed, regular, plane curve. The rotation index of $\vec{X}$ is ± 1.*

PROBLEMS

1. Consider the figure-eight curve $\vec{X}(t) = (\cos t, \sin(2t))$ with $0 \le t \le 2\pi$. Show that the rotation index is 0. [Hint: When performing an appropriate integration, reparametrize by $t = u + \frac{\pi}{2}$ and change the bounds of integration to $[-\pi, \pi]$.]

2. Show that the lemniscate given by

$$\vec{X}(t) = \left(\frac{a \cos t}{1 + \sin^2 t}, \frac{a \cos t \sin t}{1 + \sin^2 t} \right)$$

has a rotation index of 0.

3. Calculate the rotation index of the limaçon de Pascal given by

$$\vec{X}(t) = \big((1 - 2\cos t)\cos t, (1 - 2\cos t)\sin t\big).$$

4. Let $\vec{x}(t) = (\cos 3t \cos t, \cos 3t \sin t)$ be the trefoil curve. Show that $I = [0, \pi]$ is enough of a domain to make $\vec{x}$ into a closed curve. Prove that the rotation index of $\vec{x}$ is 2.

5. Let C be a closed regular plane curve. Suppose that p is a point in the plane such that there exists a ray from p that does not intersect C. Show that the winding number of C around p is 0.

6. Give an example of a curve C (by either a sketch or parametrization) and a point $\vec{p} \in \mathbb{R}^2$ such that the winding number of C around $\vec{p}$ is 2 but that every ray based at $\vec{p}$ intersects C in more than two points.

7. Prove Proposition 2.2.6.

8. The curve C pictured below has the parametrization

$$\vec{X}(t) = \big((3 + \cos(5t))\cos(3t), (3 + \cos(5t))\sin(3t)\big)$$

for $t \in [0, 2\pi]$.

   (a) Calculate explicitly the integral formula for the winding number of C around a point $\vec{p}$ as given in Proposition 2.2.8.

   (b) Use a computer algebra system to calculate this winding number for various points in the plane in all the different connected components of $\mathbb{R}^2 - C$.

   (c) Give an intuitive justification for your results.

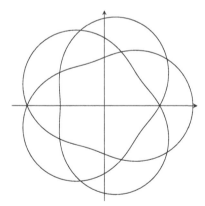

2.3   Isoperimetric Inequality

With the Jordan Curve Theorem at our disposal, we can use Green's Theorem to prove an inequality between the length of a simple closed curve and the area contained in the interior. This inequality, called the isoperimetric inequality, is another example of a global theorem since it relates quantities that take into account the entire curve at once.

Theorem 2.3.1 *Let C be a simple, closed, plane curve with length l, and let A be the area bounded by C. Then*

$$l^2 \geq 4\pi A, \tag{2.3}$$

and equality holds if and only if C is a circle.

Proof: Let L_1 and L_2 be parallel lines that are both tangents to C and such that C is contained in the strip between them. Let $\mathbb{S}^1$ be a circle that is also tangent to both L_1 and L_2 such that $\mathbb{S}^1$ does not intersect C. We call r the radius of this circle, and we set up the coordinate axes in the plane so that the origin is at the center of $\mathbb{S}^1$, the x-axis is perpendicular to L_1 (and L_2), and the y-axis is parallel to L_1 and L_2. See Figure 2.3.

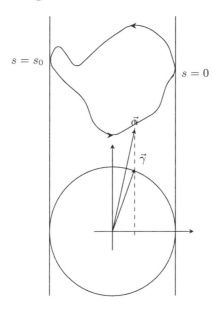

Figure 2.3: Isoperimetric inequality.

Assume that C is parametrized by arc length $\vec{\alpha}(s) = (x(s), y(s))$ and that the parametrization is positively oriented at the points of tangency with L_1 and L_2. Also, let's call $s = 0$ the parameter location for the tangency point $C \cap L_1$ and $s = s_0$ the parameter for $C \cap L_2$. We can also assume that $\mathbb{S}^1$ is parametrized by $\vec{\gamma}(s)$, where $\vec{\gamma}(s)$ is the intersection of

- the upper half of $\mathbb{S}^1$ with the vertical line through $\bar{\alpha}(s)$ if $0 \leq s \leq s_0$;

- the lower half of $\mathbb{S}^1$ with the vertical line through $\bar{\alpha}(s)$ if $s_0 \leq s \leq l$.

Notice that in this parametrization for the circle, writing $\bar{\gamma}(s) = (\bar{x}(s), \bar{y}(s))$, we have $x(s) = \bar{x}(s)$.

Let's call A the area enclosed in the curve $\mathcal{C}$, and let $\bar{A}$ be the area of $\mathbb{S}^1$. Using Green's Theorem, we have the following formulas for the area of the circle of the interior of the curve:

$$A = \int_0^l xy' \, ds.$$

We would like to apply the same formula with the parametrization $\bar{\gamma}(s) = (\bar{x}(s), \bar{y}(s))$ of the circle. However, the parametrization is in general not simple. On the other hand, using a strategy similar to that in the proof of Proposition 2.2.1, we can also use Green's Theorem and justify the claim that the area of the circle is

$$\bar{A} = \pi r^2 = -\int_0^l x' \bar{y} \, ds.$$

Adding the two areas, we get

$$A + \pi r^2 = \int_0^l xy' - x' \bar{y} \, ds \leq \int_0^l \sqrt{(xy' - x' \bar{y})^2} \, ds$$

$$\leq \int_0^l \sqrt{(x^2 + \bar{y}^2)((x')^2 + (y')^2)} \, ds = \int_0^l \sqrt{\bar{x}^2 + \bar{y}^2} \, ds = lr.$$

$$(2.4)$$

We now use the fact that the geometric mean of two positive real numbers is less than the arithmetic mean. Thus,

$$\sqrt{A \cdot \pi r^2} \leq \frac{1}{2}(A + \pi r^2) = \frac{1}{2}lr. \qquad (2.5)$$

Squaring both sides and dividing by r^2, we get $4\pi A \leq l^2$. This proves the first part of Theorem 2.3.1.

In order for equality to hold in Equation (2.3), we need equalities to hold in both the inequalities in Equation (2.4). From Equation (2.5), we conclude that $A = \pi r^2$ and $l = 2\pi r$. Furthermore, since A does not change, regardless of the direction of L_1 and L_2, neither does the distance r. We also must have

$$(xy' - x' \bar{y})^2 = (x^2 + \bar{y}^2)((x')^2 + (y')^2) = r^2,$$

which, after expanding both sides, leads to $(xx' + \bar{y}y')^2 = 0$. However, differentiating the relationship $x^2 + \bar{y}^2 = r^2$ for all parameter values s, we find that $xx' + \bar{y}\bar{y}' = 0$ for all s. Thus, for all s, we have $y'(s) = \bar{y}'(s)$, and thus $y(s) = \bar{y}(s) + D$ for some constant D. Since by construction $x(s) = \bar{x}(s)$, then $\mathcal{C}$ is a circle – a translate of $\mathbb{S}^1$ in the direction $(0, D)$. □

Since nothing in the above proof uses second derivatives, we only need the curve $\vec{\alpha}$ to be C^1, i.e., that it has a continuous first derivative. Furthermore, one can generalize the proof to apply to curves that are only piecewise C^1.

2.4   Curvature, Convexity, and the Four-Vertex Theorem

In this section, all curves are assumed to be of class C^2.

Definition 2.4.1 A subset of S of $\mathbb{R}^n$ is called *convex* if for all p and q in S, the line segment $[p, q]$ is a subset of S, i.e., lies entirely inside S. If $\mathcal{C}$ is not convex, it is called *concave*. (See Figure 2.4.)

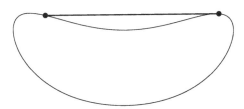

Figure 2.4: A concave curve.

With the Jordan Curve Theorem, we can affirm that a simple, closed curve has an interior. Hence, we can discuss convexity properties of a plane curve if we use the following definition.

Definition 2.4.2 A closed, regular, simple, parametrized curve $\mathcal{C}$ is called convex if the union of $\mathcal{C}$ and the interior of $\mathcal{C}$ form a convex subset of $\mathbb{R}^2$.

In the context of regular curves, it is possible to provide various characterizations for when a curve is convex based on a curve's position with respect to its tangent lines.

Proposition 2.4.3 *A regular, closed, plane curve $\mathcal{C}$ is convex if and only if it is simple and its curvature κ_g does not change sign.*

Proof: The proof involves showing that if the curvature function $\kappa_g(s)$ with respect to arc length changes sign at s_0, then there exist a value $s_0 - \varepsilon$ near and before s_0 and a value $s_0 - \varepsilon$ near and after s_0 such that the segment connecting the points corresponding to s_0 and $s_0 - \varepsilon$ and the segment connecting the points corresponding to s_0 and $s_0 + \varepsilon$ cannot be in the same connected component of $\mathbb{R}^2 - C$.

Let $\vec{\alpha} : [0, l] \to \mathbb{R}^2$ be a parametrization by arc length with positive orientation for the curve C. Let $\vec{T} : [0, l] \to \mathbb{S}^1$ be the tangential indicatrix, and let $\theta : [0, l] \to \mathbb{R}$ be the total angle function associated to the rotation index. By Equation (2.2), $\kappa_g(s) = \theta'(s)$, so the condition that $\kappa_g(s)$ does not change sign is equivalent to $\theta(s)$ being monotonic.

We first prove that convexity implies that C is simple. This follows from the definition of convexity, which assumes that C possesses an interior and, therefore, that $\mathbb{R}^2 - C$ has only two components. By the Jordan Curve Theorem, since we already assume C is regular and closed, for it to possess an interior, it must be simple.

Assume from now on that C is simple. Suppose that $\kappa_g(s)$ changes sign on $[0, l]$. At a point $s = a$, define the height function for all $s \in [0, l]$ by

$$h_a(s) = (\vec{\alpha}(s) - \vec{\alpha}(a)) \cdot \vec{U}(a).$$

This measures the height of the point $\vec{\alpha}(s)$ from the tangent line L_a to C at $\vec{\alpha}(a)$, with $\vec{U}(a)$ being considered the positive direction. The derivatives of this height function are

$$h'(s) = \vec{\alpha}'(s) \cdot \vec{U}(a) = \vec{T}(s) \cdot \vec{U}(a),$$
$$h''(s) = \kappa_g(s)\vec{U}(s) \cdot \vec{U}(a).$$

Consequently, at $s = a$, the height function satisfies $h'(a) = 0$ and $h''(a) = \kappa_g(a)$.

From the definition of differentiation, one deduces that in a neighborhood of a, say over the interval $(a - \delta, a + \delta)$, the projection $p_a : C \to L_a$ of C onto L_a is a bicontinuous bijection (i.e., both the function and its inverse are continuous). Let

$$p_a(a - \delta) = \vec{\alpha}(a) - \varepsilon_1 \vec{T}(a) \qquad \text{and} \qquad p_a(a + \delta) = \vec{\alpha}(a) + \varepsilon_2 \vec{T}(a),$$

where $\varepsilon_1, \varepsilon_2 > 0$. Define the function $g : (-\varepsilon_1, \varepsilon_2) \to [0, l]$ as follows. Let $u(s) = (\vec{\alpha}(a+s) - \vec{\alpha}(a)) \cdot \vec{T}(a)$. This corresponds to the projection of $\vec{\alpha}(a + s) - \vec{\alpha}(a)$ onto the tangent line, or in other words the $\vec{T}$ component of the curve near $\vec{\alpha}(a)$ in the frame with center $\vec{\alpha}(a)$ with ordered basis $(\vec{T}(a), \vec{U}(a))$. As discussed above, this projection is a bijection over $(-\delta, \delta)$ with its image. Let $g : (\varepsilon_1, \varepsilon_2) \to (-\delta, \delta)$ be the inverse function of $u(s)$.

Finally, define $f : (-\varepsilon_1, \varepsilon_2) \to \mathbb{R}$ by $f = h_a \circ g$. The graph of the function f placed in the frame $\left(\vec{T}(a), \vec{U}(a) \right)$ with origin $\vec{\alpha}(a)$ traces out the curve $\mathcal{C}$ in a neighborhood of $s = a$ (Figure 2.5).

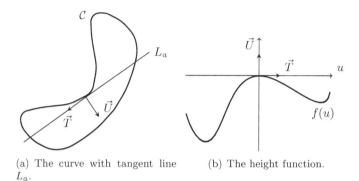

(a) The curve with tangent line L_a.   (b) The height function.

Figure 2.5: Height function $f(u)$.

The derivatives of f are

$$f'(u) = h'(g(u))g'(u),$$
$$f''(u) = h''(g(u))(g'(u))^2 + h'(g(u))g''(u).$$

Since $g(0) = a$ and $h'(a) = 0$, without knowing $g'(0)$, we deduce that $f(0) = f'(0) = 0$ and that $f''(0)$ has the same sign as $\kappa_g(a)$.

If $\kappa_g(a) < 0$, then from basic calculus we deduce that f is concave down over $(-\varepsilon_1, \varepsilon_2)$, which implies that every line segment between points on the graph of f forms a chord that is below the graph. As an example, use the segment between $(u_1, f(u_1))$ and $(u_2, f(u_2))$. Furthermore, since curve $\mathcal{C}$ has a positive orientation, the interior of $\mathcal{C}$ is in the direction of $\vec{U}(a)$, which corresponds to being above the graph of f. Thus, the line segment between the points $\vec{\alpha}(g(u_1))$ and $\vec{\alpha}(g(u_2))$ is in the exterior of the curve. Consequently, this proves that κ_g changes sign if and only if $\mathcal{C}$ is not convex and the proposition follows.   □

This proposition and a portion of the proof leads to another more geometric characterization of convex curves.

Proposition 2.4.4 *A regular, simple, closed curve $\mathcal{C}$ is convex if and only if it lies on one side of every tangent line to $\mathcal{C}$.*

Proof: Suppose we parametrize $\mathcal{C}$ by arc length with a positive orientation. From the proof of Proposition 2.4.3, one observes that at

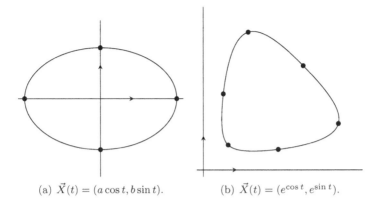

(a) $\vec{X}(t) = (a\cos t, b\sin t)$.   (b) $\vec{X}(t) = (e^{\cos t}, e^{\sin t})$.

Figure 2.6: Two closed curves and their vertices.

any point p of the curve $\mathcal{C}$ where the curvature is negative (given the positive orientation), in a neighborhood of p, the tangent line L to $\mathcal{C}$ at p is in the interior of $\mathcal{C}$. Consequently, since $\mathcal{C}$ is bounded and L is not, L must intersect the curve $\mathcal{C}$ at another point. By Proposition 2.4.3, this proves that $\mathcal{C}$ is concave if and only if there exists a tangent line that has points in the interior of $\mathcal{C}$ and the exterior of $\mathcal{C}$.   □

In Section 1.3, we gave the term vertex for a point on the curve where the curvature reaches an extremum. More precisely, for any regular parametrized curve $\vec{X} : I \to \mathbb{R}^2$, we defined a vertex to be a point $\vec{X}(t_0)$ where $\kappa_g'(t_0) = 0$. Note that since the curvature of a curve is independent of the parametrization up to a possible change of sign, vertices are independent of any parametrization of $\mathcal{C}$.

The concept of a vertex is obviously a local property of the curve, but if one were to experiment with a variety of closed curves, one would soon guess that there must be a restriction on the number of vertices. In Problem 1.3.15, the reader calculated that the noncircular ellipse has four vertices. Figure 2.6(b) shows another simple closed curve with six vertices. (In each figure, the dots indicate the vertices of the curve.)

Theorem 2.4.5 (Four-Vertex Theorem)   *Every simple, closed, regular, convex, plane curve $\mathcal{C}$ has at least four vertices.*

Proof: Let $\vec{\alpha} : [0, l] \to \mathbb{R}^2$ be a parametrization by arc length for the curve $\mathcal{C}$. Since $[0, l]$ is compact and $\kappa_g : [0, l] \to \mathbb{R}$ is continuous, then by the Extreme Value Theorem, κ_g attains both a maximum and a minimum over $[0, l]$. These values already assure that $\mathcal{C}$ has two vertices. Call these points p and q and call L the line between them. Let β be the arc along $\mathcal{C}$ from p to q, and let γ be the arc along $\mathcal{C}$ from q to p.

We claim that β and γ are contained in opposite half-planes defined by the line L. Assume one of the arcs is not contained in one of the half-planes. Then $\mathcal{C}$ meets L at another point r. By convexity, in order for the line segments $[p, q]$, $[p, r]$, and $[q, r]$ to all lie inside $\mathcal{C}$, all three points would need to have L as their tangent line to $\mathcal{C}$. By Problem 2.4.1, this is a contradiction. Assume now that the arcs β and γ are contained in half-planes but in the same half-plane. Then again by convexity, the only possibility is that one of the arcs is a line segment along L. Then along that line segment, the curvature $\kappa_g(s)$ is identically 0. However, this implies that the curvature at p and q is 0, but since p and q are extrema of κ_g, this would force $\mathcal{C}$ to be a line. This is a contradiction since $\mathcal{C}$, being closed, is bounded.

If there are no other vertices on the curve, then $\kappa_g'(s)$ does not change sign on β or on γ. If L has equation $Ax + By + C = 0$, then

$$\int_0^l (Ax + By + C)\frac{d\kappa_g}{ds}ds \tag{2.6}$$

is positive. However, this leads to a contradiction because for all real constants A, B, and C, the integral in Equation (2.6) is always 0 (Problem 2.4.2).

This proves that there is at least one other vertex. But then, since $\kappa_g'(s)$ changes sign at least three times, it must change signs a fourth time as well. Hence, $\kappa_g'(s)$ has at least four 0s. $\square$

PROBLEMS

1. A *bitangent* line L to a regular curve $\mathcal{C}$ is a line that is tangent to $\mathcal{C}$ at a minimum of two points p and q such that between p and q on $\mathcal{C}$, there are points that do not have L as a tangent line. Prove that a simple, regular curve is not convex if and only if it has a bitangent line.

2. Let $\vec{X} : [0, l] \to \mathbb{R}^2$ be a regular, closed, plane curve parametrized by arc length and write $\vec{X}(s) = (x(s), y(s))$.

   (a) Show that $x'' = -\kappa y'$ and $y'' = \kappa x'$.

   (b) Prove that

   $$\int_0^l x(s)\kappa_g'(s)\,ds = -\int_0^l \kappa_g(s)x'(s)\,ds,$$

   and do the same for $y(s)$ instead of $x(s)$.

   (c) Use the above to show that for any constants A, B, and C, we have

   $$\int_0^l (Ax + By + C)\frac{d\kappa_g}{ds}\,ds = 0.$$

3. If a closed curve C is contained inside a disk of radius r, prove that there exists a point $P \in C$ such that the curvature $\kappa_g(P)$ of C at P satisfies

$$|\kappa_g(P)| \geq \frac{1}{r}.$$

4. Let $\vec{\alpha} : I \to \mathbb{R}^2$ be a regular, closed, simple curve with positive orientation. Define the parallel curve at a distance r as

$$\vec{\beta}(t) = \vec{\alpha}(t) - r\vec{U}(t).$$

Call $l(\vec{\alpha})$ the length of the curve $\vec{\alpha}$ and $A(\vec{\alpha})$ the area of the interior of the curve. Prove the following:

(a) $\vec{\beta}$ is regular and simple if and only if $\kappa_g(t) > -\frac{1}{r}$ for all $t \in I$.

(b) If $\vec{\beta}$ is regular, then $l(\vec{\beta}) = l(\vec{\alpha}) + 2\pi r$.

(c) If $\vec{\beta}$ is regular, then $A(\vec{\beta}) = A(\vec{\alpha}) + rl(\vec{\alpha}) + \pi r^2$.

(d) At any regular point of $\vec{\beta}$, we have

$$\kappa_{\vec{\beta}}(t) = \frac{\kappa_{\vec{\alpha}}(t)}{1 + r\kappa_{\vec{\alpha}}(t)}.$$

5. Show that the limaçon $\vec{X}(t) = \big((1 - 2\cos t)\cos t, (1 - 2\cos t)\sin t\big)$ for $t \in [0, 2\pi]$ has exactly two vertices. Explain why it does not contradict the Four-Vertex Theorem.

6. Let $a > 1$ be a real number and $n \geq 2$ an integer. Consider the polar curves parametrized by $\vec{X}(t) = \big((a + \cos nt)\cos t, (a + \cos nt)\sin t\big)$ for $t \in [0, 2\pi]$. Use Exercise 1.3.14 to find an inequality involving a and n that gives a necessary and sufficient condition for the resulting curve to be convex.

7. (**CAS**) Consider the same family of curves described in the previous exercise.

(a) Use Exercise 1.3.14 to find an equation for the location of the vertices of $\vec{X}(t)$ expressed as

$$P_{a,n}(\cos nt)\sin nt = 0$$

where $P_{a,n}(x)$ is a cubic polynomial in x with parameters a and n in its coefficients.

(b) Consider now the specific curve with $a = 2$ and $n = 4$. From $\sin(4t) = 0$, this curve has vertices where $t = \pi k/4$ for any $k \in \mathbb{Z}$. Show that there are no other vertices besides these.

(c) From a picture of this curve with $a = 2$ and $n = 4$, explain the result you obtained in part (b).

8. (*) Consider the curve in Figure 2.6(b). Determine the coordinates of the vertices of $\vec{X}(t)$. [Hint: Prove that the locus of the curve is symmetric across the line $y = x$.]

9. (*) Let $[A, B]$ be a line segment in the plane, and let $l > AB$. Show that the curve C that joins A and B and has length l such that,

together with the segment $[A, B]$, bounds the largest possible area is an arc of a circle passing through A and B. (See the following figure as a reference.)

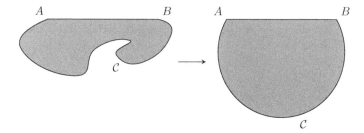

CHAPTER 3

Curves in Space: Local Properties

The local theory of space curves is similar to the theory for plane curves, but differences arise in the richer variety of configurations available for loci of curves in $\mathbb{R}^3$. In the study of plane curves, we introduced the curvature function, a function of fundamental importance that measures how much a curve deviates from being a straight line. As we shall see, in the study of space curves, we again introduce a curvature function that measures how much a space curve deviates from being linear, but we also introduce a torsion function that measures how much the curve twists away from being planar.

3.1 Definitions, Examples, and Differentiation

As in the study of plane curves, one must take some care in defining what one means by a space curve. If I is an interval of $\mathbb{R}$, to call a space curve any function $\vec{X} : I \to \mathbb{R}^3$ (or the image thereof) would allow for separate pieces or even a set of scattered points. By a curve, one typically thinks of a connected set of points, and, just as with plane curves, the desired property is continuity.

Instead of repeating the definitions provided in Section 1.1, we point out that the definition for the limit of a vector function (Definition 1.1.4), the definition for continuity of a vector function (Definition 1.1.5), Proposition 1.1.6, and Corollary 1.1.7 continue to hold for vector functions into $\mathbb{R}^n$.

Definition 3.1.1 Let I be an interval of $\mathbb{R}$. A *parametrized curve* in $\mathbb{R}^n$ is a continuous function $\vec{X} : I \to \mathbb{R}^n$. If for all $t \in I$ we have

$$\vec{X}(t) = \big(x_1(t), x_2(t), \ldots, x_n(t)\big),$$

then the functions $x_i : I \to \mathbb{R}$ for $1 \leq i \leq n$ are called the *coordinate functions* or *parametric equations* of the parametrized curve. The *locus* is the image $\vec{X}(I)$ of the parametrized curve. If $n = 3$, we call $\vec{X}$ a *space curve*.

DOI: 10.1201/9781003295341-3

In order to develop an intuition for space curves, we present a number of examples. These illustrate only some of the great variability in the shape of parametric curves but provide a short library for examples we revisit later.

 Example 3.1.2 (Lines) In space or in $\mathbb{R}^n$, a line is uniquely defined by a point $\vec{p}$ on the line and a nonzero direction vector $\vec{v}$. Parametric equations for a line are then $\vec{X} : \mathbb{R} \to \mathbb{R}^n$ with $\vec{X}(t) = \vec{v}t + \vec{p}$.

 Example 3.1.3 (Planar Curves) We define a planar curve as any space curve whose image lies in a plane. A parametrized plane curve $\vec{X} : I \to \mathbb{R}^2$ with $\vec{X}(t) = (x(t), y(t))$ can be considered as a parametrized space curve by setting $\vec{X}(t) = (x(t), y(t), 0)$. This becomes a planar curve.

More generally, recall that if $\vec{a}$ and $\vec{b}$ are linearly independent vectors of $\mathbb{R}^3$ and $\vec{p}$ is any point in $\mathbb{R}^3$, then the plane spanned by $\{\vec{a}, \vec{b}\}$ through $\vec{p}$ can be described by $\vec{p} + u\vec{a} + v\vec{b}$, where $u, v \in \mathbb{R}$. Note that a normal vector to this plane is $\vec{N} = \vec{a} \times \vec{b}$. Consequently, a planar curve in this plane will be given by a parametrized curve of the form

$$\vec{X}(t) = \vec{p} + u(t)\vec{a} + v(t)\vec{b},$$

where $u(t)$ and $v(t)$ are continuous functions $I \to \mathbb{R}$, where I is some interval of $\mathbb{R}$.

In order to properly devise parametrizations for specific curves on planes in $\mathbb{R}^3$, a judicious choice of the vectors $\vec{a}$ and $\vec{b}$ may be necessary. The usual basis vectors of $(1, 0)$ and $(0, 1)$ of $\mathbb{R}^2$ are not only linearly independent but they form an orthonormal set, both mutually perpendicular and of unit length. So, for example, a space curve of the form

$$\vec{X}(t) = \vec{p} + r(\cos t)\vec{a} + r(\sin t)\vec{b}$$

will not in general be a circle but an ellipse.

As a specific example, we can obtain the equations for a circle of radius 7 in the plane $3x - 2y + 2z = 6$ with center $\vec{p} = (2, 1, 1)$ as follows. A normal vector to the plane is $\vec{N} = (3, -2, 2)$. Two vectors that are not collinear and in this plane are $\vec{a} = (2, 3, 0)$ and $\vec{b} = (2, 0, -3)$. We use the Gram-Schmidt orthonormalization process to find an orthonormal basis of $\text{Span}(\vec{a}, \vec{b})$. The first vector in the orthonormal basis is

$$\vec{u_1} = \frac{\vec{a}}{\|\vec{a}\|} = \left(\frac{2}{\sqrt{13}}, \frac{3}{\sqrt{13}}, 0\right).$$

Then calculate

$$\vec{v_2} = \vec{b} - \text{proj}_{\vec{u_1}} \vec{b} = \vec{b} - (\vec{u_1} \cdot \vec{b})\vec{u_1} = \left(\frac{18}{13}, -\frac{12}{13}, -3\right)$$

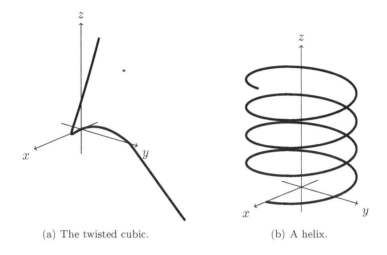

(a) The twisted cubic.   (b) A helix.

Figure 3.1: Two important space curves.

and get a second basis vector of

$$\vec{u}_2 = \frac{\vec{v}_2}{\|\vec{v}_2\|} = \left(\frac{18}{\sqrt{1729}}, -\frac{12}{\sqrt{1729}}, -\frac{39}{\sqrt{1729}} \right).$$

Finally, a parametrization for the circle we are looking for is given by

$$\vec{X}(t) = \vec{p} + (7\cos t)\vec{u}_1 + (7\sin t)\vec{u}_2.$$

Example 3.1.4 (Twisted Cubic) One of the simplest nonplanar
curves is called the twisted cubic and has the parametrization $\vec{X}(t) = (t, t^2, t^3)$ (see Figure 3.1(a)).

We might wonder why this should be considered the simplest non-planar curve. After all, could we not create a nonplanar curve with just quadratic polynomials? In fact, the answer is no. A curve $\vec{X}(t)$ with quadratic polynomials for coordinate functions can be written as

$$\vec{X}(t) = \vec{a}t^2 + \vec{b}t + \vec{c},$$

where $\vec{a}$, $\vec{b}$, and $\vec{c}$ are linearly independent vectors in $\mathbb{R}^3$. If $\vec{a}$ and $\vec{b}$ are different and nonzero, then the image of $\vec{X}(t)$ lies in the plane through $\vec{c}$ with direction vectors $\vec{a}$ and $\vec{b}$, in other words, the plane through $\vec{c}$ with normal vector $\vec{a} \times \vec{b}$.

To see clearly that the twisted cubic is not planar, consider the four points $\vec{X}(-1) = (-1, 1, -1)$, $\vec{X}(0) = (0, 0, 0)$, $\vec{X}(1) = (1, 1, 1)$, and $\vec{X}(2) = (2, 4, 8)$. The first three points lie in the plane $x = z$ but the fourth does not.

Example 3.1.5 (Cylindrical Helix) A cylindrical helix is a space curve that wraps around a circular cylinder, climbing in altitude at a constant rate. We have equations

$$\vec{X}(t) = (a\cos t, a\sin t, bt).$$

See Figure 3.1(b) for an example with $\vec{X}(t) = (2\cos t, 2\sin t, 0.2t)$.

Example 3.1.6 The coordinate functions of the parametrized curve $\vec{X}(t) = (at\cos t, at\sin t, bt)$ satisfy the algebraic equation

$$\frac{x^2}{a^2} + \frac{y^2}{a^2} - \frac{z^2}{b^2} = 0$$

for all t. Thus, the image of $\vec{X}$ lies on the circular cone described by this equation.

Example 3.1.7 Consider the parametric curve

$$\vec{X}(t) = (a\cosh t \cos t, a\cosh t \sin t, b\sinh t)$$

with $t \in \mathbb{R}$. It is not hard to see that the coordinate functions of $\vec{X}$ satisfy

$$\frac{x^2}{a^2} + \frac{y^2}{a^2} - \frac{z^2}{b^2} = 1$$

so that the image of $\vec{X}$ lies on the hyperboloid of one sheet. See Figure 3.2a for an example with

$$\vec{X}(t) = (\cosh t \cos(10t), \cosh t \sin(10t), \sinh t).$$

(Recall that hyperbolic trigonometric functions are defined as

$$\cosh t = \frac{e^t + e^{-t}}{2} \qquad \text{and} \qquad \sinh t = \frac{e^t - e^{-t}}{2}$$

and that they satisfy the relation $\cosh^2 t - \sinh^2 t = 1$ for all $t \in \mathbb{R}$.)

Example 3.1.8 (Space Cardioid) Consider the parametrization

$$\vec{X}(t) = ((1 - \cos t)\cos t, (1 - \cos t)\sin t, \sin t)$$

with $t \in [0, 2\pi]$. This curve is called the *space cardioid* (see Figure 3.2(b)). One interesting property of this curve is that it is a closed curve in $\mathbb{R}^3$. Projected onto the xy-plane it gives the cardioid, but its locus in $\mathbb{R}^3$ has no critical points. In an intuitive sense, we have stretched the cardioid out of the plane and removed its cusp.

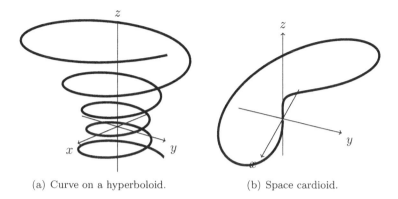

(a) Curve on a hyperboloid.   (b) Space cardioid.

Figure 3.2: Examples of parametric curves.

Using the same vocabulary as in Section 1.1, we call a *reparame-trization* of a parametrized curve $\vec{X} : I \to \mathbb{R}^n$ any other continuous function $\vec{\xi} : J \to \mathbb{R}^n$ defined by $\vec{\xi} = \vec{X} \circ g$ for some surjective function $g : J \to I$. Then the image $\mathcal{C} \subseteq \mathbb{R}^n$ of $\vec{\xi}$ is the same as the image of $\vec{X}$. When g is not surjective, the image of $\vec{\xi}$ could be a proper subset of $\mathcal{C}$, and we do not call $\vec{\xi}$ a reparametrization. In addition, the reparametrization $\vec{\xi}$ of $\vec{X}$ is

- *regular* if g is continuously differentiable and $g'(t) \neq 0$;

- *positively oriented* if it is regular and $g'(t) > 0$ for all $t \in J$;

- *negatively oriented* if it is regular and $g'(t) < 0$ for all $t \in J$.

The definition in Equation (1.2) of the derivative for a vector function was purposefully presented irrespective of the dimension. We restate the definition in an alternate form that is sometimes better suited for proofs.

Definition 3.1.9 Let I be an interval in $\mathbb{R}$, and let $t_0 \in I$. If $\vec{f} : I \to \mathbb{R}^n$ is a continuous vector function, we say that $\vec{f}$ is *differentiable* at t_0 with derivative $\vec{f}'(t_0)$ if there exists a vector function $\vec{\varepsilon}$ such that

$$\vec{f}(t_0 + h) = \vec{f}(t_0) + \vec{f}'(t_0)h + h\vec{\varepsilon}(h) \qquad \text{and} \qquad \lim_{h \to 0} \|\vec{\varepsilon}(h)\| = 0.$$

It follows as a consequence of Proposition 1.1.6 (modified to vector functions in $\mathbb{R}^n$) that a vector function $\vec{X} : I \to \mathbb{R}^n$ is differentiable at a point if and only if all its coordinate functions are differentiable at that point. Furthermore, if $\vec{X} : I \to \mathbb{R}^n$ is a continuous vector function that is differentiable at $t = t_0$ and if the coordinate functions of $\vec{X}$ are

$$\vec{X}(t) = \big(x_1(t), x_2(t), \ldots, x_n(t)\big),$$

then the derivative $\vec{X}'(t_0)$ is

$$\vec{X}'(t_0) = \big(x_1'(t_0), x_2'(t_0), \ldots, x_n'(t_0)\big).$$

As in $\mathbb{R}^2$, for any vector function $\vec{X} : I \to \mathbb{R}^n$, borrowing from the language of trajectories in mechanics, we often call $\vec{X}$ the *position function*, $\vec{X}'$ the *velocity function*, and $\vec{X}''$ the *acceleration function*. Furthermore, we define the *speed* function associated to $\vec{X}$ as the function $s' : I \to \mathbb{R}$ defined by

$$s'(t) = \|\vec{X}'(t)\|.$$

It is also a simple matter to define equations for the tangent line to a parametrized curve $\vec{X}(t)$ at $t = t_0$ as long as $\vec{X}'(t_0) \neq \vec{0}$. The parametric equations for the tangent line are

$$\vec{L}(u) = \vec{X}(t_0) + u\,\vec{X}'(t_0).$$

Proposition 3.1.10 *Proposition 1.2.3 holds for differentiable vector functions $\vec{v}$ and $\vec{w}$ from $I \subset \mathbb{R}$ into $\mathbb{R}^n$. Furthermore, suppose that $\vec{v}(t)$ and $\vec{w}(t)$ are vector functions into $\mathbb{R}^3$ that are defined on and differentiable over an interval $I \subset \mathbb{R}$. If $\vec{X}(t) = \vec{v}(t) \times \vec{w}(t)$, then $\vec{X}$ is differentiable over I and*

$$\vec{X}'(t) = \vec{v}'(t) \times \vec{w}(t) + \vec{v}(t) \times \vec{w}'(t).$$

Proof: (Left as an exercise for the reader. See Problem 3.1.14.)   □

As in Section 1.1, the problems in this section focus on the properties of vectors and vector functions in $\mathbb{R}^3$.

PROBLEMS

1. Calculate the velocity, acceleration, and speed of the twisted cubic $\vec{X}(t) = (t, t^2, t^3)$.

2. Calculate the velocity, acceleration, and speed of the space cardioid $\vec{X}(t) = ((1 - \cos t)\cos t, (1 - \cos t)\sin t, \sin t)$.

3. Calculate the velocity, acceleration, and speed of the parametrized curve $\vec{X}(t) = (\tan^{-1}(t), \sin t, \cos 2t)$.

4. Find a parametrization of the intersection of the cylinder $x^2 + z^2 = 1$ with the plane $x + 2y + z = 2$.

5. Let $\vec{a}$, $\vec{b}$, and $\vec{c}$ be three vectors in $\mathbb{R}^3$. Prove that
$$(\vec{a} \times \vec{b}) \cdot \vec{c} = (\vec{b} \times \vec{c}) \cdot \vec{a} = (\vec{c} \times \vec{a}) \cdot \vec{b}.$$

6. Let $\vec{a}$, $\vec{b}$, $\vec{c}$, and $\vec{d}$ be four vectors in $\mathbb{R}^3$. Prove that
$$(\vec{a} \times \vec{b}) \cdot (\vec{c} \times \vec{d}) = \begin{vmatrix} (\vec{a} \cdot \vec{c}) & (\vec{b} \cdot \vec{c}) \\ (\vec{a} \cdot \vec{d}) & (\vec{b} \cdot \vec{d}) \end{vmatrix}.$$

7. Let $\vec{a}, \vec{b}, \vec{c}$ be constant vectors in $\mathbb{R}^3$. Calculate the derivatives of
   (a) $\vec{X}(t) = (2t\vec{a} + t^2\vec{b}) \times (\vec{b} + (4 - t^3)\vec{c})$;
   (b) $f(t) = (2t\vec{a} + t^2\vec{b}) \cdot (\vec{b} + (4 - t^3)\vec{c})$.

8. Modify the parametrization of a helix to give the parametrization $\vec{X} : \mathbb{R} \to \mathbb{R}^3$ of a curve that lies on the cylinder $x^2 + y^2 = 1$ such that the z-component has $(0, \infty)$ as its image.

9. Give parametric equations of the tangent lines to $\vec{X}(t) = (t^2 - 1, 3t - t^3, t^4 - 4t^2)$ at $(-1, 0, 0)$ and also where $t = 2$.

10. Consider the parametrized space curve $\vec{X}(t) = (t \cos t, t \sin t, t)$ with $t \in \mathbb{R}$.
    (a) Find the equation of the tangent line where $t = \pi/3$.
    (b) Do any tangent lines of $\vec{X}(t)$ intersect the x-axis and if so, then for what values of t and at what points on the x-axis?

11. Consider the parametrization space curve $\vec{X}(t) = (\cos t, \sin, \sin 2t)$ with $t \in [0, 2\pi]$. Find the equations of the tangent lines at the points where the curve intersects the xy-plane.

12. Let $\vec{\alpha}(t)$ be a regular, plane, parametrized curve. View the xy-plane as a subset of $\mathbb{R}^3$. Let $\vec{p}$ be a fixed point in the plane and $\vec{u}$ a fixed vector. Let $\theta(t)$ be the angle $\vec{\alpha}(t) - \vec{p}$ makes with the direction $\vec{u}$. Prove that (up to a change in sign)
$$\theta'(t) = \frac{\|\vec{\alpha}'(t) \times (\vec{\alpha}(t) - \vec{p})\|}{\|\vec{\alpha}(t) - \vec{p}\|^2}.$$
Conclude that the angle function $\theta(t)$ of $\vec{\alpha}(t) - \vec{p}$ with respect to the direction $\vec{u}$ has a local extremum at a point t_0 if and only if $\vec{\alpha}'(t_0)$ is parallel to $\vec{\alpha}(t_0) - \vec{p}$.

13. Consider the space curve $\vec{X}(t) = (t \cos t, t \sin t, t)$. Find where the extremal values of the angle between $\vec{X}'$ and $\vec{X}''$ occur and determine whether they are maxima or minima.

14. Let $\vec{\alpha}(t)$ and $\vec{\beta}(t)$ be two differentiable functions $\mathbb{R} \to \mathbb{R}^3$. Prove that if $\vec{\gamma}(t) = \vec{\alpha}(t) \times \vec{\beta}(t)$, then $\vec{\gamma}'(t) = \vec{\alpha}'(t) \times \vec{\beta}(t) + \vec{\alpha}(t) \times \vec{\beta}'(t)$.

15. Consider the plane in $\mathbb{R}^3$ given by the equation $ax + by + cz + f = 0$ and a point $P = (x_0, y_0, z_0)$. Prove that the distance between the plane and the point P is
$$d = \frac{|ax_0 + by_0 + cz_0 + f|}{\sqrt{a^2 + b^2 + c^2}}.$$

16. Determine the angle of intersection between the lines $\vec{u}(t) = (2t - 1, 3t + 2, -2t + 3)$ and $\vec{v}(t) = (3t - 1, 5t + 2, 3t + 3)$.

17. Consider the two lines given by $\vec{u}(t) = (2t - 1, 3t + 2, -2t + 3)$ and $\vec{v}(t) = (3t + 1, -5t - 3, 3t - 1)$. Find the shortest distance between these two lines. [Hint: Consider the function $f(s, t) = \|\vec{u}(s) - \vec{v}(t)\|$.]

18. Consider two nonparallel lines given by the equations

$$\vec{l}_1(s) = \vec{a} + s\vec{u} \qquad \text{and} \qquad \vec{l}_2(t) = \vec{b} + t\vec{v}.$$

Prove that the distance d between these two lines is

$$d = \frac{|(\vec{u} \times \vec{v}) \cdot (\vec{b} - \vec{a})|}{\|\vec{u} \times \vec{v}\|}.$$

19. Let $\vec{X}(t)$ be any parametrized curve that is of class C^3. Prove that

$$\frac{d}{dt}\left(\vec{X}(t) \cdot (\vec{X}'(t) \times \vec{X}''(t))\right) = \vec{X}(t) \cdot (\vec{X}'(t) \times \vec{X}'''(t)).$$

[Hint: Use Problem 3.1.14.]

20. Consider a sphere of radius R and center $\vec{p}$. Prove by a geometric argument that the closest point on this sphere to a point $\vec{q} \neq \vec{p}$ is

$$\vec{p} + R\frac{\vec{q} - \vec{p}}{\|\vec{q} - \vec{p}\|}.$$

We call this closest point the projection of $\vec{q}$ to the sphere. Let $\vec{X} : I \to \mathbb{R}^3$ be a parametrized space curve that does not go through the origin. Deduce that $\vec{X}(t)/\|\vec{X}(t)\|$ parametrizes the projection of the space curve to the unit sphere.

21. Prove that no four distinct points of the twisted cubic lie in a common plane. [Hint: Prove that for all $a, b, c, d \in \mathbb{R}$,

$$\begin{vmatrix} b - a & c - a & d - a \\ b^2 - a^2 & c^2 - a^2 & d^2 - a^2 \\ b^3 - a^3 & c^3 - a^3 & d^3 - a^3 \end{vmatrix} = \begin{vmatrix} 1 & 1 & 1 & 1 \\ a & b & c & d \\ a^2 & b^2 & c^2 & d^2 \\ a^3 & b^3 & c^3 & d^3 \end{vmatrix}$$

and use the Vandermonde determinant identity.]

3.2   Curvature, Torsion, and the Frenet Frame

Let I be an interval of $\mathbb{R}$, and let $\vec{X} : I \to \mathbb{R}^3$ be a differentiable space curve. Following the setup with plane curves, we can talk about the unit tangent vector $\vec{T}(t)$ defined by

$$\vec{T}(t) = \frac{\vec{X}'(t)}{\|\vec{X}'(t)\|}. \tag{3.1}$$

The unit tangent vector is not defined at a value $t = t_0$ if $\vec{X}$ is not differentiable at t_0 or if $\vec{X}'(t_0) = 0$. This leads to the following definition (which can be generalized to curves in $\mathbb{R}^n$).

Definition 3.2.1 Let $\vec{X} : I \to \mathbb{R}^n$ be a parametrized curve. We call a point $t = t_0$ a *critical point* if $\vec{X}'(t_0)$ is not defined or if $\vec{X}'(t_0) = \vec{0}$. A point $t = t_0$ that is not critical is called a *regular point*. A parametrized curve $\vec{X} : I \to \mathbb{R}^n$ is called *regular* if it is of class C^1 and if $\vec{X}'(t) \neq \vec{0}$ for all $t \in I$.

In practice, if $\vec{X}'(t_0) = \vec{0}$ but $\vec{X}'(t) \neq \vec{0}$ for all t in some interval J around t_0, it is possible for

$$\lim_{t \to t_0} \frac{\vec{X}'(t)}{\|\vec{X}'(t)\|} \qquad (3.2)$$

to exist. In such cases, it is common to think of $\vec{T}$ as completed by continuity by calling $\vec{T}(t_0)$ the limit in Equation (3.2).

Since $\vec{T}(t)$ is a unit vector for all $t \in I$, we have $\vec{T} \cdot \vec{T} = 1$ for all $t \in I$. Therefore, if $\vec{T}$ is itself differentiable,

$$\frac{d}{dt}\left(\vec{T}(t) \cdot \vec{T}(t)\right) = 2\vec{T}'(t) \cdot \vec{T}(t) = 0.$$

Thus, the derivative of the unit tangent vector $\vec{T}'$ is perpendicular to $\vec{T}$. At any point t on the curve, where $\vec{T}'(t) \neq \vec{0}$, we define the *principal normal vector* $\vec{P}(t)$ to the curve at t to be the unit vector

$$\vec{P}(t) = \frac{\vec{T}'(t)}{\|\vec{T}'(t)\|}.$$

Again, it is still possible sometimes to complete $\vec{P}$ by continuity even at points $t = t_0$, where $\vec{T}'(t_0) = \vec{0}$. In such cases, it is common to assume that $\vec{P}(t)$ is completed by continuity wherever possible. Nonetheless, by Equation (3.1), the requirement that $\vec{T}$ be differentiable is tantamount to $\vec{X}$ being twice differentiable.

Finally, for a space curve defined over any interval J, where $\vec{T}(t)$ and $\vec{P}(t)$ exist (perhaps when completed by continuity), we complete the set $\{\vec{T}, \vec{P}\}$ to an orthonormal frame by adjoining the *binormal vector* $\vec{B}(t)$ given as

$$\vec{B} = \vec{T} \times \vec{P}.$$

Definition 3.2.2 Let $\vec{X} : I \to \mathbb{R}^3$ be a continuous space curve of class C^2 (i.e., has a continuous second derivative). To each point $\vec{X}(t)$ on the curve, we associate the *Frenet frame* as the ordered triple of vectors $(\vec{T}, \vec{P}, \vec{B})$.

Figure 3.3 illustrates the Frenet frame as it moves through $t = 0$ on the space cardioid

$$\vec{X}(t) = ((1 - \cos t) \cos t, (1 - \cos t) \sin t, \sin t). \qquad (3.3)$$

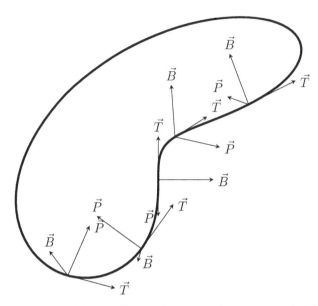

Figure 3.3: Moving Frenet frame on the space cardioid.

(In this figure, the basis vectors of the Frenet frame were scaled down by a factor of $\frac{1}{2}$ to make the picture clearer.) In this example, it is interesting to see that near $t = 0$, though the unit tangent vector doesn't change too quickly, the vectors $\vec{P}$ and $\vec{B}$ rotate quickly about the tangent line. The functions that measure how fast $\vec{T}$ (resp. $\vec{B}$) changes are called the curvature (resp. torsion) functions.

Similar to the basis $\{\vec{T}(t), \vec{U}(t)\}$ for plane curves, the Frenet frame provides a geometrically natural basis in which to study local properties of space curves. We now analyze the derivatives of a space curve in reference to the Frenet frame.

Recall that the speed $s'(t)$ is $\|\vec{X}'(t)\|$. By definition of the unit tangent vector, we have

$$\vec{X}'(t) = s'(t)\vec{T}(t).$$

If $\vec{x}$ is twice differentiable at t, then $\vec{T}'(t)$ exists. If, in addition, $\vec{P}$ is defined at t, then by definition of the principal normal vector, $\vec{T}'$ is parallel to $\vec{P}$. This allows us to define the curvature of a space curve as follows.

 Definition 3.2.3 Let $\vec{X}$ be a regular parametrized curve of class C^2. The curvature $\kappa : I \to \mathbb{R}^+$ of a space curve is

$$\kappa(t) = \frac{\|\vec{T}'(t)\|}{s'(t)}.$$

Note that at any point where $P(t)$ is defined, $\kappa(t)$ is the nonnegative number defined by

$$\vec{T}'(t) = s'(t)\kappa(t)\vec{P}(t).$$

We would like to determine how the other unit vectors of the Frenet frame behave under differentiation. Remember that $\vec{B} = \vec{T} \times \vec{P}$. Taking a derivative of this cross product, we obtain

$$\vec{B}' = \vec{T}' \times \vec{P} + \vec{T} \times \vec{P}' = \vec{T} \times \vec{P}',$$

since $\vec{T}'$ is parallel to $\vec{P}$. However, just as $\vec{T}' \perp \vec{T}$, we have the same for $\vec{P}' \perp \vec{P}$. Thus, since we are in three dimensions, we can write $\vec{P}'(t) = f(t)\vec{T}(t) + g(t)\vec{B}(t)$ for some continuous functions $f, g : I \to \mathbb{R}$. Consequently, we deduce that

$$\vec{B}' = \vec{T} \times (f(t)\vec{T} + g(t)\vec{B}) = f(t)\vec{T} \times \vec{T} + g(t)\vec{T} \times \vec{B} = -g(t)\vec{P}.$$

Thus, the derivative of the unit binormal vector is parallel to the principal normal vector.

Definition 3.2.4 Let $\vec{x} : I \to \mathbb{R}^3$ be a regular space curve of class C^2 for which the Frenet frame is defined everywhere. We define the *torsion function* $\tau : I \to \mathbb{R}$ as the unique function such that

$$\vec{B}'(t) = -s'(t)\tau(t)\vec{P}(t).$$

We now need to determine $\vec{P}'(t)$, and we already know that it has the form $f(t)\vec{T}(t) + g(t)\vec{B}(t)$. However, we can say much more without performing any specific calculations. Since $(\vec{T}, \vec{P}, \vec{B})$ form an orthonormal frame for all t, we have the following equations:

$$\vec{T} \cdot \vec{P} = 0 \quad \text{and} \quad \vec{P} \cdot \vec{B} = 0.$$

Taking derivatives with respect to t, we have

$$\vec{T}' \cdot \vec{P} + \vec{T} \cdot \vec{P}' = 0 \quad \text{and} \quad \vec{P}' \cdot \vec{B} + \vec{P} \cdot \vec{B}' = 0,$$

that is,

$$\vec{P}' \cdot \vec{T} = -\vec{T}' \cdot \vec{P} \quad \text{and} \quad \vec{P}' \cdot \vec{B} = -\vec{P} \cdot \vec{B}'.$$

Thus, we deduce that

$$\vec{P}'(t) = -s'(t)\kappa(t)\vec{T}(t) + s'(t)\tau(t)\vec{B}(t). \tag{3.4}$$

If we assume, as one does in linear algebra, that $\vec{T}$, $\vec{P}$, and $\vec{B}$ are column vectors and write $(\vec{T} \quad \vec{P} \quad \vec{B})$ as the matrix that has these vectors as columns, we can summarize Definition 3.2.3, Definition 3.2.4, and Equation (3.4) in matrix form by

$$\frac{d}{dt}\begin{pmatrix}\vec{T} & \vec{P} & \vec{B}\end{pmatrix} = \begin{pmatrix}\vec{T} & \vec{P} & \vec{B}\end{pmatrix}\begin{pmatrix} 0 & -s'\kappa & 0 \\ s'\kappa & 0 & -s'\tau \\ 0 & s'\tau & 0 \end{pmatrix}. \tag{3.5}$$

As provided, the definitions for curvature and torsion of a space curve do not particularly lend themselves to direct computations when given a particular curve. We now obtain formulas for $\kappa(t)$ and $\tau(t)$ in terms of $\vec{x}(t)$. In order for our formulas to make sense, we assume for the remainder of this section that the parametrized curve $\vec{x} : I \to \mathbb{R}^3$ is regular and of class C^3.

First, to find the curvature, we need the following derivatives:

$$\vec{X}'(t) = s'(t)\vec{T}(t), \tag{3.6}$$

$$\vec{X}''(t) = s''(t)\vec{T}(t) + (s'(t))^2\kappa(t)\vec{P}(t). \tag{3.7}$$

Taking the cross product of these two vectors, we now obtain

$$\vec{X}'(t) \times \vec{X}''(t) = (s'(t))^3\kappa(t)\vec{B}(t). \tag{3.8}$$

However, by definition of curvature for a space curve, $\kappa(t)$ is a non-negative function. Furthermore, $s'(t) = \|\vec{X}'(t)\|$, so we get

$$\kappa(t) = \frac{\|\vec{X}'(t) \times \vec{X}''(t)\|}{s'(t)^3}. \tag{3.9}$$

Secondly, to obtain the torsion function from $\vec{X}(t)$ directly, we will need to take the third derivative

$$\vec{X}'''(t) = s'''(t)\vec{T} + s''(t)s'(t)\kappa(t)\vec{P} + 2s'(t)s''(t)\kappa(t)\vec{P}$$
$$+ (s'(t))^2\kappa'(t)\vec{P} + (s'(t))^3\kappa(t)(-\kappa(t)\vec{T} + \tau(t)\vec{B}),$$

which, without writing the dependence on t explicitly, reads

$$\vec{X}''' = \left(s''' - (s')^3\kappa^2\right)\vec{T} + \left(3s''s'\kappa + (s')^2\kappa'\right)\vec{P} + (s')^3\kappa\tau\vec{B}. \tag{3.10}$$

Taking the dot product of $\vec{X}' \times \vec{X}''$ with $\vec{X}'''$ eliminates all the terms of $\vec{X}'''$ associated to $\vec{T}$ and $\vec{P}$. We get

$$(\vec{X}' \times \vec{X}'') \cdot \vec{X}''' = (s'(t))^6(\kappa(t))^2\tau(t),$$

from which we deduce a formula for $\tau(t)$ only in terms of the curve $\vec{X}(t)$ as follows:

$$\tau(t) = \frac{(\vec{X}'(t) \times \vec{X}''(t)) \cdot \vec{X}'''(t)}{\|\vec{X}'(t) \times \vec{X}''(t)\|^2}. \tag{3.11}$$

Example 3.2.5 (Helices) We wish to calculate the curvature and torsion of the circular helix $\vec{X}(t) = (a\cos t, a\sin t, bt)$. We need the following:

$$\vec{X}'(t) = (-a\sin t, a\cos t, b),$$
$$s'(t) = \sqrt{a^2\sin^2 t + a^2\cos^2 t + b^2} = \sqrt{a^2 + b^2},$$
$$\vec{X}''(t) = (-a\cos t, -a\sin t, 0),$$
$$\vec{X}'(t) \times \vec{X}''(t) = (ab\sin t, -ab\cos t, a^2),$$
$$\vec{X}'''(t) = (a\sin t, -a\cos t, 0).$$

Note that one can easily parametrize the helix by arc length since $s'(t) = \sqrt{a^2 + b^2}$ is a constant function, so $s(t) = t\sqrt{a^2 + b^2}$. Thus, the parametrization by arc length for the circular helix is

$$\vec{X}(s) = \left(a\cos\left(\frac{s}{c}\right), a\sin\left(\frac{s}{c}\right), \frac{b}{c}s\right), \qquad \text{where } c = \sqrt{a^2 + b^2}.$$

We now calculate the curvature and torsion as

$$\kappa(t) = \frac{\|\vec{X}' \times \vec{X}''\|}{(s')^3} = \frac{|a|\sqrt{a^2 + b^2}}{(a^2 + b^2)^{3/2}} = \frac{|a|}{a^2 + b^2},$$
$$\tau(t) = \frac{(\vec{X}' \times \vec{X}'') \cdot \vec{X}'''}{\|\vec{X}' \times \vec{X}''\|^2} = \frac{a^2 b}{a^2(a^2 + b^2)} = \frac{b}{a^2 + b^2}.$$

Consequently, we find that this circular helix has constant curvature and constant torsion.

The circular helix, however, is a particular case of a larger class of curves simply called *helices*. These are defined by requiring that the unit tangent make a constant angle with a fixed line in space. Thus, $\vec{X}(t)$ is a helix if and only if for some unit vector $\vec{u}$,

$$\vec{T} \cdot \vec{u} = \cos\alpha = \text{const.} \tag{3.12}$$

Taking a derivative of Equation (3.12), we obtain

$$\vec{P} \cdot \vec{u} = 0.$$

Hence for all t, $\vec{u}$ is in the plane determined by $\vec{T}$ and $\vec{B}$, so we can write

$$\vec{u} = \cos\alpha\vec{T} + \sin\alpha\vec{B} \text{ or } \vec{u} = \cos\alpha\vec{T} - \sin\alpha\vec{B}$$

for all t. Taking the derivative, we obtain

$$0 = s'\kappa\cos\alpha\vec{P} - s'\tau\sin\alpha\vec{P},$$

which implies that

$$\frac{\kappa}{\tau} = \tan\alpha,$$

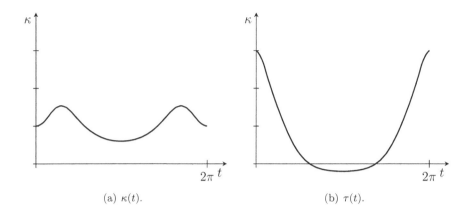

(a) $\kappa(t)$.   (b) $\tau(t)$.

Figure 3.4: Curvature and torsion of the space cardioid.

and thus, for any helix, the ratio of curvature to torsion is a constant. This ratio $\frac{\kappa}{\tau} = \frac{a}{b}$ is called the *pitch* of the helix.

One can follow the above discussion in reverse and also conclude that the converse is true. Therefore, a curve is a helix if and only if the ratio of curvature to torsion is a constant.

Example 3.2.6 (Space Cardioid) We consider the space cardioid again as a follow-up to Figure 3.3. Figure 3.3 shows that in the vicinity of $t = 0$, the Frenet frame twists quickly about the tangent line, even while the tangent line does not move much. This indicates that near 0, $\kappa(t)$ is not large, while $\tau(t)$ is relatively large.

We leave the precise calculation of the curvature and torsion functions to the space cardioid as an exercise for the reader but plot their graphs in Figure 3.4. The graphs of $\kappa(t)$ and $\tau(t)$ justify the intuition provided by Figure 3.3 concerning how the Frenet frame moves through $t = 0$. In particular, the torsion function has a high peak at $t = 0$, which indicates that the Frenet frame rotates quickly about the tangent line.

In some proofs that we will encounter later, it is often useful to assume that a curve is parametrized by arc length. In this case, in all of the above formulas, one has $s' = 1$ and $s'' = 0$ as functions. The transformation properties of the Frenet frame then read

$$\frac{d}{ds} \begin{pmatrix} \vec{T} & \vec{P} & \vec{B} \end{pmatrix} = \begin{pmatrix} \vec{T} & \vec{P} & \vec{B} \end{pmatrix} \begin{pmatrix} 0 & -\kappa & 0 \\ \kappa & 0 & -\tau \\ 0 & \tau & 0 \end{pmatrix}. \qquad (3.13)$$

If $\vec{X}(s)$ is parametrized by arc length and is of class C^3, then Equation

(3.7) gives us $\vec{X}''(s) = \kappa(s)\vec{P}(s)$. Hence, the curvature is given by

$$\kappa(s) = \|\vec{X}''(s)\|.$$

Furthermore, at any point where $\kappa(s) \neq 0$, the torsion function is

$$\tau(s) = \frac{(\vec{X}'(s) \times \vec{X}''(s)) \cdot \vec{X}'''(s)}{\|\vec{X}''(s)\|^2}.$$

A key property of the curvature and torsion functions of a space curve is summarized in the following proposition.

Proposition 3.2.7 *Let $\vec{X} : I \to \mathbb{R}^3$ be a regular parametric curve.*

1. *Suppose that $\vec{X}$ is of class C^2. If the curvature $\kappa(t)$ is identically 0, then the locus of $\vec{X}$ is a line segment.*

2. *Suppose that $\vec{X}$ is of class C^3. If the torsion $\tau(t)$ is identically 0, then the locus of $\vec{X}$ lies in a plane.*

Proof: (Left as an exercise for the reader. See Problem 3.2.14.) $\square$

PROBLEMS

In Problems 3.2.1 through 3.2.5, calculate both the curvature function and the torsion function of the given space curve

1. The twisted cubic: $\vec{X}(t) = (t, t^2, t^3)$ for $t \in \mathbb{R}$.

2. The space cardioid: $\vec{X}(t) = ((1 - \cos t)\cos t, (1 - \cos t)\sin t, \sin t)$ for $t \in [0, 2\pi]$.

3. The curve $\vec{X}(t) = (t, f(t), g(t))$, where f and g are functions of class C^3.

4. The curve $\vec{X} = (a(t - \sin t), a(1 - \cos t), bt)$ for $t \in \mathbb{R}$.

5. The curve $\vec{X}(t) = (t, \frac{1+t}{t}, \frac{1-t^2}{t})$ for $t \in \mathbb{R}^{>0}$.

6. Consider the parametrized curve $\vec{X}(t) = (\cos(t), \sin(t), \sin(2t))$.
   (a) Calculate the curvature and torsion of $\vec{X}(t)$.
   (b) Prove that the curvature is never 0.
   (c) Find the exact locations of the vertices, i.e., where $\kappa'(t) = 0$.

7. Fix a positive real number r. Consider the space curve given by the intersection of the cylinder $(x - r)^2 + y^2 = r^2$ and the sphere $x^2 + y^2 + z^2 = (2r)^2$. Find the parametric equations of this space curve; then calculate its curvature and torsion functions.

8. In Chapter 5, we will encounter a surface called a torus. Curves that lie on torus are often knotted. Define a *torus knot* as a curve parametrized by

$$\vec{X}(t) = \big((a + b\cos(qt))\cos(pt), (a + b\cos(qt))\sin(pt), b\sin(qt)\big)$$

where a and b are real numbers, with $a > b > 0$, and p and q are relatively prime positive integers. Let $a = 2$ and $b = 1$. Calculate the curvature function of the corresponding torus knot.

9. Let $\vec{x}$ be a parametrized curve, and let $\vec{\xi} = \vec{x} \circ g$ be a regular reparametrization of $\vec{X}$.

   (a) Prove that the curvature function κ is unchanged under a regular reparametrization, i.e., that $\kappa_{\vec{x}}(g(u)) = \kappa_{\vec{\xi}}(u)$.
   (b) Prove that the torsion function τ is unchanged under a positively oriented reparametrization and becomes $-\tau$ under a negatively oriented reparametrization.

10. Consider the curve $\vec{X}(t) = (a\sin t\cos t, a\sin^2 t, a\cos t)$.

    (a) Prove that the locus of $\vec{X}$ lies on a sphere.
    (b) Calculate the curvature and torsion functions of $\vec{X}(t)$.

11. If $\vec{X}(t)$ is a parametrization for a planar curve, then for some fixed vectors $\vec{a}$, $\vec{b}$, and $\vec{c}$, with $\vec{b}$ and $\vec{c}$ not collinear, and for some real functions $f(t)$ and $g(t)$, we can write

$$\vec{X}(t) = \vec{a} + f(t)\vec{b} + g(t)\vec{c}.$$

Prove that for a planar curve, its torsion function is identically 0.

12. Consider all space curves of the form $\vec{X}(t) = (\cos t, \sin t, f(t))$, where $f : \mathbb{R} \to \mathbb{R}$ is a smooth function. Find all functions f such that $\vec{X}(t)$ has torsion identically 0.

13. Determine the vector functions $\vec{T}(t)$, $\vec{P}(t)$, and $\vec{B}(t)$ of the Frenet frame for the twisted cubic.

14. (**ODE**) Prove Proposition 3.2.7.

15. Similar to the case of plane curves, we define the evolute to a space curve $\vec{X}(t)$ as the curve

$$\vec{\gamma}(t) = \vec{X}(t) + \frac{1}{\kappa(t)}\vec{P}(t).$$

Prove that the evolute to the circular helix $\vec{X}(t) = (a\cos t, a\sin t, bt)$ is another circular helix. Find the pitch of this new helix.

16. Consider the circular cone with equation

$$\frac{x^2}{a^2} + \frac{y^2}{a^2} = \frac{z^2}{b^2}.$$

    (a) Find parametric equations for a curve that lies on this cone such that the tangent makes a constant angle with the z-axis.
    (b) Prove that they project to logarithmic spirals on the xy-plane.

(c) Find the curvature and torsion functions.

17. Let $\vec{\alpha} : I \to \mathbb{R}^3$ be a parametrized regular curve with $\kappa(t) \neq 0$ and $\tau(t) \neq 0$ for $t \in I$. The curve $\vec{\alpha}$ is called a *Bertrand curve* if there exists another curve $\vec{\beta} : I \to \mathbb{R}^3$ such that the principal normal lines to $\vec{\alpha}$ and $\vec{\beta}$ are equal at all $t \in I$. The curve $\vec{\beta}$ is called the Bertrand mate of $\vec{\alpha}$.

(a) Prove that we can write $\vec{\beta}(t) = \vec{\alpha}(t) + r\vec{P}(t)$ for some constant r.

(b) Prove that $\vec{\alpha}$ is a Bertrand curve if and only if there exists a linear relation

$$a\kappa(t) + b\tau(t) = 1 \qquad \text{for all } t \in I,$$

where a and b are nonzero constants and $\kappa(t)$ and $\tau(t)$ are the curvature and the torsion of $\vec{\alpha}$ respectively.

(c) Prove that a curve $\vec{\alpha}$ has more than one Bertrand mate if and only if $\vec{\alpha}$ is a circular helix.

3.3 Osculating Plane and Osculating Sphere

As with plane curves, if $\vec{X}(t)$ is a regular space curve of class C^2, one can talk about osculating circles to $\vec{X}(t)$ at a point $t = t_0$. Recall that an osculating circle to a curve at a point is a circle with contact of order 2 at that point. Following the proof of Proposition 1.4.4, one determines that at any point $t = t_0$, where $\kappa(t_0) \neq 0$, the osculating circle exists and a parametric formula for it is

$$\vec{\gamma}(t) = \vec{X}(t_0) + \frac{1}{\kappa(t_0)}\vec{P}(t_0) + \frac{1}{\kappa(t_0)}\Big((\sin t)\vec{T}(t_0) - (\cos t)\vec{P}(t_0)\Big). \quad (3.14)$$

Even without reference to osculating circles, given any parametric curve $\vec{X} : I \to \mathbb{R}^3$, the second-order Taylor approximation to $\vec{X}$ at $t = t_0$ is a planar curve with contact of order 2. Furthermore, if $\vec{T}'(t_0) \neq \vec{0}$, this second-order approximation $\vec{f}$ of $\vec{X}$ is

$$\vec{f}(t) = \vec{X}(t_0) + (t - t_0)\vec{X}'(t_0) + \frac{1}{2}(t - t_0)^2\vec{X}''(t_0)$$

$$= \vec{X}(t_0) + \Big(s'(t_0)(t - t_0) + \frac{1}{2}s''(t_0)(t - t_0)^2\Big)\vec{T}(t_0)$$

$$+ \frac{1}{2}s'(t_0)^2\kappa(t_0)(t - t_0)^2\vec{P}(t_0).$$

The vector function $\vec{f}$ is a planar curve that lies in the plane that goes through the point $\vec{X}(t_0)$ and has $\vec{T}(t_0)$ and $\vec{P}(t_0)$ as direction vectors. (If $\kappa(t_0) = 0$, then $\vec{P}(t_0)$ is not strictly defined and might not even be defined by completing by continuity. In this case, the second-order approximation $\vec{f}$ to $\vec{X}$ at t_0 lies on a line.) This leads to the following definition.

Definition 3.3.1 Let $\vec{X} : I \to \mathbb{R}^3$ be a parametrized curve, and let $t_0 \in I$. Suppose that $\vec{X}$ is of class C^2 over an open interval containing t_0 and that $\kappa(t_0) \neq 0$. The *osculating plane* to $\vec{X}$ at $t = t_0$ is the plane through $\vec{X}(t_0)$ spanned by $\vec{T}(t_0)$ and $\vec{P}(t_0)$. In other words, the osculating plane is the set of points $\vec{u} \in \mathbb{R}^3$ such that

$$\vec{B}(t_0) \cdot (\vec{u} - \vec{X}(t_0)) = 0.$$

Note that from (3.8), $\vec{B}(t_0)$ is parallel to $\vec{X}'(t_0) \times \vec{X}''(t_0)$, so the osculating plane also has the equation $(\vec{X}'(t_0) \times \vec{X}''(t_0)) \cdot (\vec{u} - \vec{X}(t_0)) = 0$ in points $\vec{u} \in \mathbb{R}^3$.

We introduced the notion of order of contact between two curves in Section 1.4, but we can also talk about the order of contact between a curve C and a surface Σ by defining this latter notion as the order of contact between C and C_2, where C_2 is the orthogonal projection of C onto Σ.

Proposition 3.3.2 *Let $\vec{X} : I \to \mathbb{R}^3$ be a regular parametrized curve of class C^2. The osculating plane to $\vec{X}$ at $t = t_0$ is the unique plane in $\mathbb{R}^3$ of contact order 2 or greater. Furthermore, assuming $\vec{X}$ is of class C^3, the osculating plane has contact order 3 or greater if and only if $\tau(t_0) = 0$ or $\kappa(t_0) = 0$.*

Proof: Set $A = \vec{X}(t_0)$ and reparametrize $\vec{X}$ by arc length so that $s = 0$ corresponds to point A. The orthogonal distance $f(s)$ between $\vec{X}(s)$ and the osculation plane $\mathcal{P}$ is

$$f(s) = |\vec{B}(0) \cdot (\vec{X}(s) - \vec{X}(0))|.$$

Using the Taylor approximation of $\vec{X}(s)$ near $s = 0$; Equations (3.6), (3.7), and (3.10); and the fact that $\vec{B} \cdot \vec{T} = \vec{B} \cdot \vec{P} = 0$, we deduce that

$$f(s) = \left| \frac{1}{6}\kappa(0)\tau(0)s^3 + \text{ higher order terms} \right|.$$

Thus,

$$\lim_{s \to 0} \frac{f(s)}{s^3} = \frac{|\kappa(0)\tau(0)|}{6}$$

and the proposition follows.   □

We can now interpret the sign of the torsion function $\tau(t)$ of a parametrized curve in terms of the curve's position with respect to

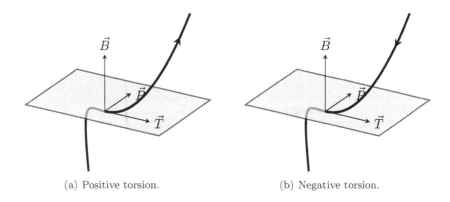

(a) Positive torsion.   (b) Negative torsion.

Figure 3.5: Torsion and the osculating plane.

its osculating plane at a point. In fact, $\tau(t_0) > 0$ at $\vec{x}(t_0)$ when the curve comes up through the osculating plane (where the binormal $\vec{B}$ defines the up direction) and $\tau(t_0) < 0$ when the curve goes down through the osculating plane (see Figure 3.5).

The osculating plane along with two other planes form what is called the moving trihedron, which consists of the coordinate planes in the Frenet frame. The plane through $\vec{X}(t_0)$ and spanned by the principal normal and binormal is called the *normal plane* and is the set of points $\vec{u}$ that satisfy

$$\vec{T}(t_0) \cdot (\vec{u} - \vec{X}(t_0)) = 0.$$

The plane through the tangent and binormal is called the *rectifying plane* and is the set of points $\vec{u}$ that satisfy

$$\vec{P}(t_0) \cdot (\vec{u} - \vec{X}(t_0)) = 0.$$

Figure 3.6 shows the osculating plane, the normal plane, and the rectifying plane together at a point on the space cardioid.

We now apply the theory of order of contact from Section 1.4 to find the *osculating sphere* – a sphere that has order of contact 3 or higher to a curve at a point.

Suppose that a sphere has center $\vec{c}$ and radius r so that its points $\vec{Z}$ satisfy the equation

$$\|\vec{Z} - \vec{c}\|^2 = r^2.$$

Consider a curve $\mathcal{C}$ parametrized by arc length by $\vec{X} : I \to \mathbb{R}^3$. The distance $f(s)$ between the point $\vec{X}(s)$ on the curve and the sphere is

$$f(s) = \big| \, \|\vec{X}(s) - \vec{c}\| - r \big|.$$

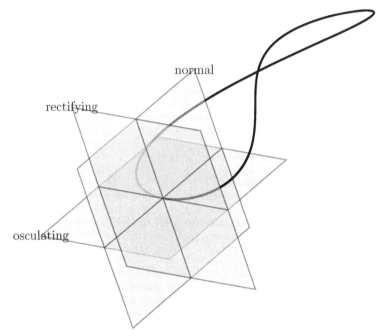

normal

rectifying

osculating

Figure 3.6: Moving trihedron.

Since derivatives of $G(t) = \sqrt{g(t)}$ are equal to 0 if and only if $g'(t) = 0$, then the derivatives of $f(s)$ are equal to 0 if and only if the derivatives of

$$h(s) = (\vec{X}(s) - \vec{c}) \cdot (\vec{X}(s) - \vec{c})$$

are equal to 0. The first three derivatives of $h(s)$ lead to

$$h'(s) = 0 \iff (\vec{X} - \vec{c}) \cdot \vec{T} = 0,$$
$$h''(s) = 0 \iff (\vec{X} - \vec{c}) \cdot \kappa \vec{P} + 1 = 0,$$
$$h'''(s) = 0 \iff (\vec{X} - \vec{c}) \cdot (\kappa' \vec{P} - \kappa^2 \vec{T} + \kappa \tau \vec{B}) = 0.$$

Consequently, at any point $\vec{X}(s_0)$ on the curve such that $\kappa(s_0) \neq 0$ and $\tau(s_0) \neq 0$, the first three derivatives can be equal to 0 if we have

$$(\vec{X} - \vec{c}) \cdot \vec{T} = 0, \quad (\vec{X} - \vec{c}) \cdot \vec{P} = -\frac{1}{\kappa}, \quad (\vec{X} - \vec{c}) \cdot \vec{B} = \frac{\kappa'}{\kappa^2 \tau}. \quad (3.15)$$

The equations in (3.15) give a decomposition of the center of the osculating sphere to a curve $\vec{X}(s)$ at the point $s = s_0$. Isolating $\vec{c}$, the center is

$$\vec{c} = \vec{X}(s_0) + R(s_0)\vec{P}(s_0) + \frac{R'(s_0)}{\tau(s_0)}\vec{B}(s_0) \quad (3.16)$$

where $R(s) = \frac{1}{\kappa(s)}$. The radius of the osculating sphere is

$$r = \sqrt{R(s_0)^2 + \left(\frac{R'(s_0)}{\tau(s_0)}\right)^2}.$$

That the first three derivatives of the distance function $f(s)$ are 0 implies that the curve $\mathcal{C}$ and the osculating sphere have contact of order 3.

If $\tau(s_0) = 0$, then $h'''(s)$ may still be 0 as long as $\kappa'(s_0) = 0$ at the same time. In that case, the curve admits a one-parameter family of osculating spheres at $\vec{x}(s_0)$, where the $\vec{B}$ component of the center $\vec{c}$ can be anything. This discussion leads to the following proposition.

Proposition 3.3.3 *Let $\vec{X} : I \to \mathbb{R}^3$ be a regular parametrized curve of class C^3. Let $t_0 \in I$ be a point on the curve where $\kappa(t_0) \neq 0$. The curve $\vec{X}$ admits an osculating sphere at $t = t_0$ if either (1) $\tau(t_0) \neq 0$ or (2) $\tau(t_0) = 0$ and $\kappa'(t_0) = 0$. Define $R(t) = \frac{1}{\kappa(t)}$. If $\tau(t_0) \neq 0$, then at t_0 the curve admits a unique osculating sphere with center $\vec{c}$ and radius r, where*

$$\vec{c} = \vec{X}(t_0) + R(t_0)\vec{P}(t_0) + \frac{R'(t_0)}{s'(t_0)\tau(t_0)}\vec{B}(t_0)$$

$$and \quad r = \sqrt{R(t_0)^2 + \left(\frac{R'(t_0)}{s'(t_0)\tau(t_0)}\right)^2}.$$

If $\tau(t_0) = 0$ and $\kappa'(t_0) = 0$, then at t_0 the curve admits as an osculating sphere any sphere with center $\vec{c}$ and radius r where

$$\vec{c} = \vec{X}(t_0) + R(t_0)\vec{P}(t_0) + c_B\vec{B}(t_0)$$

and $r = \sqrt{R(t_0)^2 + c_B^2}$, where c_B is any real number.

Proof: The only matter to address beyond the previous discussion is to see how the various quantities in Equation (3.16) change under a reparametrization.

Let $J \subset \mathbb{R}$ be an interval, and let $f : J \to I$ be of class C^3 such that f' doesn't change sign. Then $\vec{\xi} = \vec{x} \circ f$ is a regular reparametrization of $\vec{X}$. It is not hard to check that

$$\vec{T}_\xi = \operatorname{sign}(f')\vec{T}, \quad \vec{P}_\xi = \vec{P}, \quad \vec{B}_\xi = \operatorname{sign}(f')\vec{B}.$$

Also, in Problem 3.2.9, the reader showed that

$$\kappa_\xi = \kappa, \quad \text{and} \quad \tau_\xi = \operatorname{sign}(f')\tau.$$

Consequently, we have $R_\xi = R$ and also that $\frac{1}{\tau}\vec{B}$ is invariant under any regular reparametrization.

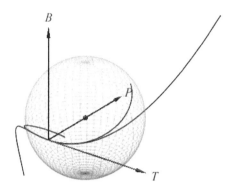

Figure 3.7: Osculating sphere.

On the other hand, if $t = f(u)$ so that $\vec{\xi}(u) = \vec{X}(t)$, then $R_\xi(u) = R(f(u))$ and therefore, $R'_\xi(u) = R'(f(u))f'(u) = R'(t)f'(u)$. In particular, $dR/ds = (1/s'(t))(dR/dt)$.

The proposition then follows from the prior discussion. □

Figure 3.7 illustrates an example of a curve and its osculating sphere at a point. The figure also shows the Frenet frame at the point in question, along with the orthogonal projection of the curve onto the sphere.

PROBLEMS

1. Find the osculating plane to the space cardioid, parametrized by $\vec{X}(t) = ((1 - \cos t) \cos t, (1 - \cos t) \sin t, \sin t)$, at $t = \pi$.

2. Find the osculating circle to twisted cubic $\vec{X}(t) = (t, t^2, t^3)$ at $t = 1$.

3. Consider the curve $\vec{X}(t) = (\cos t, \sin t, \sin 2t)$ for $t \in [0, 2\pi]$. Note that this curve lies on the cylinder $x^2 + y^2 = 1$.

   (a) Find equations for the osculating plane at t_0.

   (b) For what values of t is the osculating plane tangent to the cylinder?

4. Calculate the osculating sphere of the twisted cubic $\vec{X}(t) = (t, t^2, t^3)$ at (a) $(0, 0, 0)$ and at (b) $(1, 1, 1)$.

5. Consider the helix parametrized by $\vec{X}(t) = (a \cos t, a \sin t, bt)$. This is a helix around the z-axis and with pitch $\kappa/\tau = a/b$. Let $\vec{Y}(t)$ be the locus of the centers of the osculating sphere to this helix. Prove that $\vec{Y}(t)$ again parametrizes a helix but with a pitch that is negative the reciprocal of the pitch of the original helix.

6. Let $\vec{X} : I \to \mathbb{R}^3$ be a parametrized curve, and let $t \in I$ be a fixed point where $\kappa(t) \neq 0$. Define $\pi : \mathbb{R}^3 \to \mathbb{R}^2$ as the orthogonal projection of $\mathbb{R}^3$ onto the osculating plane to $\vec{X}$ at t. Define $\gamma = \pi \circ \vec{X}$ as the

orthogonal projection of the space curve $\vec{X}$ into the osculating plane. Prove that the curvature $\kappa(t)$ of $\vec{X}$ is equal to the curvature $\kappa_g(t)$ of the plane curve $\vec{\gamma}$.

7. Consider a generic curve on the unit cylinder parametrized by $\vec{X}(t) = (\cos t, \sin t, f(t))$ for some smooth function $f : \mathbb{R} \to \mathbb{R}$.

   (a) Calculate $\kappa(t)$, $\tau(t)$, $\vec{T}(t)$, $\vec{P}(t)$ and $\vec{B}(t)$ for this curve.

   (b) Deduce that the torsion is zero precisely when $f'(t) + f'''(t) = 0$.

   (c) Deduce that the osculating plane is never tangent to the cylinder $x^2 + y^2 = 1$.

8. Prove that if all the normal planes to a curve $\mathcal{C}$ pass through a fixed point $\vec{p}$, then $\mathcal{C}$ lies on a sphere of center $\vec{p}$.

9. Let $\vec{X} : I \to \mathbb{R}^3$ parametrize a generalized helix $\mathcal{C}$. Recall that this means that $\kappa(t)/\tau(t)$ is a constant. Let $\vec{Y}(t)$ trace out the centers of the osculating spheres to $\mathcal{C}$ at $\vec{X}(t)$. Call $\vec{T}_Y$, $\vec{P}_Y$ and $\vec{B}_Y$ the unit tangent, the principal normal and the binormal vectors to $\vec{Y}(t)$. Assuming that $\tau > 0$, prove that

$$\vec{T}_Y = \vec{B}, \quad \vec{P}_Y = -\vec{P}, \quad \text{and} \quad \vec{B}_Y = \vec{T}.$$

10. Suppose that $\vec{X}(s)$ parametrizes by arclength a regular space curve $\mathcal{C}$ and suppose that $P = \vec{X}(s_0)$. Suppose also that there exists some positive $h > 0$ such that all the points of $\vec{X}(s_0 + t)$ with $t \in [-h, h]$ lie on the same side of the osculating plane of $\mathcal{C}$ at P. Prove that $\tau(s_0) = 0$.

11. Determine a condition where the osculating circle has contact of order 3 or higher.

12. Let C be a regular curve with a parametrization $\vec{X} : I \to \mathbb{R}^3$ of class C^3. Define the *osculating helix* to C at a point $P = \vec{X}(t_0)$ as the unique helix that goes through P and has curvature $\kappa(t_0)$ and $\tau(t_0)$.

   (a) Find the parametric equations of the osculating helix.

   (b) Prove that when $\kappa'(t_0) = 0$ the osculating helix has contact of order 3 with C at P.

13. (*) Let $\vec{X} : I \to \mathbb{R}^3$ be the parametrization by arc length of a curve $\mathcal{C}$. Let $s \in I$ and let $h \in \mathbb{R}$ such that the interval $[s - h, s + h] \subseteq I$. Assume that $\vec{X}(s - h)$, $\vec{X}(s)$, and $\vec{X}(s + h)$ are not collinear and consider the unique circle $\Gamma_{s,h}$ in space that goes through these three points (the *circumcircle*). Prove that the limiting position of this circle as $h \to 0$ is the osculating circle to $\mathcal{C}$ at the point $\vec{X}(s)$. [Hint: Work first with three non-collinear points in space A, B, and C with O as the origin. Show that perpendicular bisector of the segment $\overline{AB}$ in the plane defined by A, B, and C has a parametrization

$$\vec{\gamma}(t) = \overrightarrow{OA} + \frac{1}{2}\overrightarrow{AB} + t\left((\overrightarrow{AB} \times \overrightarrow{AC}) \times \overrightarrow{AB}\right).$$

Find the same for the perpendicular bisector to $\overline{AC}$. Determine that the center P of the circle occurs on $\vec{\gamma}(t)$ where $t = (\overrightarrow{AC} \cdot \overrightarrow{BC})/(2\|\overrightarrow{AB} \times \overrightarrow{AC}\|^2)$. Then set $A = \vec{X}(s)$, $B = \vec{X}(s + h)$ and $C = \vec{X}(s - h)$ and let $h \to 0$.]

14. Let C be a curve that lies on a sphere of radius r and suppose that its curvature and torsion functions with respect to arc length are $\kappa(s)$ and $\tau(s)$. Since the curve lies on a sphere, this sphere is its osculating sphere at all points on C. Thus, by Proposition 3.3.3, if $R(s) = 1/\kappa(s)$, then $R(s)^2 + \left(\frac{R'(s)}{\tau(s)}\right)^2$ is a constant function equation to r^2. Prove the converse: that if C is a curve with $\kappa(s)$ and $\tau(s)$ such that $R(s)^2 + \left(\frac{R'(s)}{\tau(s)}\right)^2$ is constant, then C lies on a sphere. [Hint: Prove that the center of the osculating sphere does not move.]

3.4   Natural Equations

In Section 1.5, we showed that the curvature (in terms of arc length) function uniquely specifies a regular curve up to its location and orientation in $\mathbb{R}^2$. For space curves, we must introduce the torsion function for a measurement of how much the curve twists away from being planar. Since we know how $\vec{T}$, $\vec{P}$, and $\vec{B}$ change with respect to t, we can express all higher derivatives $\vec{X}^{(n)}(t)$ of $\vec{X}(t)$ in the Frenet frame in terms of $s'(t)$, $\kappa(t)$, and $\tau(t)$ and their derivatives. This leads one to surmise that a curve is to some degree determined uniquely by its curvature and torsion functions. The following theorem shows that this is indeed the case.

Theorem 3.4.1 (Fundamental Theorem of Space Curves)

Given functions $\kappa(s) \geq 0$ and $\tau(s)$ continuously differentiable over some interval $J \subseteq \mathbb{R}$ containing 0, there exists an open interval I containing 0 and a regular vector function $\vec{X} : I \to \mathbb{R}^3$ that parametrizes its locus by arc length, with $\kappa(s)$ and $\tau(s)$ as its curvature and torsion functions, respectively. Furthermore, any two curves C_1 and C_2 with curvature function $\kappa(s)$ and torsion function $\tau(s)$ can be mapped onto one another by a rigid motion of $\mathbb{R}^3$.

Proof: Let $\kappa(s)$ and $\tau(s)$ be functions defined over an interval $J \subset \mathbb{R}$ with $\kappa(s) \geq 0$. Consider the following system of 12 linear first-order differential equations:

$$
\begin{aligned}
x_1'(s) &= t_1(s), & p_1'(s) &= -\kappa(s)t_1(s) + \tau(s)b_1(s), \\
x_2'(s) &= t_2(s), & p_2'(s) &= -\kappa(s)t_2(s) + \tau(s)b_2(s), \\
x_3'(s) &= t_3(s), & p_3'(s) &= -\kappa(s)t_3(s) + \tau(s)b_3(s), \\
t_1'(s) &= \kappa(s)p_1(s), & b_1'(s) &= -\tau(s)p_1(s), \\
t_2'(s) &= \kappa(s)p_2(s), & b_2'(s) &= -\tau(s)p_2(s), \\
t_3'(s) &= \kappa(s)p_3(s), & b_3'(s) &= -\tau(s)p_3(s)
\end{aligned}
\tag{3.17}
$$

where $x_i(s)$, $t_i(s)$, $p_i(s)$, and $b_i(s)$, with $i = 1, 2, 3$, are unknown functions. With the stated conditions on $\kappa(s)$ and $\tau(s)$, according to the existence and uniqueness theorem for first-order systems of differential equations (see [2, Section 31.8]), there exists a solution to the above system defined for s in a neighborhood of 0. Furthermore, there exists a unique solution with specified initial conditions

$$x_1(0) = x_{10},\ x_2(0) = x_{20},\ x_3(0) = x_{30},$$
$$t_1(0) = t_{10},\ t_2(0) = t_{20},\ t_3(0) = t_{30},$$
$$p_1(0) = p_{10},\ p_2(0) = p_{20},\ p_3(0) = p_{30},$$
$$b_1(0) = b_{10},\ b_2(0) = b_{20},\ b_3(0) = b_{30}.$$

Define the two matrices of functions $A(s)$ and $M(s)$ equal to

$$\begin{pmatrix} 0 & -\kappa(s) & 0 \\ \kappa(s) & 0 & -\tau(s) \\ 0 & \tau(s) & 0 \end{pmatrix} \text{ and } \begin{pmatrix} t_1(s) & p_1(s) & b_1(s) \\ t_2(s) & p_2(s) & b_2(s) \\ t_3(s) & p_3(s) & b_3(s) \end{pmatrix}$$

respectively. Recall from (3.13) that $M'(s) = M(s)A(s)$. It is possible to show that since $A(s)$ is antisymmetric, the function $f(s) = M(s)^T M(s)$ is constant as a matrix of functions. Therefore, any solution to (3.17) is such that $M(s)^T M(s)$ remains constant. In particular, if we choose initial conditions such that

$$\begin{pmatrix} t_{10} \\ t_{20} \\ t_{30} \end{pmatrix}, \qquad \begin{pmatrix} p_{10} \\ p_{20} \\ p_{30} \end{pmatrix}, \qquad \begin{pmatrix} b_{10} \\ b_{20} \\ b_{30} \end{pmatrix} \qquad (3.18)$$

form an orthonormal basis of $\mathbb{R}^3$, then solutions to (3.17) are such that, for all s in the domain of the solution, the vectors

$$\begin{pmatrix} t_1(s) \\ t_2(s) \\ t_3(s) \end{pmatrix}, \qquad \begin{pmatrix} p_1(s) \\ p_2(s) \\ p_3(s) \end{pmatrix}, \qquad \begin{pmatrix} b_1(s) \\ b_2(s) \\ b_3(s) \end{pmatrix} \qquad (3.19)$$

form an orthonormal basis.

We have shown that with a choice of initial conditions such that the vectors in (3.18) form an orthonormal basis, the corresponding solution to (3.17) is such that $\vec{X}(s) = (x_1(s), x_2(s), x_3(s))$ is a regular curve of class C^3 with curvature $\kappa(s)$ and torsion $\tau(s)$. In addition, the vector functions in (3.19) are the $\vec{T}$, $\vec{P}$, and $\vec{B}$ vectors of the Frenet frame associated to $\vec{X}(s)$. This proves existence.

The existence and uniqueness theorem of systems of differential equations states that a solution is unique once (the right number of) initial conditions are specified. However, we have imposed the additional condition that the vectors in (3.18) form an orthonormal set. Any different choice for the initial conditions in (3.18) corresponds to a rotation in $\mathbb{R}^3$. Also, two different choices of initial conditions

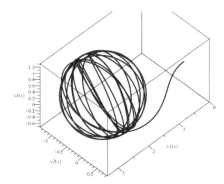

Figure 3.8: A ball of yarn by natural equations.

x_{10}, x_{20}, x_{30} correspond to a translation in $\mathbb{R}^3$. Therefore, different allowed initial conditions correspond to a rigid motion of the locus of $\vec{X}(s)$ in $\mathbb{R}^3$. This proves the theorem.   □

Because of the Fundamental Theorem of Space Curves, the pair of functions $\kappa(s)$ and $\tau(s)$, where $\kappa(s)$ is a positive function, are called the natural equations of a curve. Taken as a pair, they define the curve uniquely up to a rigid motion in the plane.

Just as in Section 1.5, we strongly encourage the reader to access the software that accompanies this book, as described in the preface. The applets contained in the workbook implement a differential equations solver and plot a space curve given any curvature function and torsion function.

Figure 3.8 is generated from the functions $\kappa(s) = (1 - s^2)/(1 + s^2)$ and $\tau(s) = \frac{1}{5}\sin(s)$, but only for $s \geq 0$. Note that

$$\lim_{s \to \infty} \kappa(s) = 1,$$

so as s gets large, the space curve has approximately constant curvature $\kappa = 1$ but with a torsion that oscillates between $-\frac{1}{5}$ and $\frac{1}{5}$. From Proposition 3.3.3, we expect that as s grows, the curve would lie more and more on a sphere of radius 1.

PROBLEMS

1. Assume that $\kappa(s)$ and $\tau(s)$ are of class C^∞ over an interval I. Use the Taylor expansion of $\vec{X}(s)$ to prove Theorem 3.4.1.

2. Prove that if $\kappa(s)$ and $\tau(s)$ are nonzero constant functions, then the resulting curve must be a helix.

CHAPTER 4

Curves in Space: Global Properties

Paralleling our presentation of curves in the plane, we now turn from local properties of space curves to global properties. As before, global properties of curves are properties that involve the curve as a whole as opposed to properties that are defined in the neighborhood of a point on the curve.

The Jordan Curve Theorem does not apply to curves in $\mathbb{R}^3$, so Green's Theorem, the isoperimetric inequality, and theorems connecting curvature and convexity do not have an equivalent for space curves. On the other hand, curves in space exhibit new types of global properties, in particular, knottedness and linking.

4.1 Basic Properties

Definition 4.1.1 A parametrized space curve $\mathcal{C}$ is called *closed* if there exists a parametrization $\vec{X} : [a, b] \to \mathbb{R}^3$ of $\mathcal{C}$ such that $\vec{X}(a) = \vec{X}(b)$. A closed curve is of class C^k if, in addition, all the (one-sided) derivatives of $\vec{X}$ at a and at b are equal of order $i = 0, 1, \ldots, k$; in other words, if as one-sided derivatives $\vec{X}'(a) = \vec{X}'(b)$, $\vec{X}''(a) = \vec{X}''(b)$, and so on up to $\vec{X}^{(k)}(a) = \vec{X}^{(k)}(b)$. A space curve $\mathcal{C}$ is called *simple* if it has no self-intersections, and a closed curve is called simple if it has no self-intersections except at the endpoints.

As discussed in Section 2.1, a closed space curve can be understood as a function $f : \mathbb{S}^1 \to \mathbb{R}^3$ that is continuous as a function between topological spaces (see [24, Appendix A]). Again, to say that a closed curve is simple is tantamount to saying that as a continuous function $f : \mathbb{S}^1 \to \mathbb{R}^3$, it is injective.

Our first result has an equivalent in the theory of plane curves.

Proposition 4.1.2 *If a regular curve is closed, then it is bounded.*

Proof: (Left as an exercise for the reader.) □

DOI: 10.1201/9781003295341-4

Stokes' Theorem presents a global property of curves in space in that it relates a quantity calculated along the curve with a quantity that depends on any surface that has that curve as a boundary. However, the theorem involves a vector field in $\mathbb{R}^3$, and it does not address "geometric" properties of the curve, by which we mean properties that are independent of the curve's location and orientation in space. Therefore, we simply state Stokes' Theorem as an example of a global theorem and trust the reader has seen it in a previous calculus course.

Theorem 4.1.3 (Stokes' Theorem) *Let $\mathcal{S}$ be an oriented, piecewise regular surface bounded by a closed, piecewise regular curve $\mathcal{C}$. Let $\vec{F} : \mathbb{R}^3 \to \mathbb{R}^3$ be a vector field over $\mathbb{R}^3$ that is of class C^1 on $\mathcal{S}$. Supposing that $\mathcal{C}$ is oriented according to the right-hand rule,*

$$\int_{\mathcal{C}} \vec{F} \cdot d\vec{s} = \iint_{\mathcal{S}} (\vec{\nabla} \times \vec{F}) \cdot d\vec{S}. \tag{4.1}$$

If $\mathcal{S}$ has no boundary curve, then the line integral on the left is taken as 0.

Note as a reminder that the line element $d\vec{s}$ stands for $\vec{\gamma}'(t)dt$, where $\vec{\gamma}(t)$ is a parametrization of $\mathcal{C}$, and $d\vec{S}$ stands for $\vec{n}dA$ where $\vec{n}$ is the unit normal and dA is the surface element at a point on the surface. If a surface is parametrized by a vector function $\vec{X}(u, v)$ into $\mathbb{R}^3$, then $d\vec{S} = \vec{X}_u \times \vec{X}_v \, du\, dv$.

The following is an interesting corollary to Stokes' Theorem.

Corollary 4.1.4 *Let $\mathcal{C}$ be a simple, regular, closed space curve parametrized by $\vec{\gamma} : I \to \mathbb{R}^3$. There exists no function $f : \mathbb{R}^3 \to \mathbb{R}$ of class C^2 such that the gradient satisfies $\vec{\nabla} f = \vec{T}(t)$ at all points of the curve.*

Proof: Recall that for all functions $f : \mathbb{R}^3 \to \mathbb{R}$ of class C^2, the curl of the gradient is identically 0, namely, $\vec{\nabla} \times \vec{\nabla} f = \vec{0}$. If $\mathcal{S}$ is any orientable piecewise regular surface that has $\mathcal{C}$ as a boundary, then

$$\iint_{\mathcal{S}} (\vec{\nabla} \times \vec{\nabla} f) \cdot d\vec{S} = \iint_{\mathcal{S}} \vec{0} \cdot d\vec{S} = \vec{0}.$$

If f did satisfy $\vec{\nabla} f = \vec{T}(t)$ at all points of the curve $\mathcal{C}$, then

$$\int_{\mathcal{C}} \vec{\nabla} f \cdot d\vec{s} = \int_{I} \vec{T}(t) \cdot \vec{T}(t)s'(t)dt = \int_{C} ds,$$

which is the length of the curve. By Stokes' Theorem, this leads to a contradiction, assuming the curve has length greater than 0.   □

Example 4.1.5 Stokes' Theorem is a very powerful theorem but, as this example illustrates, it is important to check that the conditions are satisfied. Consider the vector field $\vec{F}(x, y, z)$ defined by

$$\vec{F}(x, y, z) = \left(0, -\frac{z}{y^2 + z^2}, \frac{y}{y^2 + z^2} \right)$$

and the curve $\mathcal{C}$ parametrized by $\vec{X}(t) = (\cos(2t), \cos t, \sin t)$. The curl of the vector field $\vec{F}$ is $\nabla \times \vec{F} = (0, 0, 0)$ so the flux through any surface $\mathcal{S}$ over which $\vec{F}$ is C^1 is

$$\iint_{\mathcal{S}} \nabla \times \vec{F} \cdot d\vec{S} = 0.$$

On the other hand, the vector field is defined along $\mathcal{C}$ and the circulation of $\vec{F}$ along $\mathcal{C}$ is

$$\int_{\mathcal{C}} \vec{F} \cdot d\vec{s} = 2\pi.$$

This may seem like a contradiction because there certainly exists a surface $\mathcal{S}$ whose boundary is $\mathcal{C}$. However, we cannot forget to observe that $\vec{F}$ is well-defined on all of $\mathbb{R}^3$ except for the x-axis. The reason this example does not give a contradiction to Stokes' Theorem is because there is no surface $\mathcal{S}$ whose boundary is $\mathcal{C}$ and which does not intersect the x-axis.

Intuitively speaking, the curve $\mathcal{C}$ and the x-axis are "linked," in the sense that there is no way to move and deform one curve continuously in such a way as to move $\mathcal{C}$ and the x-axis into two separate half-spaces without making the curves intersect. We discuss linked curves more in Section 4.3.

PROBLEMS

1. Verify the flux and circulation calculations in Example 4.1.5.

2. Let $\vec{\alpha} : I \to \mathbb{R}^3$ be a regular, closed space curve, and let $\vec{p}$ be a point. Prove that a point t_0 of maximum distance on the curve away from $\vec{p}$ is such that $\vec{\alpha}(t_0) - \vec{p}$ is in the normal plane to the curve at $\vec{\alpha}(t_0)$.

3. The diameter of a closed curve $\vec{\alpha}$ is the maximum of the function

$$f(t, u) = \|\vec{\alpha}(t) - \vec{\alpha}(u)\|.$$

Use the second derivative test in multivariable calculus to show that if (t_0, u_0) gives a diameter of a space curve, then the following hold:

(a) The vector $\vec{\alpha}(u_0) - \vec{\alpha}(t_0)$ is in the intersection of the normal planes of $\alpha(t_0)$ and $\alpha(u_0)$.

(b) The vector $\vec{\alpha}(u_0) - \vec{\alpha}(t_0)$ is on the side of the rectifying plane of $\vec{\alpha}(t_0)$ that makes an acute angle with $\vec{P}(t_0)$ (and similarly for $\alpha(u_0)$).

(c) The diameter is greater than or equal to $\max\{1/\kappa(t_0), 1/\kappa(u_0)\}$.

[Hint: For simplicity, assume that $\vec{\alpha}$ is parametrized by arc length. Also, the extrema of $f(t, u)$ occur at and have the same properties as the extrema of $g(t, u) = f(t, u)^2$.]

4. Prove Proposition 4.1.2.

5. By using the vector field $\vec{F}(x, y, z) = (0, x, 0)$, Stokes' Theorem establishes the familiar Green's Theorem formula for area of the interior of a curve $\mathcal{C}$ in the xy-plane:

$$ A = \int_{\mathcal{C}} (0, x, 0) \cdot d\vec{s} = \int_{\mathcal{C}} x \, dy. $$

Suppose that we consider a curve $\mathcal{C}$ on the sphere of radius R and centered at the origin. Prove that there does not exist a vector field $\vec{F}(x, y, z)$ in $\mathbb{R}^3$ that could be used to calculate the area of the "interior" of the curve as a line integral. [Hint: Recall relations among vector differential operators.]

4.2   Indicatrices and Total Curvature

 Definition 4.2.1 Given a space curve $\vec{X} : I \rightarrow \mathbb{R}^3$, define the *tangent*, *principal*, and *binormal indicatrices* respectively, as the loci of the space curves on the unit sphere given by $\vec{T}(t)$, $\vec{P}(t)$, and $\vec{B}(t)$, as defined in Section 3.2.

Since the vectors of the Frenet frame have a length of 1, the indicatrices are curves on the unit sphere $\mathbb{S}^2$ in $\mathbb{R}^3$. In contrast to the tangent indicatrix for plane curves, there are more possibilities for curves on the sphere than for curves on a circle. Therefore, results such as the theorem on the rotation index or results about the winding number do not have an immediate equivalent for curves in space.

Example 4.2.2 (Lines) The tangent indicatrix of a line is a single point $\vec{v}/\|\vec{v}\|$ on the sphere, where the line is parametrized by

$$ \vec{X}(t) = \vec{v}t + \vec{p}. $$

Example 4.2.3 (Helices) Consider a helix around an axis L through the origin. The tangent indicatrix of this helix is the circle on the unit sphere $\mathbb{S}^2$, given as the intersection of $\mathbb{S}^2$ with the plane perpendicular to L at a distance of $1/\sqrt{1 + (\kappa/\tau)^2}$ from the origin, where $\frac{\kappa}{\tau}$ is the pitch of the helix. Notice that this result is true for general helices (see Example 3.2.5) and not just circular helices.

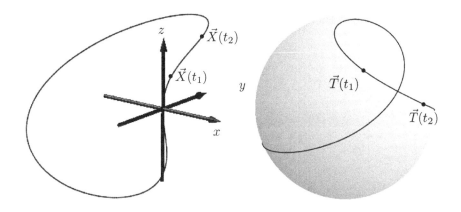

Figure 4.1: The space cardioid and its tangent indicatrix.

Example 4.2.4 (Space Cardioid) The space cardioid, given by the parametrization

$$\vec{X}(t) = \big((1 - \cos t)\cos t, (1 - \cos t)\sin t, \sin t\big), \qquad t \in [0, 2\pi],$$

is a closed curve. The tangent indicatrix is again a closed curve and is shown in Figure 4.1. The figure shows that the tangent indicatrix has a double point. In Problem 4.2.4, the reader is invited to show this and a few other interesting properties of the tangent indicatrix of the space cardioid.

Studying Figure 4.1, the reader may have some difficulty in visualizing that the placement of the two points $\vec{X}(t_1)$ and $\vec{X}(t_2)$ along with their corresponding unit tangent vectors is correct. All the applets provided in the associated MAPLE workbook are valuable, but here in particular, the corresponding app helps intuition immensely.

Definition 4.2.5 The *total curvature* of a closed curve C parametrized by $\vec{X} : [a, b] \to \mathbb{R}^3$ is

$$\int_C \kappa \, ds = \int_a^b \kappa(t) s'(t) \, dt.$$

This is a nonnegative real number since $\kappa(t) \geq 0$ for space curves.

Though we cannot define the concept of a rotation index, Fenchel's Theorem gives a lower bound for the total curvature of a space curve. Since $\vec{T}' = s'\kappa\vec{P}$, the speed of the tangent indicatrix is $s'(t)\kappa(t)$. Therefore, the total curvature of $\vec{X}$ is the length of the tangent indicatrix. One can also note that the tangent indicatrix has a critical point where $\kappa(t) = 0$.

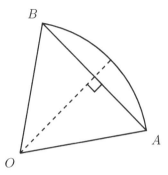

Figure 4.2: Spherical distance.

Theorem 4.2.6 (Fenchel's Theorem) *The total curvature of a regular closed space curve C is greater than or equal to 2π. It is equal to 2π if and only if C is a convex plane curve.*

Before we prove Fenchel's Theorem, we need to discuss the concept of distance between points on the unit sphere $\mathbb{S}^2$. We can define the distance between two points p and q on $\mathbb{S}^2$ as

$$d(p,q) = \inf\{\text{length}(\Gamma) \mid \Gamma \text{ is a curve connecting } p \text{ and } q\}.$$

Let O be the center of the unit sphere. It is not difficult to show that the path connecting p and q of shortest distance is an arc between p and q on the circle defined by the intersection of the sphere and the plane containing O, p, and q. (We will obtain this result in Example 8.4.8 when studying geodesics, but it is possible to prove this claim without the techniques of geodesics.) Then the spherical distance between two points on the unit sphere is

$$d(p,q) = \cos^{-1}(p \cdot q),$$

where we view p and q as vectors in $\mathbb{R}^3$.

We will denote by AB the Euclidean distance between two points A and B as elements in $\mathbb{R}^3$. If $A, B \in \mathbb{S}^2$, we use the notation $\overline{AB}$ to denote the distance between A and B on the sphere. Since OAB forms an isosceles triangle (see Figure 4.2), we see that the spherical and Euclidean distance are related via

$$AB = 2\sin\left(\frac{\overline{AB}}{2}\right) \quad \text{and} \quad \overline{AB} = 2\sin^{-1}\left(\frac{AB}{2}\right).$$

Lemma 4.2.7 (Horn's Lemma) *Let Γ be a regular closed curve on the unit sphere $\mathbb{S}^2$. If Γ has length less than 2π, then there exists a great circle C on the sphere such that Γ does not intersect C.*

Proof: The proof we give to this lemma is due to R. A. Horn [20].

Let P and Q be two points of Γ that divide the curve into two arcs of equal length $L/2$. The points P and Q separate Γ into two curves Γ_1 and Γ_2 of equal length, each with endpoints of P and Q. Since $L < 2\pi$, the spherical distance between P and Q is less than the length of Γ_1, which is strictly less than π. Consequently, P and Q are not antipodal (i.e., the segment $[P, Q]$ is not a diameter of the sphere) and therefore there exists a unique point M on the minor arc from P to Q midway between P and Q. We claim that Γ does not meet the equator $\mathcal{C}$ that has M as the north pole, so the curve lies in the hemisphere centered at M.

To show that Γ_1 does not meet the equator, consider a copy Γ_1' of Γ_1 rotated one half turn about M. The curve Γ_1' is a curve from Q to P of length $L/2$. Define the closed curve Γ'' as the curve that follows Γ_1 from P and Q and then follows Γ_1' from Q back to P. The curve Γ'' has length L. Assume Γ_1 intersects $\mathcal{C}$; then so would Γ_1'. Hence, Γ'' would contain antipodal points R and R'. However, the spherical distance $\overline{RR'} = \pi$ so then both of the two arcs between R to R' along Γ'' would have an arclength greater than or equal to π. This contradicts the fact that the arclength of Γ'' is strictly less than 2π. We conclude that Γ_1 does not intersect $\mathcal{C}$.

Thus any curve with length less than 2π lies in an open hemisphere. $\qquad\square$

We are now in a position to prove Fenchel's Theorem.

Proof of Theorem 4.2.6: Let $\vec{\gamma} : [a, b] \to \mathbb{R}^3$ be a parametrization for the regular closed curve C, and let p be any point on the unit sphere. Consider the function $g(t) = p \cdot \vec{\gamma}(t)$. Since $[a, b]$ is a closed and bounded interval and $g(t)$ is continuous, then it attains a maximum and minimum value in $[a, b]$. This value occurs where

$$g'(t) = p \cdot \vec{\gamma}'(t) = 0.$$

We remark that the set of points $\vec{x} \in \mathbb{S}^2$ such that $p \cdot \vec{x} = 0$ is the great circle on $\mathbb{S}^2$ that is on the plane perpendicular to the line (Op), where O is the center of the sphere. Since $\vec{\gamma}'(t)$ and $\vec{T}(t)$ are collinear and since p was chosen arbitrarily, we conclude that the tangent indicatrix intersects every great circle on $\mathbb{S}^2$. Consequently, by Lemma 4.2.7, the length of the tangent indicatrix is greater than or equal to 2π. Thus, since the length of the tangent indicatrix is the total curvature,

$$\int_a^b \kappa(t) s'(t)\, dt = \int_C \kappa\, ds \geq 2\pi.$$

To prove the second part of the theorem, first note that if $\vec{\gamma}(t)$ traces out a convex plane curve, then $\kappa(t) = |\kappa_g(t)|$ for $\vec{\gamma}$ as a plane curve. Furthermore, by Propositions 2.4.3 and 2.2.11, the total curvature is 2π. Therefore, to finish proving the theorem, we only need to prove the converse.

Suppose that the total curvature of $\vec{\gamma}$ is 2π. The length of the tangent indicatrix is therefore 2π. Furthermore, since by the above reasoning the tangent indicatrix must intersect every great circle, the tangent indicatrix must itself be a great circle. Thus $\vec{T}(t)$, $\vec{T}'(t)$, and $\vec{T}''(t)$ are coplanar and so $\tau(t) = 0$. Thus, by Proposition 3.2.7, $\vec{\gamma}(t)$ is planar. In this case, it is not hard to check that, $\kappa(t) = |\kappa_g(t)|$, where $\kappa_g(t)$ is the curvature of $\vec{\gamma}(t)$ as a plane curve. Thus,

$$2\pi = \int_C |\kappa_g|\, ds$$

is the total distance that $\vec{T}(t)$ travels on the unit circle, and this is the length of the unit circle. Consequently, if $t_1, t_2 \in [a, b)$, with $\vec{T}(t_1) = \vec{T}(t_2)$, then $\vec{T}(t)$ is constant over $[t_1, t_2]$ because, otherwise, the total distance $\vec{T}(t)$ travels on the unit circle would exceed 2π by at least

$$\int_{t_1}^{t_2} |\kappa_g(t)|\, ds.$$

We conclude then that $\vec{\gamma}$ has no bitangent lines, and by Problem 2.4.1, we conclude that C is a convex plane curve.   □

As an immediate corollary, we obtain the following result.

Corollary 4.2.8 *If a regular, closed space curve has a curvature function κ that satisfies*

$$\kappa \le \frac{1}{R},$$

then C has length greater than or equal to $2\pi R$.

Proof: Suppose that a regular, closed space curve has a curvature function $\kappa(s)$, given in terms of arc length, that satisfies the given hypothesis. Then if L is the length of the curve, we have

$$L = \int_0^L ds \ge \int_0^L R\kappa(s)\, ds = R \int_0^L \kappa(s)\, ds \ge 2\pi R,$$

where the last inequality follows from Fenchel's Theorem.   □

Though it is outside the scope of our current techniques, we wish to state here Jacobi's Theorem since it is a global theorem on space curves. The proof follows as an application of the Gauss-Bonnet Theorem (Problem 8.3.11).

Theorem 4.2.9 (Jacobi's Theorem) *Let* $\vec{\alpha} : I \to \mathbb{R}^3$ *be a closed, regular parametrized curve whose curvature is never 0. Suppose that the principal normal indicatrix* $\vec{P} : I \to \mathbb{S}^2$ *is simple. Then the locus* $\vec{P}(I)$ *of the principal normal indicatrix separates the sphere into two regions of equal area.*

The reader should note that Jacobi's Theorem is quite profound in the following sense. Given a parametrized curve $\vec{\gamma} : I \to \mathbb{R}^3$ such that $\|\vec{\gamma}(t)\| = 1$ for all $t \in I$, the problem of calculating the surface area of the sphere lying on one side or the other of the locus of this curve is not a tractable problem.

We now present Crofton's Theorem, which we will use in the next section to prove a theorem by Fary and Milnor on the total curvature of a knot.

Let O be the center of the unit sphere $\mathbb{S}^2$. Each great circle $\mathcal{C}$ is uniquely defined by a line L through the origin of the sphere as the intersection between $\mathbb{S}^2$ and the plane through O perpendicular to L. Furthermore, L is defined by two "poles," the points of intersection of L with $\mathbb{S}^2$. However, there exists a bijective correspondence between oriented great circles and points on $\mathbb{S}^2$ by associating to $\mathcal{C}$ the pole of L that is in the direction on L that is positive in the sense of the right-hand rule of motion on $\mathcal{C}$.

Consider a set Z of oriented great circles on $\mathbb{S}^2$. We call the *measure* $m(Z)$ of the set Z the area of the region traced out on $\mathbb{S}^2$ by the positive poles of the oriented circles in Z.

Theorem 4.2.10 (Crofton's Theorem) *Let* Γ *be a curve of class* C^1 *on* $\mathbb{S}^2$. *The measure of the great circles of* $\mathbb{S}^2$ *that meet* Γ *is equal to four times the length of* Γ.

Proof: Suppose that $\vec{e}_1 : [0, L] \to \mathbb{S}^2$ parametrizes Γ by arc length. Complete $\{\vec{e}_1(s)\}$ to form an orthonormal basis $\{\vec{e}_1(s), \vec{e}_2(s), \vec{e}_3(s)\}$ so that $\vec{e}_i(s)$ for $i = 2, 3$ are of class C^1. Without loss of generality, we can construct $\vec{e}_2$ and $\vec{e}_3$ so that

$$\det(\vec{e}_1, \vec{e}_2, \vec{e}_3) = 1$$

for all $s \in [0, L]$. Using the same reasoning that established Equation (3.5) or the result of Problem 9.1.12, we have

$$\frac{d}{ds} \begin{pmatrix} \vec{e}_1 & \vec{e}_2 & \vec{e}_3 \end{pmatrix} = \begin{pmatrix} \vec{e}_1 & \vec{e}_2 & \vec{e}_3 \end{pmatrix} \begin{pmatrix} 0 & a_2 & a_3 \\ -a_2 & 0 & a_1 \\ -a_3 & -a_1 & 0 \end{pmatrix} \qquad (4.2)$$

for some continuous functions $a_i : [0, L] \to \mathbb{R}$. Furthermore, since $\vec{e}_1$ is parametrized by arc length we know that it has unit speed, so

$a_2(s)^2 + a_3(s)^2 = 1$. Consequently, there exists a function $\alpha : [0, L] \to \mathbb{R}$ of class C^1 such that

$$\vec{e}_1'(s) = \cos(\alpha(s))\vec{e}_2 + \sin(\alpha(s))\vec{e}_3.$$

The set of oriented great circles that meet Γ at $\vec{e}_1(s)$ is parametrized by its positive poles

$$(\cos\theta)\vec{e}_2(s) + (\sin\theta)\vec{e}_3(s)$$

for $\theta \in [0, 2\pi]$. Therefore, the region traced out by the positive poles of oriented great circles that meet Γ is parametrized by

$$\vec{Y}(s, \theta) = (\cos\theta)\vec{e}_2(s) + (\sin\theta)\vec{e}_3(s) \qquad \text{for } (s, \theta) \in [0, L] \times [0, 2\pi].$$

We wish to determine the area element $|dA|$ for the set of poles traced out by the set of oriented great circles meeting Γ. However,

$$|dA| = \|\vec{Y}_s \times \vec{Y}_\theta\| \, d\theta \, ds,$$

so after some calculation and using Equation (4.2), one obtains

$$|dA| = |a_2(s)\cos\theta + a_3(s)\sin\theta| \, d\theta \, ds = |\cos(\alpha(s) - \theta)| \, d\theta \, ds.$$

Call $\mathcal{C}_{s,\theta}$ the oriented great circle with positive pole $\vec{Y}(\theta, s)$ and denote by $n(\mathcal{C}_{s,\theta})$ the number of points in $\mathcal{C}_{s,\theta} \cap \Gamma$. Then the measure m of oriented great circles in $\mathbb{S}^2$ that meet Γ is

$$m = \iint n(\mathcal{C}_{s,\theta})|dA| = \int_0^L \int_0^{2\pi} |\cos(\alpha(s) - \theta)| \, d\theta \, ds. \qquad (4.3)$$

However, for all fixed α_0, we have

$$\int_0^{2\pi} |\cos(\alpha_0 - \theta)| \, d\theta = 4,$$

so we conclude that $m = 4L$, which establishes the theorem. $\square$

PROBLEMS

1. Consider the twisted cubic $\vec{X}(t) = (t, t^2, t^3)$ with $t \in \mathbb{R}$.

   (a) Prove that the closure of the tangential indicatrix of the twisted cubic is a closed curve, in particular, that

   $$\lim_{t \to \infty} \vec{T}(t) = \lim_{t \to -\infty} \vec{T}(t) = (0, 0, 1).$$

   (b) Prove that the tangential indicatrix does not have a corner at $(0, 0, 1)$, namely, that

   $$\lim_{t \to \infty} \vec{T}'(t) = \lim_{t \to -\infty} \vec{T}'(t).$$

(c) Prove directly that the total curvature of the twisted cubic is bounded.

2. (**CAS**) Consider the modified twisted cubic $\vec{X}(t) = (3t, 3t^2, 4t^3)$. Calculate the total curvature, allowing for $t \to -\infty$ and $t \to \infty$.

3. Consider the helix $\vec{X}(t) = (a\cos t, a\sin t, bt)$ for $t \in \mathbb{R}$. Determine the locus of the tangent indicatrix of $\vec{X}(t)$.

4. Figure 4.1 appears to show that the tangent indicatrix of the space cardioid has a double point.
   (a) Find the value t_0 of t at which the double point occurs.
   (b) Show that the antipodal point to $\vec{T}(t_0)$, i.e., $-\vec{T}(t_0)$ is also on the tangent indicatrix of the space cardioid.
   (c) Show that the curve obtained by projecting the tangent indicatrix onto the plane perpendicular to $\vec{T}(t_0)$ through the origin has two cusps.

5. Calculate directly the total curvature of the space cardioid

$$\vec{X}(t) = \big((1 - \cos t)\cos t, (1 - \cos t)\sin t, \sin t\big) \qquad \text{for } t \in [0, 2\pi].$$

6. Prove Fenchel's Theorem as a corollary to Crofton's Theorem.

7. (*) Let $\vec{\alpha} : [0, L] \to \mathbb{R}^3$ be a regular closed curve (parametrized by arc length) whose image lies on a sphere. Suppose also that $\kappa(s)$ is never 0. Prove that

$$\int_0^L \tau(s)\, ds = 0.$$

4.3 Knots and Links

The reader should be forewarned that the study of knots and links is a vast and fruitful area that one usually considers as a subbranch of topology. In this section, we only have space to give a cursory introduction to the concept of a knotted curve in $\mathbb{R}^3$ or two linked curves in $\mathbb{R}^3$. However, the property of being knotted or linked is a global property of a curve or curves (since it depends on the curve as a whole), which motivates us to briefly discuss these topics in this chapter. This section presents two main theorems: the Fary-Milnor Theorem, which shows how the property of knottedness imposes a condition on the total curvature of a curve, and Gauss's formula for the linking number of two curves.

4.3.1 Knots

Intuitively speaking, a knot is a simple closed curve in $\mathbb{R}^3$ that cannot be deformed into a circle without breaking the curve and reconnecting it. In other words, a knot cannot be deformed into a circle by a

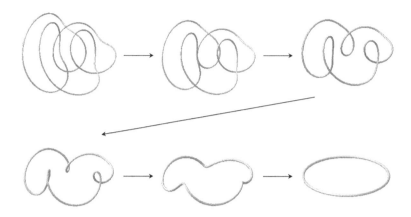

Figure 4.3: An unknotting homotopy in $\mathbb{R}^3$.

continuous process without passing through a stage where it is not a simple curve.

The following gives a precise definition to the above intuition.

Definition 4.3.1 A simple closed curve Γ in $\mathbb{R}^3$ is called *unknotted* if there exists a continuous function $H : \mathbb{S}^1 \times [0,1] \to \mathbb{R}^3$ such that $H(\mathbb{S}^1 \times \{0\}) = \Gamma$ and $H(\mathbb{S}^1 \times \{1\}) = \mathbb{S}^1$ and such that $\Gamma_t = H(\mathbb{S}^1 \times \{t\})$ is a simple, closed space curve. If there does not exist such a function H, the curve Γ is called *knotted*.

The function H described in the above definition is called a *homotopy* in $\mathbb{R}^3$ between Γ and a circle $\mathbb{S}^1$. Figure 4.3 illustrates four intermediate stages of a homotopy between a space curve and a circle.

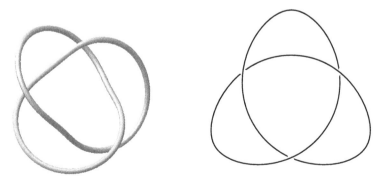

Figure 4.4: A trefoil knot and its diagram.

Figure 4.4 shows the trefoil knot realized as a curve in space, along with a two-dimensional diagram. As it is somewhat tedious to plot general knotted curves, even with the assistance of a computer algebra system, one often uses a diagram that shows the "crossings," i.e., which part of the curve passes above the other whenever the diagram would intersect in the given perspective. (The interested reader is encouraged to read [23] for an advanced introduction to knot theory.)

The following theorem gives a necessary relationship between a knotted curve and the curvature of a space curve.

Theorem 4.3.2 (Fary-Milnor Theorem) *The total curvature of a knot is greater than or equal to 4π.*

Proof: Let $\vec{x} : [0, L] \to \mathbb{R}^3$ be a closed, regular space curve parametrized by arc length, and let $\vec{T} : [0, L] \to \mathbb{S}^2$ be the tangent indicatrix. Call C the image of $\vec{x}$ in $\mathbb{R}^3$ and Γ the image of $\vec{T}$ on the sphere.

Recall the notations in the proof of Crofton's Theorem (Theorem 4.2.10): namely the function $\vec{Y}(s, \theta)$, such for any given s, parametrizes all unit vectors perpendicular to $\vec{T}(s)$, and also the oriented great circle $\mathcal{C}_{s,\theta}$ associated to $\vec{Y}(s, \theta)$. Recall also that $n(\mathcal{C}_{s,\theta})$ is the number of times that $\mathcal{C}_{s,\theta}$ intersects the curve Γ (see Figure 4.5). By part of the proof of Crofton's Theorem, $n(\mathcal{C}_{s,\theta}) > 0$ over the domain $(s, \theta) \in [0, L] \times [0, 2\pi]$, and $n(\mathcal{C}_{s,\theta})$ is even.

Suppose that the total curvature of the curve $\vec{x}$, which is also the length of Γ, is less than 4π. By Equation (4.3) in the proof of Crofton's Theorem,

$$\iint n(\mathcal{C}_{s,\theta}) |dA| < 16\pi.$$

Since the area of the sphere is 4π, there exists (s_0, θ_0) such that $n(\mathcal{C}_{s_0,\theta_0}) = 2$.

Define a height function $h(s) = \vec{Y}(s_0, \theta_0) \cdot \vec{x}(s)$. This is the height of the curve $\vec{x}(s)$ above the plane that goes through the origin and is perpendicular to $\vec{Y}(s_0, \theta_0)$. Then the derivative is $h'(s) = \vec{Y}(s_0, \theta_0) \cdot \vec{T}(s)$ so that the critical points of $h(s)$ occur where Γ intersects $\mathcal{C}_{s_0,\theta_0}$. Since $n(\mathcal{C}_{s_0,\theta_0}) = 2$, the height function $h(s)$ has exactly two critical points, which must be one maximum and one minimum. Call s_1 the maximum point and s_2 the minimum. The points $\vec{x}(s_1)$ and $\vec{x}(s_2)$ divide C into two curves over which the height function $h(s)$ is monotonic, one increasing and the other decreasing. Consequently, any plane perpendicular to $\vec{Y}(s_0, \theta_0)$ that intersects $\vec{x}$ between the points $\vec{x}(s_1)$ and $\vec{x}(s_2)$ of extremal height function, intersects $\vec{x}$ in exactly two points. This fact allows us to construct a homotopy between C and a circle, which we now describe.

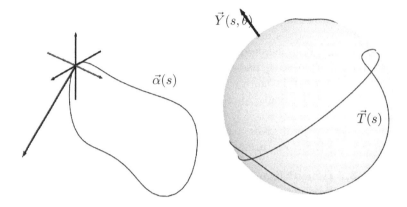

$\vec{Y}(s,\theta)$

$\vec{\alpha}(s)$

$\vec{T}(s)$

Figure 4.5: Proof of the Fary-Milnor Theorem.

Call v the height parameter $h(s)$ for a point on the curve $\vec{x}(s)$. Call $v_{\max} = h(s_1)$ and $v_{\min} = h(s_2)$. Every plane perpendicular to $\vec{Y}(s_0, \theta_0)$ is determined by a parameter v as the unique plane perpendicular to $\vec{Y}(s_0, \theta_0)$ and going through the point $v\vec{Y}(s_0, \theta_0)$. According to the discussion in the previous paragraph, the plane $\mathcal{P}_v$ perpendicular to $\vec{Y}(s_0, \theta_0)$ at height v intersects C in two points. Therefore, C can be expressed as the union of the locus of two continuous curves $\vec{\gamma}_1, \vec{\gamma}_2 : [v_{\min}, v_{\max}] \longrightarrow \mathbb{R}^3$ such that $\mathcal{P}_v \cap C = \{\vec{\gamma}_1(v), \vec{\gamma}_2(v)\}$. Furthermore, these two curves intersect only at the end points $\vec{\gamma}_1(v_{\max}) = \vec{\gamma}_2(v_{\max}) = \vec{x}(s_1)$ and $\vec{\gamma}_1(v_{\min}) = \vec{\gamma}_2(v_{\min}) = \vec{x}(s_2)$.

The required homotopy involves differentially rotating and stretching the two halves of the curve parametrized respectively by $\vec{\gamma}_1$ and $\vec{\gamma}_2$ into two halves of a circle. Call $v_0 = \frac{1}{2}(v_{\min} + v_{\max})$ and $R = \frac{1}{2}(v_{\max} - v_{\min})$. Complete $\{\vec{Y}(s_0, \theta_0)\}$ to an orthonormal basis $\mathcal{B} = \{\vec{Y}(s_0, \theta_0), \vec{e}_2, \vec{e}_3\}$. We construct a homotopy between C and the circle of center $\vec{x}(s_2) + v_0\vec{Y}(s_0, \theta_0)$, of radius R in the plane through this center and spanned by $\{\vec{Y}(s_0, \theta_0), \vec{e}_2\}$.

By the construction of $\vec{\gamma}_1$ and $\vec{\gamma}_2$, in $\mathcal{B}$-coordinates we can write

$$[\vec{\gamma}_1(v) - \vec{x}(s_2)]_{\mathcal{B}} = \begin{pmatrix} v \\ a_1(v)\cos(\beta_1(v)) \\ a_1(v)\sin(\beta_1(v)) \end{pmatrix}$$

and

$$[\vec{\gamma}_2(v) - \vec{x}(s_2)]_{\mathcal{B}} = \begin{pmatrix} v \\ a_2(v)\cos(\beta_2(v)) \\ a_2(v)\sin(\beta_2(v)) \end{pmatrix},$$

where $a_1, a_2, \beta_1, \beta_2 : [v_{\min}, v_{\max}] \to \mathbb{R}$ are continuous functions that satisfy (i) $a_1(v_{\min}) = a_2(v_{\min}) = 0$, (ii) $a_1(v), a_2(v) \geq 0$, and (iii) $\beta_1(v) < \beta_2(v)$ for all $v \in [v_{\min}, v_{\max}]$. Let N be an integer such that $2\pi N > \beta_2(v)$ for all v.

For $i = 1, 2$, define functions $G_{i,1}$, $G_{i,2}$, and G_i by, expressed in $\mathcal{B}$-coordinates,

$$G_{i,1}(v,t) = \vec{x}(s_2) + \begin{pmatrix} v \\ a_i(v)\cos((1-2t)\beta_i(v) + 2t \cdot 2\pi(N + i - 1)) \\ a_i(v)\cos((1-2t)\beta_i(v) + 2t \cdot 2\pi(N + i - 1)) \end{pmatrix}$$

for $(v,t) \in [v_{\min}, v_{\max}] \times [0, 1/2]$,

$$G_{i,1}(v,t) = \vec{x}(s_2) + \begin{pmatrix} v \\ ((2-2t)a_i(v) + (2t-1)\sqrt{R^2 - (v_0 - v)^2})(-1)^{i-1} \\ 0 \end{pmatrix}$$

for $(v,t) \in [v_{\min}, v_{\max}] \times [1/2, 1]$, and finally

$$G_i(v,t) = \begin{cases} G_{i,1}(v,t) & \text{for } 0 \leq t \leq 1/2 \\ G_{i,2}(v,t) & \text{for } 1/2 \leq t \leq 1 \end{cases}$$

for $(v,t) \in [v_{\min}, v_{\max}] \times [0, 1]$. We note that $G_{i,1}$ first continuously rotates $\vec{\gamma}_i(v) - \vec{x}(s_2)$ so that when projected onto a plane spanned by $\{\vec{e}_2, \vec{e}_3\}$ the point $G_{1,1}(v, 1/2)$ (resp. $G_{2,1}(v, 1/2)$) lies on the positive (resp. negative) $\vec{e}_2$ ray. Then, for $1/2 \leq t \leq 1$, the function $G_{i,2}(v,t)$ stretches the resulting curve after $G_{i,1}$ to half circle.

In order to construct a complete homotopy from C to a circle, we define $H : [0, 2\pi] \times [0, 1] \to \mathbb{R}^3$ by

$$H(u,t) = \begin{cases} G_1(v_0 - R\cos u, t) & \text{for } 0 \leq u \leq \pi \\ G_2(v_0 - R\cos u, t) & \text{for } \pi \leq u \leq 2\pi. \end{cases}$$

This reparametrization of v ensures that for the angle u with $0 \leq u\pi$ we use the homotopy of associated to γ_1, and that for $\pi \leq u \leq 2\pi$, we come back along γ_2.

We now leave it as an exercise to the reader to complete the proof that $H(u,t)$ is a homotopy between C and a circle of radius R (Problem 4.3.8).

By Definition 4.3.1, the existence of H as defined above allows us to conclude that C is unknotted. Therefore, if C is a knot, we conclude that the total curvature of C is greater than or equal to 4π. □

4.3.2 Links

A simple, closed space curve in itself resembles a circle. Knottedness is a property that concerns how a simple closed curve "sits" in the

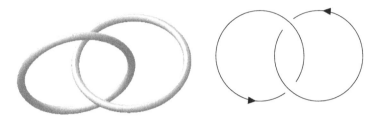

Figure 4.6: Linked curves; linking number $+1$.

ambient space. The technical language for this scenario is that any simple closed curve is homeomorphic to a circle, but knottedness concerns how the curve is embedded in $\mathbb{R}^3$. In a similar way, the notion of linking between two simple closed curves is a notion that considers how the curves are embedded in space in relation to each other.

Intuitively speaking, we would like the linking number between two simple closed curves C_1 and C_2 in $\mathbb{R}^3$ to be the minimum number of crossings required in order to continuously move the curves in space (including deformations that do not break the curves) until they are separated by a plane in $\mathbb{R}^3$. We cannot use this intuition as a definition since counting the number of times it takes to break one curve in order to pass through another is too vague.

Figure 4.6 illustrates a basic scenario of linked curves. The figure on the left shows a rendering of two thick circles in space that are linked. Our intuitive picture leads us to think of this curve as having a single link. Since it is difficult to accurately sketch curves in $\mathbb{R}^3$ and to see which curve is in front of which, it is common to use a diagram to depict the curves. The diagram on the right is obtained by projecting the curves into a plane (perpendicular to some vector $\vec{n}$), careful to show the white space along the curve diagram to communicate which curve is above (in reference to $\vec{n}$) the other at the intersections that occur when projecting. The diagram is called the *link diagram*.

In order to give a definition of the linking number, we need to give orientations to the curves. The standard definition of the linking number of two curves uses the link diagram along with the sign associated to each crossing, as shown below:

$$\nearrow\!\!\!\!\nwarrow \quad -1 \qquad \nwarrow\!\!\!\!\nearrow \quad +1$$

The arrows indicate the orientation on the respective curves at a crossing point. In Figure 4.6, the picture on the left makes no reference to orientation, but the diagram on the right imposes an orientation by the depicted arrows.

Definition 4.3.3 The linking number $\text{link}(C_1, C_2)$ of two simple closed oriented curves C_1 and C_2 is half the sum of the signs of all the crossings in the diagram of the pair (C_1, C_2).

For example, in the oriented diagram on the right in Figure 4.6, both crossings have sign $+1$. Thus half the sum of the signs of the crossings is $+1$.

One of the surprising results about linking is Gauss's formula for the linking number between two curves. Though linking and the linking number are obviously global properties of the curves, Gauss's formula calculates the linking number in terms of the parametrizations of the curves, thereby connecting the global property to local properties.

Theorem 4.3.4 (Gauss's Linking Formula) *Let C_1 and C_2 be two closed, regular space curves parametrized by $\vec{\alpha} : I \to \mathbb{R}^3$ and $\vec{\beta} : J \to \mathbb{R}^3$, respectively. The linking number between C_1 and C_2 is*

$$\text{link}(C_1, C_2) = \frac{1}{4\pi} \int_I \int_J \frac{\det(\vec{\alpha}(u) - \vec{\beta}(v), \vec{\alpha}'(u), \vec{\beta}'(v))}{\|\vec{\alpha}(u) - \vec{\beta}(v)\|^3} \, du \, dv. \quad (4.4)$$

Though we cannot give a proof for this formula at this time, we briefly sketch a few of the concepts and techniques that go into it.

In Section 2.2, we introduced the notion of degree for a continuous map between circles, $f : \mathbb{S}^1 \to \mathbb{S}^1$. Intuitively speaking, the degree of f counts how many times f covers its codomain and with what orientation. The degree of f takes into account when f might double back. For example, if $q \in \mathbb{S}^1$ has six preimages from f and, for four of the preimages, f passes through q with the same orientation as on the domain and, for the other two preimages, f passes through q in an opposite orientation, then the degree of f is $4 - 2 = 2$.

Similarly (and leaving many of the technical details for later), one can define the degree of a continuous map between spheres $f : \mathbb{S}^2 \to \mathbb{S}^2$ as how often f covers $\mathbb{S}^2$. Again, one should note the difficulty inherent in making this definition precise since one must take into account a form of orientation and doubling back, i.e., whether f folds back over itself over some region of the codomain. In the same way, one can define the degree of a continuous map $F : S \to \mathbb{S}^2$, where S is a regular surface without boundary. (We give the definition for a regular surface in Chapter 5.) The conditions that S is regular and has no boundary guarantee that F generically covers $\mathbb{S}^2$ by the same amount at all its points, so long as we take into account orientation and assign a negative covering value to F when F covers $\mathbb{S}^2$ negatively.

As discussed in Definition 2.2.7 and in the following, it is more appropriate to view the parametrization of a simple closed curve C

as a continuous function $\vec{x} : \mathbb{S}^1 \to \mathbb{R}^3$. So consider two simple closed curves with parametrizations $\vec{\alpha} : \mathbb{S}^1 \to \mathbb{R}^3$ and $\vec{\beta} : \mathbb{S}^1 \to \mathbb{R}^3$. The function

$$\Psi : \mathbb{S}^1 \times \mathbb{S}^1 \to \mathbb{S}^2 \quad \text{given by} \quad \Psi(t, u) = \frac{\vec{\beta}(u) - \vec{\alpha}(t)}{\|\vec{\beta}(u) - \vec{\alpha}(t)\|}$$

defines a continuous function from the torus $\mathbb{S}^1 \times \mathbb{S}^1$ to the unit sphere $\mathbb{S}^2$. The technical definition of the linking number of two curves is the degree of $\Psi(t, u)$. The proof of Theorem 4.3.4 consists of calculating the degree of $\Psi(t, u)$.

Equation (4.4) is difficult to use in general, and in the exercises, we often content ourselves with using a computer algebra system to calculuate the integrals. Some basic facts about linking are not difficult to show with the appropriate topological background but are difficult using Equation (4.4). For example, if two simple closed curves can be separated by a plane, then $\text{link}(C_1, C_2) = 0$. This fact is simple once one has a few facts about the degrees of maps to $\mathbb{S}^2$, but using Gauss's formula to show the same result is a difficult problem.

PROBLEMS

1. Let C_1 and C_2 be two simple, closed, oriented space curves. Suppose that C_2^- is the same locus as C_2 but with the opposite orientation. Prove that $\text{link}(C_1, C_2^-) = -\text{link}(C_1, C_2)$.

2. Let C_1 and C_2 be two simple, closed, oriented space curves. Prove that $\text{link}(C_1, C_2) = \text{link}(C_2, C_1)$.

3. Consider the two linked curves depicted in Figure 4.7. Impose an orientation on the curves, sketch the link diagram, and then calculate the linking number.

4. (**CAS**) Consider the trefoil knot parametrized by

$$\vec{\alpha}(t) = \left((3 + \cos 3t) \cos 2t, (3 + \cos 3t) \sin 2t, \sin 3t\right).$$

Calculate the total curvature of this trefoil knot. Show how to create another simple closed curve that is homotopic to this trefoil knot, with a total curvature of $4\pi + \varepsilon$ for any $\varepsilon > 0$.

5. Let C_1 and C_2 be the two circles in $\mathbb{R}^3$ parametrized by

$$\vec{\alpha}(t) = (\cos t, \sin t, 0), \qquad \text{for } t \in [0, 2\pi],$$
$$\vec{\beta}(t) = (1 - \cos t, 0, \sin t), \quad \text{for } t \in [0, 2\pi].$$

   (a) Using Equation (4.4), write an integral that gives the linking number of these two curves.

   (b) Use a computer algebra system to calculate this linking number.

   (c) (*) Calculate the integral explicitly.

Figure 4.7: An interesting link.

6. A rigid motion in $\mathbb{R}^3$ is a transformation $F : \mathbb{R}^3 \to \mathbb{R}^3$ defined by $F(\vec{x}) = A\vec{x} + \vec{b}$, where A is an orthogonal matrix. Prove from Equation (4.4) that the linking number is a geometric invariant, i.e., that if a rigid motion is applied to two simple closed curves C_1 and C_2, then their linking number does not change.

7. Consider the two simple closed curves

$$\vec{\alpha}(t) = (3\cos t, 3\sin t, 0), \qquad \text{for } t \in [0, 2\pi],$$

$$\vec{\beta}(t) = \big((3 + \cos(nt))\cos t, (3 + \cos(nt))\sin t, \sin(nt)\big), \text{ for } t \in [0, 2\pi].$$

(a) Give the link diagram of these two space curves, including the orientation, and show that the linking number of these two curves is n.

(b) Gauss's formula in Equation (4.4) is quite difficult to use, but, using a computer algebra system, give support for the above answer.

8. This exercise finishes the proof of the Fary-Milnor Theorem. Consider the function $H(u, t)$ defined in the proof of the Fary-Milnor Theorem.

(a) Show that the locus of $H(u, 0)$ is the curve C.

(b) Show that the locus of $H(u, 1)$ is a circle of center $\vec{x}(s_2) + v_0 \vec{Y}(s_0, \theta_0)$ and radius R.

(c) Prove that, for all $t \in [0, 1]$, $H(u, t_0)$ is a simple closed curve.

(d) Use the previous parts of the exercise to conclude that $H(u, t)$ is a homotopy between C and a circle.

CHAPTER 5

Regular Surfaces

5.1 Parametrized Surfaces

There are many approaches that one can take to introduce surfaces. Some texts immediately build the formalism of differentiable manifolds, some texts first encounter surfaces in $\mathbb{R}^3$ as the solution set to an algebraic equation $F(x, y, z) = 0$ with three variables, and other texts present surfaces as the images of vector functions of two variables. In this section, we introduce the last of these three options and occasionally refer to the connection with the surfaces as solution sets to algebraic equations. Only later will we show why one requires more technical definitions to arrive at a workable definition that matches what one typically means by a "surface."

We imitate the definition of parametrized curves in $\mathbb{R}^n$ and begin our study of surfaces (yet to be defined) by considering continuous functions $\vec{X} : U \to \mathbb{R}^3$, where U is a subset of $\mathbb{R}^2$. Below are some examples of such functions whose images in $\mathbb{R}^3$ are likely to appear in a multivariable calculus course.

Example 5.1.1 (Planes) If $\vec{a}$ and $\vec{b}$ are linearly independent vectors in $\mathbb{R}^3$, then the plane through the point $\vec{p}$ and parallel to the vectors $\vec{a}$ and $\vec{b}$ can be expressed as the image of the following function:

$$\vec{X}(u, v) = \vec{p} + u\vec{a} + v\vec{b}$$

for $(u, v) \in \mathbb{R}^2$. In other words, we can write

$$\vec{X}(u, v) = (p_1 + a_1 u + b_1 v, p_2 + a_2 u + b_2 v, p_3 + a_3 u + b_3 v)$$

for constants p_i, a_i, b_i, with $1 \leq i \leq 3$ that make (a_1, a_2, a_3) and (b_1, b_2, b_3) noncollinear vectors.

Example 5.1.2 (Graphs of Functions) If $z = f(x, y)$ is a real function of two variables defined over some set $U \subset \mathbb{R}^2$, one can ob-

DOI: 10.1201/9781003295341-5

tain the graph of this function as the image of a function $\vec{X} : U \to \mathbb{R}^3$ by setting

$$\vec{X}(u, v) = (u, v, f(u, v)).$$

 Example 5.1.3 (Spheres) One can obtain a sphere of radius R centered at the origin as the image of

$$\vec{X}(u, v) = (R \cos u \sin v, R \sin u \sin v, R \cos v),$$

with $(u, v) \in [0, 2\pi] \times [0, \pi]$. This expression should not appear too mysterious since it merely puts the equation of spherical coordinates $r = R$ into Cartesian coordinates, with $u = \theta$ and $v = \phi$. This method of parametrizing the sphere is the astronomical one, where v measures the colatitude, i.e., the angle down from the north pole. (Unfortunately, there is no uniformity in the parametrization used for spheres in calculus books.)

 Example 5.1.4 (Conics) Besides getting the sphere, one could also obtain an ellipsoid as the image of some vector function $\vec{X} : U \to \mathbb{R}^2$ by modifying the coefficients in front of the x, y, and z coordinate functions of the parametrization for the sphere. In particular, the parametrization

$$\vec{X}(u, v) = (a \cos u \sin v, b \sin u \sin v, c \cos v),$$

with $(u, v) \in [0, 2\pi] \times [0, \pi]$ has for image the ellipsoid of equation

$$\frac{x^2}{a^2} + \frac{y^2}{b^2} + \frac{z^2}{c^2} = 1.$$

One can obtain all of the other conic surfaces as follows. The circular cone

$$\frac{x^2}{a^2} + \frac{y^2}{a^2} = \frac{z^2}{b^2}$$

is the image of

$$\vec{X}(u, v) = (au \cos v, au \sin v, bu),$$

with $(u, v) \in \mathbb{R} \times [0, 2\pi]$. The function

$$\vec{X}(u, v) = (\cosh u \cos v, \cosh u \sin v, \sinh u),$$

with $(u, v) \in \mathbb{R} \times [0, 2\pi]$, traces out the hyperboloid of one sheet. (See Figure 5.1 for two perspectives of a hyperboloid of one sheet.) For the hyperboloid of two sheets, we need two separate functions

$$\vec{X}(u, v) = (\sinh u \cos v, \sinh u \sin v, \cosh u)$$

and

$$\vec{X}(u, v) = (\sinh u \cos v, \sinh u \sin v, -\cosh u).$$

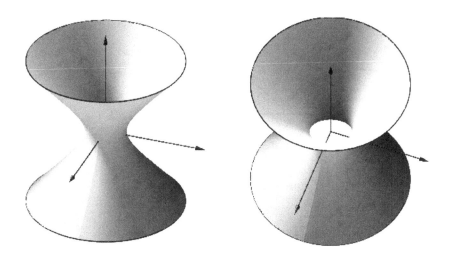

Figure 5.1: Hyperboloid of one sheet.

Example 5.1.5 (Surfaces of Revolution) A *surface of revolution*
is a set S in $\mathbb{R}^3$ obtained by rotating, in $\mathbb{R}^3$, a regular plane curve C
in a plane $\mathcal{P}$ about a line L in $\mathcal{P}$ that does not meet C. The curve C
is called the *generating curve*, and the line l is called the *rotation axis*.

The circles described by the rotation locus of the points of C are
called the *parallels* of S, and the various positions of C are called the
meridians of S. A surface of revolution can be parametrized naturally
by the parameter on the curve C and the angle of rotation u about
the axis.

Let $\vec{x} : I \to \mathbb{R}^2$ be a parametric curve in the xy-plane, with $\vec{x}(t) =$
$(x(t), y(t))$, and suppose we want to find the surface of revolution of
this curve about either the x- or y-axis. About the y-axis, we take
$\vec{X}(t, u) = (x(t) \cos u, y(t), x(t) \sin u)$, and about the x-axis, we would
take $\vec{X}(t, u) = (x(t), y(t) \cos u, y(t) \sin u)$. Both of these options have
domains of $I \times [0, 2\pi]$. (See Figure 5.2 for an example of a surface of
revolution about the x-axis based on a curtate cycloid.)

Looking more closely at Example 5.1.3, one should note that
though the function $\vec{X}(u, v)$ maps onto the sphere of radius R centered at the origin, this function exposes two potential problems with
the intuitive approach. The first problem is that $\vec{X}$ is not injective.
In particular, for any point (x, y, z) on this sphere with $y = 0$ and
$x > 0$, we have

$$v = \cos^{-1}\left(\frac{z}{R}\right) \qquad \text{and} \qquad u = 0 \text{ or } 2\pi,$$

and so there exist two preimages for such points. Worse still, if we

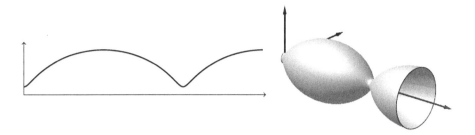

Figure 5.2: Surface of revolution.

consider the north and south poles $(0, 0, 1)$ and $(0, 0, -1)$, the preimages are the points $(u, 0)$ and (u, π), respectively, for all $u \in [0, 2\pi]$. With the given parametrization, the arc on the sphere defined by $u = 0$ is the set of points with more than one preimage. This latter remark leads to the second problem because the points on this arc have no particular geometric significance, and one would wish to avoid any formulation of a surface that confers special properties on some points that are geometrically ordinary.

Before addressing these and other issues necessary to be able to "do calculus" on a surface, we need to define the class of subsets of $\mathbb{R}^3$ that one can even hope to study with differential geometry. The primary criterion for such a subset is that one must be able to describe its points using continuous functions.

Definition 5.1.6 A subset $S \subseteq \mathbb{R}^3$ is called a *parametrized surface* if for each $p \in S$, there exists an open set $U \subset \mathbb{R}^2$, an open neighborhood V of p in $\mathbb{R}^3$, and a continuous function $\vec{X} : U \to \mathbb{R}^3$ such that $\vec{X}(U) = V \cap S$. Each such $\vec{X}$ is called a *parametrization* of a neighborhood of S. We call a parametrized surface *of class C^r* if it can be covered by parametrizations $\vec{X}$ of class C^r.

Note that in contrast to space curves, where we use intervals as the domain for parametrizations, this definition uses open sets. Furthermore, this definition does not insist that S be the image of a single parametrization but, rather that it be covered by images of parametrizations.

For any such function $\vec{X} : U \to V \cap S$, where $V \cap S$ is an open neighborhood of p in S, we call $\vec{X}$ a parametrization of the *coordinate patch* $V \cap S$. (See Figure 5.3 for an illustration of the relation of the domain $U \subset \mathbb{R}^2$ and the coordinate patch on the surface $V \cap S$.)

Definition 5.1.6 states that, for every point $p \in S$, there is an open neighborhood of p on S that is the image of some vector function.

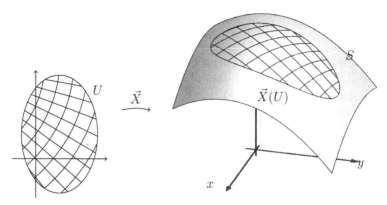

Figure 5.3: Coordinate patch.

With the tools of curves in the plane or space, one can study properties of a surface S by considering various families of curves on the surface. If one wishes to study S near a point p, one could look for common properties of all the curves on S passing through p.

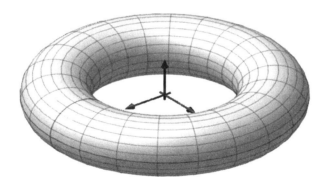

Figure 5.4: Coordinate lines on a torus.

There are two natural classes of curves on any given parametrized surface S that arise naturally. First, we can consider what are generically called the *coordinate lines* of the surface S. A coordinate line on S is the image of a space curve defined by fixing one of the variables in a particular parametrization of a coordinate patch of S. Namely, if $\vec{X} : U \to \mathbb{R}^3$ parametrizes a patch of S, then a curve of the form

$$\vec{\gamma}_1(u) = \vec{X}(u, v_0), \quad \text{respectively} \quad \vec{\gamma}_2(v) = \vec{X}(u_0, v),$$

is called a coordinate line of the variable u, and respectively for v. (See Figure 5.4 for an example of coordinate lines of a torus parametrized by $\vec{X}(u,v) = \big((3+\cos v)\cos u, (3+\cos v)\sin u, \sin v\big)$ for $(u,v) \in [0,2\pi] \times [0,2\pi].$)

Example 5.1.7 With the surfaces of revolution described in Example 5.1.5, the coordinate lines of t and u, respectively, are precisely the parallels and the meridians.

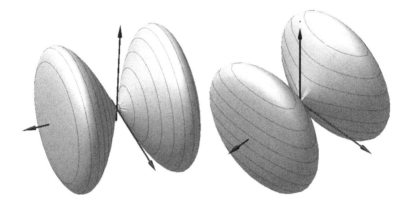

Figure 5.5: Slices of a surface of revolution.

Another family of curves we might study are the *slices* of S, by which we mean the intersection of S with a family of parallel planes. As an example, Figure 5.5 shows the parametric surface

$$\vec{X}(u,v) = (\cos v, \sin 2v \cos u, \sin 2v \sin u),$$

with domain $[0,2\pi] \times [0,\pi]$ along with two sets of slices. Though any set of parallel planes can produce a set of slices on S, we often consider the intersection of S with planes of the form $x = a$, $y = b$, or $z = c$. In the figure on the left, we sliced the surface with planes $x = a$, which are perpendicular to the axis of revolution. These slices result in a set of circles. If we fix $x = a$, with $-1 \leq a \leq 1$, we get $\cos v = a$, and there exists a value $v_0 \in [0,\pi]$ satisfying this equation. Then the slice of $\vec{X}(u,v)$ intersecting the $x = a$ plane gives a circle parallel to the yz-plane, given by

$$(x,y,z) = (\cos(v_0), \sin(2v_0)\cos u, \sin(2v_0)\sin u).$$

On the other hand, if we consider planes perpendicular to the z-axis, that is planes of the form $z = c$, the $z = c$ slice corresponds to the set of points $(\cos v, \sin(2v)\cos u)$ where $(u,v) \in [0,2\pi] \times [0,\pi]$ satisfy $\sin(2v)\sin(u) = c$. This defines a curve implicitly. In this case, the best we can do is to solve for $\sin u = c/\sin(2v)$ and then get a parametrization for the slices by

$$(\cos v, \pm\sqrt{\sin^2(2v) - c^2}, c).$$

These are shown on the right in Figure 5.5.

PROBLEMS

1. Find a parametrization $\vec{X}$ for the cylinder $\{(x, y, z) \in \mathbb{R}^3 \mid x^2 + y^2 = 1\}$. Can the domain of this parametrization be chosen so that $\vec{X}$ is bijective onto the cylinder? Explain why or why not.

2. Prove that the following functions provide parametrizations of the hyperbolic paraboloid $x^2 - y^2 = z$. Provide appropriate domains of definitions. Which ones are injective? Which ones are surjective onto the surface?

    (a) $\vec{X}(u, v) = (v \cosh u, v \sinh u, v^2)$.
    (b) $\vec{X}(u, v) = ((u + v), (u - v), 4uv)$.
    (c) $\vec{X}(u, v) = (uv, u(1 - v), u^2(2v - 1))$.

3. Let $\mathcal{P}$ be the plane through the origin with normal vector $\vec{n} = (a, b, c)$ and let $F = (x_F, y_F, z_F)$ be a point not on $\mathcal{P}$. Let S be the set of points M in $\mathbb{R}^3$ such that the distance QM is equal to the distance between M and the plane $\mathcal{P}$. Find a parametrization for the surface S.

4. Let $\vec{\alpha} : I \to \mathbb{R}^2$ be a regular, plane curve, and let L be a line that does not intersect the locus of $\vec{\alpha}$. Suppose that L goes through the point $\vec{p}$ and has direction vector $\vec{u}$ so that $\vec{p} + t\vec{u}$ parametrizes L. Find a formula for the parametrization of the surface of revolution obtained by rotating $\vec{\alpha}(I)$ about L.

5. Let $\vec{\alpha}, \vec{\beta} : I \to \mathbb{R}^3$ be two parametrized space curves with the same domain. Define the *secant surface* by the parametrization $\vec{X} : I \times \mathbb{R} \to \mathbb{R}^3$ with

$$\vec{X}(t, u) = (1 - u)\vec{\alpha}(t) + u\vec{\beta}(t).$$

    (a) Let $\vec{\alpha}(t) = (2\cos t, \sin t, 1)$ and $\vec{\beta}(t) = (\cos t, 2\sin t, -1)$. Prove that every slice by a constant u coordinate is an ellipse and give the eccentricity of the ellipse as a function of u. (Recall that the eccentricity of an ellipse with half-axes $a > b > 0$ is $e = \sqrt{1 - \frac{b^2}{a^2}}$.)

    (b) Now let $\vec{\alpha}(t) = (\cos t, \sin t, 1)$ and $\vec{\beta}(t) = (-\sin t, \cos t, -1)$. Prove that every slice by a constant u coordinate is a circle and give the radius as a function of u.

6. Prove that on the unit sphere parametrized by

$$\vec{X}(\theta, \phi) = (\cos\theta \sin\phi, \sin\theta \sin\phi, \cos\phi),$$

    the curves given by

$$(\theta, \phi) = (\ln t, 2\tan^{-1} t)$$

intersect every meridian with an angle of $\frac{\pi}{4}$. (Note that a meridian is a curve such that $\theta = $ const. A curve on the sphere that intersects the meridians at a constant angle is called a *loxodrome*.)

7. Consider a curve in the plane $\vec{\alpha}(t) = (x(t), y(t))$ and the surface of revolution obtained by rotating the image of $\vec{\alpha}$ about the y-axis. This surface of revolution is parametrized by

$$\vec{X}(t, u) = (x(t)\cos(u), y(t), x(t)\sin(u)).$$

   (a) Consider u-coordinate lines, i.e., the space curves

$$\gamma_1(t) = \big(x(t)\cos(u_0), y(t), x(t)\sin(u_0)\big),$$

   where u_0 is fixed. Calculate the space curvature and torsion of γ_1.

   (b) Repeat the above question with the t-coordinate lines, i.e., the space curves $\gamma_2(u) = \big(x(t_0)\cos(u), y(t_0), x(t_0)\sin(u)\big)$, where t_0 is fixed.

8. Suppose that C is a curve in the xy-plane defined by the equation $F(x, y) = 0$.

   (a) Using the previous exercise, show that the surface of revolution of C about the y-axis satisfies the equation $F(\sqrt{x^2 + z^2}, y) = 0$.

   (b) Show that the surface of revolution about the x-axis of the curve $\vec{\alpha}$ from the previous exercise is parametrized by $\vec{Y}(t, u) = (x(t), y(t)\cos u, y(t)\sin u)$.

   (c) Show that the surface of revolution of C about the x-axis satisfies the equation $F(x, \sqrt{y^2 + z^2}) = 0$.

9. Consider the set of points $S = \{(x, y, z) \in \mathbb{R}^3 \mid x^4 + y^4 + z^4 = 1\}$. Modify the usual parametrization for a sphere to find parametrizations that cover S.

5.2   Tangent Planes; The Differential

In the local theory of curves, we called a curve C regular at a point p if there is a parametrization $\vec{x}(t)$ of C near $p = \vec{x}(t_0)$ such that $\vec{x}'(t_0)$ exists and $\vec{x}'(t_0) \neq \vec{0}$. In Section 1.2, we saw that this definition is tantamount to requiring that the unit tangent vector

$$\lim_{t \to t_0} \frac{\vec{x}'(t)}{\|\vec{x}'(t)\|},$$

at least given in the limit, exists. Along with the requirement that the curve be simple, regularity means that there exists a tangent line to C at p. Imitating this latter geometric property, we will eventually call a point $p \in S$ regular if one can define a tangent plane to p at S. However, we must delay a precise definition for a regular surface until after we discuss the tangent vectors to a surface S at p.

5.2.1 Tangent Planes

Definition 5.2.1 Let S be a parametrized surface and p a point in S.
Consider the set of space curves $\vec{\gamma} : (-\varepsilon, \varepsilon) \to \mathbb{R}^3$ such that $\vec{\gamma}(0) = p$,
and the image of $\vec{\gamma}$ lies entirely in S. A *tangent vector* to S at p is
any vector in $\mathbb{R}^3$ that can be expressed as $\vec{\gamma}'(0)$, where $\vec{\gamma}$ is such a
space curve.

From the view point of intuition, we often consider tangent vectors
to S at p to be based at p. The set of tangent vectors, as a subset of
$\mathbb{R}^3$, gives an approximation of the behavior of S at p. By rescaling
the parameter t, we notice that the set of tangent vectors to S at a
point p contains $\vec{0}$ and is closed under scalar multiplication. Hence,
the set of tangent vectors will be a (usually infinite) union of lines,
and, as we shall see, it is often, but not always, a plane.

Example 5.2.2 As an example of a set of tangent vectors that is not
a plane, consider the cone with equation

$$x^2 + y^2 - z^2 = 0,$$

with parametrization

$$\vec{X}(u, v) = (v \cos u, v \sin u, v),$$

for $(u, v) \in [0, 2\pi] \times \mathbb{R}$ (see Figure 5.6). We note that $\vec{X}(u, 0) = (0, 0, 0)$ for all $u \in [0, 2\pi]$ and consider the point $p = \vec{X}(u, 0) = (0, 0, 0)$. A curve on the cone through p has a parametric equation
of the form $\vec{\gamma}(t) = \vec{X}(u(t), v(t))$, where $v(0) = 0$ and u is any real in
$[0, 2\pi]$. For such a curve, using the multivariable chain rule, we have

$$\vec{\gamma}'(t) = \frac{\partial \vec{X}}{\partial u}\frac{du}{dt} + \frac{\partial \vec{X}}{\partial v}\frac{dv}{dt} = u'(t)\frac{\partial \vec{X}}{\partial u} + v'(t)\frac{\partial \vec{X}}{\partial v},$$

and since $\vec{X}_u(u, 0) = (0, 0, 0)$,

$$\vec{\gamma}'(0) = v'(0)\,(\cos u, \sin u, 1).$$

Since $v'(0)$ and u can be any real values, we determine that the union
of the endpoints the set of tangent vectors to the cone at $(0, 0, 0)$ is
precisely the cone itself.

Definition 5.2.3 Let S be a parametrized surface and p a point in
S. If the set of tangent vectors to S at p forms a two-dimensional
subspace of $\mathbb{R}^3$, we call this subspace the *tangent space* to S at p and
denote it by $T_p S$. If $T_p S$ exists, we call the *tangent plane* the set of
points

$$\{p + \vec{v} \mid \vec{v} \in T_p S\}.$$

$\vec{\gamma}(t)$

$\vec{\gamma}'(0)$

Figure 5.6: Double cone.

We wish to find a condition that determines when the set of tangent vectors is a two-dimensional vector subspace.

Let p be a point on a surface S, and let $\vec{X} : U \to \mathbb{R}^3$ be a parametrization of a neighborhood $V \cap S$ of p for some $U \subseteq \mathbb{R}^2$. Suppose that $p = \vec{X}(u_0, v_0)$ and consider the coordinate lines $\vec{\gamma}_1(t) = \vec{X}(u_0 + t, v_0)$ and $\vec{\gamma}_2(t) = \vec{X}(u_0, v_0 + t)$ through p. Both $\vec{\gamma}_1$ and $\vec{\gamma}_2$ lie on the surface and, by the chain rule,

$$\vec{\gamma}_1'(0) = \frac{\partial \vec{X}}{\partial u}(u_0, v_0) \quad \text{and} \quad \vec{\gamma}_2'(0) = \frac{\partial \vec{X}}{\partial v}(u_0, v_0).$$

Furthermore, for any curve $\vec{\gamma}(t)$ on the surface with $\vec{\gamma}(0) = p$, we can write

$$\vec{\gamma}(t) = \vec{X}(u(t), v(t))$$

for some functions $u(t)$ and $v(t)$, with $u(0) = u_0$ and $v(0) = v_0$. By the chain rule,

$$\vec{\gamma}'(t) = u'(t)\frac{\partial \vec{X}}{\partial u}(u(t), v(t)) + v'(t)\frac{\partial \vec{X}}{\partial v}(u(t), v(t)),$$

so at p,

$$\vec{\gamma}'(0) = u'(0)\frac{\partial \vec{X}}{\partial u}(u_0, v_0) + v'(0)\frac{\partial \vec{X}}{\partial v}(u_0, v_0) = u'(0)\vec{\gamma}_1'(0) + v'(0)\vec{\gamma}_2'(0).$$

Thus, $\vec{\gamma}'(0)$ is a linear combination of $\frac{\partial \vec{X}}{\partial u}(u_0, v_0)$ and $\frac{\partial \vec{X}}{\partial v}(u_0, v_0)$.

Definition 5.1.6 does not assume that the parametrization $\vec{X}$ is injective. Therefore, the set of tangent vectors is a plane if and only if the union of all linear subspaces

$$\text{Span}\Big(\frac{\partial \vec{X}}{\partial u}(u_0, v_0), \frac{\partial \vec{X}}{\partial v}(u_0, v_0)\Big),$$

where $\vec{X}(u_0, v_0) = p$ is a plane. In particular, if there exists only one (u_0, v_0) such that $\vec{X}(u_0, v_0) = p$, then a tangent plane exists at p if $\frac{\partial \vec{X}}{\partial u}(u_0, v_0)$ and $\frac{\partial \vec{X}}{\partial v}(u_0, v_0)$ are not collinear.

To simplify notations, we often use the following abbreviated notation for partial derivatives:

$$\vec{X}_u(u, v) = \frac{\partial \vec{X}}{\partial u}(u, v) \qquad \text{and} \qquad \vec{X}_v(u, v) = \frac{\partial \vec{X}}{\partial v}(u, v).$$

We then write succinctly that if $\{(u_0, v_0)\} = \vec{X}^{-1}(p)$, then S has a tangent plane at p if and only if $\vec{X}_u(u_0, v_0) \times \vec{X}_v(u_0, v_0) \neq \vec{0}$. Furthermore, the tangent plane to S at p is the unique plane through p with normal vector $\vec{X}_u(u_0, v_0) \times \vec{X}_v(u_0, v_0) \neq \vec{0}$. This leads us to a formula for the tangent plane.

Proposition 5.2.4 *Let S be a surface parametrized near a point p* *by $\vec{X} : U \to \mathbb{R}^3$. Suppose also that $\vec{X}$ is injective at p (i.e., p has only one preimage) and that $p = \vec{X}(u_0, v_0)$. Then the set of tangent vectors forms a two-dimensional subspace of $\mathbb{R}^3$ if and only if $\vec{X}_u(u_0, v_0) \times \vec{X}_v(u_0, v_0) \neq \vec{0}$. In this case, the tangent plane to S at p satisfies the following equation for position vectors $\vec{x} = (x, y, z)$:*

$$(\vec{x} - \vec{X}(u_0, v_0)) \cdot (\vec{X}_u(u_0, v_0) \times \vec{X}_v(u_0, v_0)) = 0. \qquad (5.1)$$

Through an abuse of language, even if $\vec{X}$ is not injective at p and $\vec{X}_u(u_0, v_0) \times \vec{X}_v(u_0, v_0) \neq 0$ for some $(u_0, v_0) \in U$, we will call Equation (5.1) the equation of the tangent plane to $\vec{X}$ at $q = (u_0, v_0)$. This is an abuse of language since if $p = \vec{X}(u_0, v_0)$ and S is the image of $\vec{X}$, then p might have more than one preimage and, therefore, the set of tangent vectors to S of p might not be a plane but be a union of planes. Problem 5.2.18 gives an example of a parametrized surface that has points where the set of tangent vectors is the union of two planes.

Example 5.2.5 Consider the function $\vec{X} : \mathbb{R}^2 \to \mathbb{R}^3$ defined by $\vec{X}(u, v) = (u, v, 4 + 3u^2 - u^4 - 2v^2)$. We calculate

$$\vec{X}_u \times \vec{X}_v = (1, 0, 6u - 4u^3) \times (0, 1, -4v)$$
$$= (-6u + 4u^3, 4v, 1).$$

Figure 5.7 shows the surface along with the tangent plane at the point $\vec{X}(-1.4, -1)$. At this specific point, we have $p = \vec{X}(-1.4, -1) = (-1.4, -1, 4.0384)$ and $\vec{X}_u \times \vec{X}_v(-1.4, -1) = (-2.576, -4, 1)$. Since

Figure 5.7: A tangent plane.

$\vec{X}$ is in fact the graph of an injective two-variable function, by Proposition 5.2.4, the tangent plane to $\vec{X}$ at p is

$$-2.576(x + 1.4) - 4(y + 1) + (z - 4.0384) = 0$$
$$\Longleftrightarrow -2.576x - 4y + z = 11.6448.$$

 Example 5.2.6 As a second example, consider the hyperboloid of one sheet given by the parametrization

$$\vec{X}(u, v) = (\cosh u \cos v, \cosh u \sin v, \sinh u).$$

A calculation produces

$$\vec{X}_u \times \vec{X}_v = (-\cosh^2 u \cos v, -\cosh^2 u \sin v, \cosh u \sinh u).$$

Then at the point $p = \vec{X}(1, 0) = (\cosh 1, 0, \sinh 1)$, the tangent plane exists, and its equation is

$$-\cosh^2 1(x - \cosh 1) + \cosh 1 \sinh 1(z - \sinh 1) = 0$$
$$\Longleftrightarrow (\cosh 1)x - (\sinh 1)z = 1.$$

Figure 5.8 shows two perspectives of this hyperboloid along with its tangent plane at p. It is interesting to contrast this picture with Figure 5.7, in which the tangent plane does not intersect the surface except at p. The shape of the hyperboloid forces it to intersect its tangent plane at all points.

It is clear that $(\vec{X}_u \times \vec{X}_v)(q)$ is a normal vector to the tangent plane to S at p. This motivates the following definition.

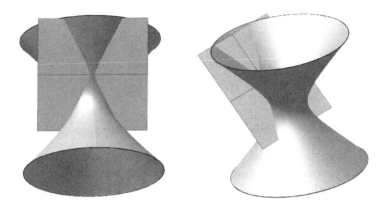

Figure 5.8: A tangent plane on a hyperboloid of one sheet.

Definition 5.2.7 The *unit normal vector* to a parametrized surface S at p, associated to a given parametrization $\vec{X}$, is

$$\vec{N}(q) = \frac{\vec{X}_u \times \vec{X}_v}{\|\vec{X}_u \times \vec{X}_v\|}(q). \tag{5.2}$$

Since the tangent space T_pS is two-dimension, its perpendicular space $(T_pS)^\perp$ is one dimensional. Hence, there are precisely two possibilities for the unit normal vector to a surface at a given point. In Section 5.4, we discuss how the unit normal vector depends on the parametrization and implications for orientability.

5.2.2 The Differential

It is possible to summarize Proposition 5.2.4 in another way that is more convenient when we discuss parametrized surfaces in higher dimensions. To do so, we need the concept of a differential.

If $\vec{F}$ is a function from an open set $U \subset \mathbb{R}^n$ to $\mathbb{R}^m$, we write $\vec{F} = (F_1, F_2, \ldots, F_m)^T$ and think of $\vec{F}$ as a column vector of functions F_i, each in n variables. For all $\vec{a} \in U$, we say that F is *differentiable* at $\vec{a}$ if each F_i is differentiable at $\vec{a}$. Furthermore, we define the *differential* of $\vec{F}$ at $\vec{a}$, written $d\vec{F}_{\vec{a}}$, as the linear transformation $\mathbb{R}^n \to \mathbb{R}^m$ that sends the jth standard vector $\vec{e}_j$ to $\dfrac{\partial \vec{F}}{\partial x_j}(\vec{a})$. We denote by $[d\vec{F}_{\vec{a}}]$ the matrix of $d\vec{F}_{\vec{a}}$ with respect to the standard bases of $\mathbb{R}^n$ and $\mathbb{R}^m$, which we can write explicitly as

$$[d\vec{F}_{\vec{a}}] = \left(\begin{matrix} \dfrac{\partial \vec{F}}{\partial x_1} & \dfrac{\partial \vec{F}}{\partial x_2} & \cdots & \dfrac{\partial \vec{F}}{\partial x_n} \end{matrix}\right) = \begin{pmatrix} \dfrac{\partial F_1}{\partial x_1} & \dfrac{\partial F_1}{\partial x_2} & \cdots & \dfrac{\partial F_1}{\partial x_n} \\ \dfrac{\partial F_2}{\partial x_1} & \dfrac{\partial F_2}{\partial x_2} & \cdots & \dfrac{\partial F_2}{\partial x_n} \\ \vdots & \vdots & \ddots & \vdots \\ \dfrac{\partial F_m}{\partial x_1} & \dfrac{\partial F_m}{\partial x_2} & \cdots & \dfrac{\partial F_m}{\partial x_n} \end{pmatrix}.$$

$$(5.3)$$

By a slight abuse of notation, we will sometimes write $[d\vec{F}]$ or even $d\vec{F}$, without the subscript, to mean the matrix of functions with $\partial \vec{F}/\partial x_j$ as the jth column.

We can now restate Proposition 5.2.4 using the differential, and hence borrowing from linear algebraic concepts.

Corollary 5.2.8 *Let S be a surface, where $p \in S$, and suppose that in an open neighborhood of p, the surface S is parametrized by $\vec{X}$: $U \to \mathbb{R}^3$. The tangent space (and tangent plane) to S at p exists if $\vec{X}^{-1}(p) = \{q\}$ (a singleton set), the differential $d\vec{X}_q$ exists, and $d\vec{X}_q$ has maximal rank. Furthermore, the tangent space is given by $T_pS = \mathrm{Im}(d\vec{X}_q)$.*

Proof: Note that with respect to the standard bases in $\mathbb{R}^2$ and $\mathbb{R}^3$, if $q = (u_0, v_0)$, the matrix of $d\vec{X}_q$ is the 3×2 matrix

$$[d\vec{X}_q] = \begin{pmatrix} \dfrac{\partial X_1}{\partial u}(q) & \dfrac{\partial X_1}{\partial v}(q) \\ \dfrac{\partial X_2}{\partial u}(q) & \dfrac{\partial X_2}{\partial v}(q) \\ \dfrac{\partial X_3}{\partial u}(q) & \dfrac{\partial X_3}{\partial v}(q) \end{pmatrix}, \qquad (5.4)$$

where we have written $\vec{X}(u, v) = \big(X_1(u, v), X_2(u, v), X_3(u, v)\big)$.

First note that for the tangent plane to exist, the partial derivatives of $\vec{X}$ must exist, so the differential exists. From Proposition 5.2.4, we know that the surface is a plane if $\vec{X}^{-1}(p)$ is a singleton set, which we'll denote by $\{q\} \subset U$, and $\vec{X}_u(q) \times \vec{X}_v(q)$ exists and is nonzero. This is equivalent to saying that $\vec{X}_u(q)$ and $\vec{X}_v(q)$ are linearly independent, which means that $d\vec{X}_q$ has maximal rank. $\square$

The condition in Corollary 5.2.8 that requires $d\vec{X}_q$ to have maximal rank occurs often enough that we give it a name.

Definition 5.2.9 *Let $m, n \geq 1$ and let F be a function from an open set $U \subset \mathbb{R}^n$ to $\mathbb{R}^m$. A point $q \in U$ is called a critical point if dF_q does not exist or does not have maximal rank, i.e., $\mathrm{rank}\, dF_q < \min(m, n)$. If q is a critical point of F, we call $F(q) \in \mathbb{R}^m$ a critical value of F.*

If $p \in \mathbb{R}^m$ is not a critical value of F (even if p is not in the image of F), then we call p a *regular value* of F.

Note that if $n = 1$, then F is a parametrized curve. According to this definition, a point t_0 is critical if $[dF_{t_0}] = F'(t_0) = \vec{0}$. This is the same definition as before. Furthermore, if instead we have $m = 1$, then F is a real-valued multivariable function. Then according to this definition, $q = (q_1, q_2, \ldots, q_n)$ is a critical point when

$$[dF_q] = \begin{pmatrix} \frac{\partial F}{\partial x_1}(q) & \frac{\partial F}{\partial x_2}(q) & \cdots & \frac{\partial F}{\partial x_n}(q) \end{pmatrix} = \begin{pmatrix} 0 & 0 & \cdots & 0 \end{pmatrix}.$$

This condition is the same as original defined for critical points of multivariable. Hence, Definition 5.2.9 directly generalizes all previous definitions of critical points.

In practice, to determine if a point of a function $F : \mathbb{R}^n \to \mathbb{R}^m$ is a critical point, we can do two things. If $m = n$, then dF_q has less than maximal rank if dF_q is a singular linear transformation. Hence, q is a critical point of F is and only if $\det(dF_q) = 0$. Now suppose $m \neq n$ and without loss of generality, suppose $n < m$ so that the matrix of dF_q has more rows than columns. Then dF_q has less than maximal rank if and only if the columns are linearly independent. From a result in linear algebra, this condition is also equivalent to all the maximal minors being 0.

Example 5.2.10 Consider the function $F : \mathbb{R}^2 \to \mathbb{R}^3$ defined by $F(x, y) = (x^2 + y, 2y^2 - 3x, -2xy)$. The matrix of the differential at every point is

$$[dF] = \begin{pmatrix} 2x & 1 \\ -3 & 4y \\ -2y & -2x \end{pmatrix}.$$

Setting the three minors equal to 0 gives

$$\begin{cases} 8xy + 3 = 0 \\ -4x^2 + 2y = 0 \\ 6x + 8y^2 = 0. \end{cases}$$

The second equation gives $y = 2x^2$ and the remaining two equations become $16x^3 + 3 = 0$ and $2x(3 + 16x^3) = 0$. So all three minors are zero, and hence the function has one critical point, at

$$\left(-\frac{1}{2}\sqrt[3]{\frac{3}{2}}, \frac{1}{2}\sqrt[3]{\frac{9}{4}} \right).$$

Definition 5.2.11 Let S be a parametrized surface in $\mathbb{R}^3$. A point $p \in S$ is called a *singular point* of S if no tangent plane exists at p.

In light of Corollary 5.2.8 and Definition 5.2.9 a point p of a surface S is a singular point if every parametrization $\vec{X} : U \to \mathbb{R}^3$ of an open neighborhood of p on S is a critical value of $\vec{X}$ or $\vec{X}^{-1}(p)$ is not a singleton.

PROBLEMS

1. Consider the parametrized surface $\vec{X} : [0, 2\pi] \times \mathbb{R} \to \mathbb{R}^3$ given by

$$\vec{X}(u, v) = ((v^2 + 1) \cos u, (v^2 + 1) \sin u, v).$$

Find the equation of the tangent plane to this surface at $p = (2, 0, 1)$.

2. Calculate the equation of the tangent plane to the torus

$$\vec{X}(u, v) = ((2 + \cos v) \cos u, (2 + \cos v) \sin u, \sin v)$$

at the point $(u, v) = \left(\frac{\pi}{6}, \frac{\pi}{3}\right)$.

3. Consider the parametrized surface S given by

$$\vec{X}(u, v) = (v^3 \cos u, v^3 \sin u, v)$$

with $(u, v) \in [0, \pi] \times \mathbb{R}$. Prove that the set of tangent vectors at $p = (0, 0, 0)$ is a line.

4. Show that the equation of the tangent plane at (x_0, y_0, z_0) at a regular surface given by $f(x, y, z) = 0$ is

$$f_x(x_0, y_0, z_0)(x - x_0) + f_y(x_0, y_0, z_0)(y - y_0) + f_z(x_0, y_0, z_0)(z - z_0) = 0.$$

5. Use the previous exercise to determine the tangent planes to $x^2 + y^2 - z^2 = 1$ at the points $(x, y, 0)$ and show that they are all parallel to the z-axis.

In Problems 5.2.6 through 5.2.10 (a) determine the matrix of the differential for the given function, and (b) find the critical points of the function.

6. $F : \mathbb{R}^2 \to \mathbb{R}^2$ with $F(x, y) = (x^2 y, 2x - y^2)$.

7. $F : \mathbb{R}^2 \to \mathbb{R}^2$ with $F(s, t) = (e^{2s} \cos t, e^s \sin t)$.

8. $F : \mathbb{R}^2 \to \mathbb{R}^3$ with $F(x, y) = (3xy + 2, 2y^2 - 1, 3x^2 + x + y)$.

9. $F : \mathbb{R}^2 \to \mathbb{R}^3$ with $F(t, u) = (t, \cosh t \cos u, \cosh t \sin u)$.

10. $F : \mathbb{R}^3 - \{(0, 0, w) \mid w \in \mathbb{R}\} \to \mathbb{R}^3$ with

$$F(u, v, w) = \left(\frac{uv \cos w}{(u^2 + v^2)^2}, \frac{uv \sin w}{(u^2 + v^2)^2}, \frac{u^2 - v^2}{2(u^2 + v^2)^2}\right).$$

[The function parameters (u, v, w) with $(x, y, z) = F(u, v, w)$ are called *cardioidal coordinates*.]

11. Let $D \subseteq \mathbb{R}^2$ and let $f : D \to \mathbb{R}$ be a differentiable function. Calculate the unit normal vector $\vec{N}(u, v)$ of a function graph associated to the parametrization $\vec{X}(u, v) = (u, v, f(u, v))$.

12. Let f and g be real functions such that $f(v) > 0$ and $g'(v) \neq 0$. Consider a parametrized surface given by

$$\vec{X}(u,v) = \big(f(v)\cos u, f(v)\sin u, g(v)\big).$$

Show that all the normal lines to this surface pass through the z-axis. [Hint: See Problem 5.2.14.]

13. Let $0 < r < R$ be radii and consider the torus parametrized by

$$\vec{X}(u,v) = ((R + r\cos v)\cos u, (R + r\cos v)\sin u, r\sin v)$$

with $(u,v) \in [0, 2\pi] \times [0, 2\pi]$.

(a) Calculate the associated unit normal vector function $\vec{N}(u,v)$.

(b) Consider the circle parametrized by $\vec{\gamma}(u) = (R\cos u, R\sin u, 0)$. Show that the outward-pointing unit normal on the torus at a point with coordinates (u_0, v_0) is the unit vector in the direction of $\vec{X}(u_0, v_0) - \vec{\gamma}(u_0)$.

14. *Surfaces of Revolution.* Consider the surface of revolution S with parametrization $\vec{X}(u,v) = (f(v)\cos u, f(v)\sin u, g(v))$, where the pair $(f(v), g(v))$ defines a regular plane curve (in the xz-plane).

(a) Calculate the unit normal vector $\vec{N}(u,v)$ associated to $\vec{X}(u,v)$.

(b) Prove that $d\vec{X}_{(u,v)}$ has maximal rank if and only if for all v in the domain of f, $f(v) \neq 0$ and $g'(v) \neq 0$ or $f(v) \neq 0$ and $f'(v) \neq 0$.

15. *Tangential Surfaces.* Let $\vec{\alpha} : I \to \mathbb{R}^3$ be a regular parametrized curve with curvature $\kappa(t) \neq 0$. We call the tangential surface to $\vec{\alpha}$ the parametrized surface

$$\vec{X}(t,u) = \vec{\alpha}(t) + u\vec{\alpha}'(t), \qquad \text{with } t \in I \text{ and } u \neq 0.$$

(a) Prove that for any fixed $t_0 \in I$, along any curve $\vec{X}(t_0, u)$, the tangent planes are all equal.

(b) Prove that $q = (t_0, u_0)$ is a critical point of $\vec{X}$ if and only if $u_0 = 0$ or $\kappa(t_0) = 0$.

16. Consider the parametrized $\vec{X}(u,v) = (u, v^3, -v^2)$. Prove that for any point on the surface $p = \vec{X}(q)$ such that $q = (u, 0)$, $d\vec{X}_q$ does not have maximal rank.

17. Consider $\vec{X}(u,v) = (v\cos u, v\sin u, e^{-v})$ for $(u,v) \in [0, 2\pi] \times [0, \infty)$. Find all points where $\vec{X}$ is not regular.

18. Consider the surface parametrized by $\vec{X}(u,v) = (uv, u^2 - v^2, u^3 - v^3)$. Prove that $\vec{X}$ intersects itself along the ray $(x, 0, 0)$, with $x \geq 0$. Find the points (u, v) that map to this ray. Use these to prove that the tangent space to $\vec{X}$ at any point on the open ray $(x, 0, 0)$, with $x > 0$, is the union of two planes. (See Figure 5.10 for the plot of a portion of this surface near $(0, 0, 0)$.)

5.3 Regular Surfaces

In Section 5.2, we belabored the issue of whether or not one can define a tangent plane to S at a point p. We did this because, in an intuitive sense, if a tangent plane does not exist locally near p, then the surface S does not "resemble" a two-dimensional plane and thus does not look like what we would expect in a surface. Just as with curves we introduced the notion of a regular curve (i.e., a curve $\vec{x} : I \to \mathbb{R}^n$ such that $\vec{x}'(t) \neq \vec{0}$ for all $t \in I$), so with surfaces, we would like to define a class of surfaces that is smooth enough so that we can "do calculus" on it. This is the class of regular surfaces.

Definition 5.3.1 A subset $S \subseteq \mathbb{R}^3$ is a *regular surface* if for each $p \in S$, there exists an open set $U \subseteq \mathbb{R}^2$, an open set $V \subseteq \mathbb{R}^3$ such that $p \in V$, and a surjective continuous function $\vec{X} : U \to V \cap S$ such that

1. $\vec{X}$ is of class C^1: the coordinate functions $x(u,v)$, $y(u,v)$, and $z(u,v)$ of $\vec{X}(u,v)$ have continuous partial derivatives;

2. $\vec{X}$ is a homeomorphism: $\vec{X}$ is continuous and has an inverse $\vec{X}^{-1} : V \cap S \to U$ such that $\vec{X}^{-1}$ is continuous;

3. $\vec{X}$ satisfies the regularity condition: for each $(u,v) \in U$, the differential $d\vec{X}_{(u,v)} : \mathbb{R}^2 \to \mathbb{R}^3$ is a one-to-one linear transformation.

The coordinates for the domain U of the parametrization $\vec{X}$ form what we call a *system of coordinates* (in a neighborhood) of p, and the neighborhood $V \cap S$ of p in S is called a *coordinate neighborhood*. We note that a regular surface is defined as a subset of $\mathbb{R}^3$ and not as a function from a subset of $\mathbb{R}^2$ to $\mathbb{R}^3$. However, we may view a regular surface as the union of the images of an appropriate set of systems of coordinates.

Definition 5.3.2 A regular surface is called of class C^r if all the coordinate functions of $\vec{X}$ are of class C^r, i.e., all partial derivatives of order up to r exist and are continuous. A regular surface is called *smooth* if it is of class C^∞, i.e., if all partial derivatives of any order of all the coordinate functions of $\vec{X}$ exist and are continuous.

Definition 5.3.1 is arguably far more involved than the definition for a regular curve in space. We spend the rest of the section unpacking this definition and showing why these conditions are what we want for giving a precise definition to a surface as a subset of $\mathbb{R}^3$ such that at every point the set of tangent vectors is a plane.

First, we remind the reader of the concept of open sets in $\mathbb{R}^n$. Open sets in $\mathbb{R}^n$ are built on the concept of an open ball:

$$B_r(p) = \{q \in \mathbb{R}^n \,|\, d(p, q) < \varepsilon\},$$

where $d(p, q)$ is the Euclidean distance between p and q, also denoted by $\|p - q\|$. A set $A \subseteq \mathbb{R}^n$ is called *open* if for all $p \in A$, there exists a positive real number $\varepsilon > 0$ such that $B_\varepsilon(p) \subseteq A$. Intuitively speaking, open sets do not contain any boundary points, or edge. However, in analysis and geometry, using open sets has a different benefit. The definition implies that at every point p of an open set A, it is possible to move, at least a little $\varepsilon > 0$, in every direction and remain in A. So from an intuitive perspective, around every point of A, the set A appears fully n-dimensional.

We illustrate Definition 5.3.1 with a first example before discussing the three conditions for a regular surface.

Example 5.3.3 We can prove directly that the unit sphere $\mathbb{S}^2$ is a regular surface in a few different ways. We'll use rectangular coordinates first. Consider a point $p = (x, y, z) \in \mathbb{S}^2$, and let $U = \{(u, v) \,|\, u^2 + v^2 < 1\}$. If $z > 0$, then the mapping $\vec{X}_{(1)} : U \to \mathbb{R}^3$ defined by $(u, v, \sqrt{1 - u^2 - v^2})$ is clearly a bijection between U and $\mathbb{S}^2 \cap \{(x, y, z)|z > 0\}$. It is an easy exercise to check that $\vec{X}_{(1)}$ is of class C^1. For Condition 2, $\vec{X}_{(1)}$ is also a homeomorphism: it is clearly continuous; its inverse $\vec{X}_{(1)}^{-1}$, which is simply the vertical projection of the upper unit sphere onto $\mathbb{R}^2$, is a restriction of a linear transformation, so is continuous. For Condition 3, we calculate the differential

$$d\vec{X}_{(1)} = \begin{pmatrix} 1 & 0 \\ 0 & 1 \\ -\dfrac{u}{\sqrt{1 - u^2 - v^2}} & -\dfrac{v}{\sqrt{1 - u^2 - v^2}} \end{pmatrix}.$$

For all $q = (u, v) \in U$, the entries in this matrix are well defined and produce a matrix that defines a one-to-one linear transformation. Consequently, $[d\vec{X}_{(1)q}]$ is of maximal rank.

The mapping $\vec{X}_{(1)} : U \to \mathbb{R}^3$ so far only tells us that the open upper half of the sphere is a regular surface, but we wish to show that the whole sphere is a regular surface.

To cover $\mathbb{S}^2$, we use the following six parametrizations, each defined over the open unit disk U:

$$\vec{X}_{(1)}(u, v) = (u, v, \sqrt{1 - u^2 - v^2}), \quad \vec{X}_{(2)}(u, v) = (u, v, -\sqrt{1 - u^2 - v^2}),$$

$$\vec{X}_{(3)}(u, v) = (u, \sqrt{1 - u^2 - v^2}, v), \quad \vec{X}_{(4)}(u, v) = (u, -\sqrt{1 - u^2 - v^2}, v),$$

$$\vec{X}_{(5)}(u, v) = (\sqrt{1 - u^2 - v^2}, u, v), \quad \vec{X}_{(6)}(u, v) = (-\sqrt{1 - u^2 - v^2}, u, v).$$

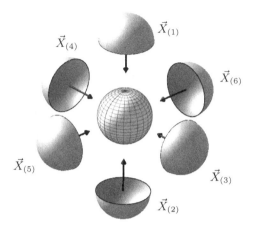

$\vec{X}_{(4)}$ $\vec{X}_{(1)}$ $\vec{X}_{(6)}$ $\vec{X}_{(5)}$ $\vec{X}_{(3)}$ $\vec{X}_{(2)}$

Figure 5.9: Six coordinate patches on the sphere.

We note that these parametrizations cover the portion of the sphere that satisfy respectively $z > 0$, $z < 0$, $y > 0$, $y < 0$, $x > 0$, and $x < 0$. These define six coordinate patches such that for all $p \in \mathbb{S}^2$, p is in at least one of these coordinate patches (see Figure 5.9 for an illustration of the six patches covering the sphere). Since the domain for each parametrization is open, all six patches are necessary.

We now look in more detail at the three conditions in Definition 5.3.1, with a view toward understanding why we need of them. Recall that our intuitive goal is to define a class of surfaces S as subsets of $\mathbb{R}^3$ that is smooth enough so that we can "do calculus" on it.

Condition (1) simply requires that around each point p of S we can parametrize S by a function $\vec{X} : U \to \mathbb{R}^3$ that is differential. The differentiability condition is expected since we anticipate that "doing calculus" on S will require us to take derivatives of $\vec{X}$. The condition that $\vec{X}$ be differentiable eliminates the possibility of corners or folds. A cube, for example, is not a regular surface because for whatever parametrization is used in the neighborhood of an edge where two faces meet, at least one of the partial derivatives will not exist.

Condition (2) requires that the parametrization $\vec{X}$ be a homeomorphism. This condition is likely the least intuitive. Though the concept of a homeomorphism has a broader definition in topology, we define it here for subsets of $\mathbb{R}^n$.

Definition 5.3.4 Let $A \subseteq \mathbb{R}^n$ and $B \subseteq \mathbb{R}^m$ be subsets. A function $f : A \to B$ is called a *homeomorphism* if (1) f is bijective, (2) f is continuous, and (3) f^{-1} is continuous.

The definition of continuity here is that of multivariable functions, restricted to the subsets A and B. Using the notion of open balls, we can restate the usual definition of continuity given in calculus courses as follows. By definition, the function $f : A \to B$ is continuous at $q \in A$ means that for all real $\varepsilon > 0$, there exist a positive real $\delta > 0$ such that $f(B_\delta(q)) \subseteq B_\varepsilon(f(q))$. If f^{-1} is continuous at $p = f(q)$, then the reverse relationship holds, namely for all $\varepsilon > 0$, there exists $\delta > 0$ such that $f^{-1}(B_\delta(p)) \subseteq B_\varepsilon(q)$.

As an example, we consider parametrizations of the unit circle $\mathbb{S}^1$ in the plane. The most natural parametrization is $f : [0, 2\pi] \to \mathbb{R}^2$ with $f(\theta) = (\cos\theta, \sin\theta)$. As stated, f is not a bijection. We need to restrict both the codomain to $\mathbb{S}^1$, and the domain to $[0, 2\pi)$. So we define

$$\gamma : [0, 2\pi) \to \mathbb{S}^1 \quad \gamma(\theta) = (\cos\theta, \sin\theta). \tag{5.5}$$

Both component functions of γ are continuous, so γ is continuous. Furthermore, it is a bijection. Its inverse function is $\gamma^{-1} : \mathbb{S}^1 \to [0, 2\pi)$, with its inputs expressed in (x, y) coordinates as

$$\gamma^{-1}(x, y) = \begin{cases} \tan^{-1}\left(\frac{y}{x}\right) & \text{if } x > 0 \text{ and } y > 0 \\ \pi/2 & \text{if } x = 0 \text{ and } y = 1 \\ \tan-1\left(\frac{y}{x}\right) + \pi & \text{if } x < 0 \\ 3\pi/2 & \text{if } x = 0 \text{and } y = -1 \\ \tan^{-1}\left(\frac{y}{x}\right) + 2\pi & \text{if } x > 0 \text{ and } y < 0. \end{cases} \tag{5.6}$$

However, γ is not a homeomorphism. No matter how small ε is, there exists no $\delta > 0$ such that $\gamma^{-1}(B_\delta(p)) \subseteq B_\varepsilon(q)$. In the figure below, the gray disk around p illustrates some $B_\delta(p)$; the pre-image $\gamma^{-1}(B_\delta(p))$ contains elements near 0 but also some near 2π. This shows that γ^{-1} is not continuous.

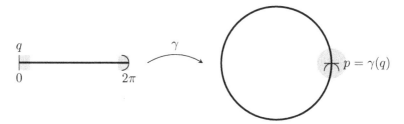

This example shows that the particular γ is not a homeomorphism. However, a more general argument shows that there exists no homeomorphism between an interval of $\mathbb{R}$ and the unit circle.

We now consider the concept of homeomorphism for coordinate charts and see what this means for regular surfaces.

Figure 5.10: Not a bijection.

Since we assume that in the neighborhood of each point $p \in S$ there is a parametrization $\vec{X} : U \to V \cap S$ that is a bijection, we already eliminate a parametrized surface that intersects itself or degenerates to a curve. Figure 5.10 shows the image of a parametrized surface $\vec{X}$ that intersects itself along a ray.

The tangent surface to this surface at any point along this ray is the union of two planes and not a single plane. Problem 5.2.18 shows the explicit parametrization $\vec{X}(u, v)$ for this surface; since it is not a bijection, it could not serve as a coordinate chart for the locus set S. However, simply requiring that $\vec{X}$ be a bijection will not suit our purposes as we will now illustrate.

Requiring not only that each patch $\vec{X} : U \to V \cap S$ of a parametrized surface be a homeomorphism means that if $\vec{X}(q_1)$ and $\vec{X}(q_2)$ are arbitrarily close in $\mathbb{R}^3$, then q_1 and q_2 are arbitrarily close in some coordinate patch. Not unlike the issue with an interval not being homeomorphic to the circle, this eliminates situations similar to that depicted in Figure 5.11, where an open strip is twisted back onto itself so that it does not intersect itself but that the distance between one end and another part in the middle of the strip is 0. Take a point p on the segment where the edge of the open strip comes back and "touches" the strip. Then for any ball $B_\delta(p)$, there are some points in $f^{-1}(B_\delta(p))$ that are not inside a small ball $B_\varepsilon(q)$. Hence $\vec{X}$ is not a homeomorphism.

Finally, Condition (3) Definition 5.3.1 is necessary because, even with the first two conditions satisfied, it is still possible to create a parametrized surface $\vec{X}$ that has a point p such that the tangent space to the surface at p is not a plane but a line. (See Problem 5.3.3 for an example of a surface that satisfies the first two conditions for a regular surface but not the third.)

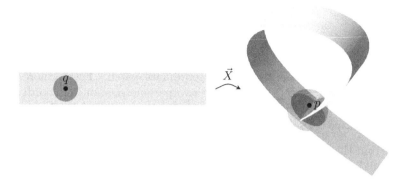

Figure 5.11: Not a homeomorphism.

Example 5.3.5 Example 5.3.3 showed the unit sphere to be a regular surface using six coordinate patches and rectangular coordinates. Carefully using two colatitude–longitude parametrizations, we get another way to show that the unit sphere $\mathbb{S}^2$ is a regular surface.

Let H_1 be the closed half-plane in $\mathbb{R}^3$ given by

$$H_1 = \{(x, y, z) \in \mathbb{R}^3 \mid x \geq 0 \text{ and } y = 0\}.$$

Let $U = (0, 2\pi) \times (0, \pi)$ and define $\vec{X}_{(1)} : U \to \mathbb{S}^2 - H_1$ the function by

$$\vec{X}_{(1)}(u, v) = (\cos u \sin v, \sin u \sin v, \cos v).$$

It is clear that $\vec{X}_{(1)}$ is a bijection and is continuous. To see that $\vec{X}_{(1)}$ is a homeomorphism, we can explicitly find the inverse function and show that it is continuous. Suppose that $(x, y, z) = \vec{X}_{(1)}(u, v)$. Then since $v \in (0, \pi)$, we have $z = \arccos(v)$. Furthermore, since $\sin v > 0$ for $v \in (0, \pi)$, we have $\sin v = \sqrt{1 - z^2}$. Then we have

$$\vec{X}_{(1)}^{-1}(x, y, z) = \left(\gamma^{-1} \left(\frac{x}{\sqrt{1 - z^2}}, \frac{y}{\sqrt{1 - z^2}} \right), \arccos v \right),$$

where γ is the function defined in (5.5). The functions γ^{-1} and $\arccos v$ are continuous. Furthermore, the functions $x/\sqrt{1 - z^2}$ and $y/\sqrt{1 - z^2}$ are defined and continuous on $\mathbb{S}^2 - H_1$. The composition of continuous functions is continuous, so $\vec{X}_{(1)}^{-1}$ is continuous, and therefore $\vec{X}_{(1)}$ is a homeomorphism. It also satisfies regularity condition.

The coordinate patch defined by $\vec{X}_{(1)}$ covers the unit sphere except for where it is intersected by the half plane H_1. To cover the rest, consider the half-plane

$$H_2 = \{(x, y, z) \in \mathbb{R}^3 \mid x \leq 0 \text{ and } z = 0\},$$

and the function $\vec{X}_{(2)} : U \to \mathbb{S}^2 - H_2$ defined by

$$\vec{X}_{(2)}(u, v) = (-\cos u \sin v, \cos v, -\sin u \sin v).$$

Using the same discussion as for the first coordinate patch, $\vec{X}_{(2)}$ satisfies all three conditions for a coordinate patch. Furthermore, $(\mathbb{S}^2 - H_1) \cup (\mathbb{S}^2 - H_2) = \mathbb{S}^2$, so the two parametrizations of coordinate patches cover the whole sphere, explicitly showing the $\mathbb{S}^2$ is a regular surface.

From the above explanations of the criteria for a regular surface, it would seem hard to determine whether or not a given surface is regular. However, the next two propositions show under what conditions function graphs and surfaces given as solutions to one equation are regular.

Proposition 5.3.6 *Let $U \subset \mathbb{R}^2$ be open. Then if a function $f : U \to \mathbb{R}$ is differentiable, the subset $S = \{(x, y, x) \in \mathbb{R}^3 \mid (x, y, z) = (u, v, f(u, v))\}$ is a regular surface.*

Proof: The inverse function $F^{-1} : S \to U$ is the projection onto the xy-plane. F^{-1} is continuous, and since f is differentiable, F is also continuous. Thus, F is a homeomorphism between U and S. The regularity condition is clearly satisfied because

$$dF_{(u,v)} = \begin{pmatrix} 1 & 0 \\ 0 & 1 \\ f_u & f_v \end{pmatrix}$$

is always one-to-one regardless of the values of f_u and f_v. $\qquad \Box$

Another class of surfaces in $\mathbb{R}^3$ one studies arises as level surfaces of functions $f : \mathbb{R}^3 \to \mathbb{R}$, that is, as points that satisfy the equation $f(x, y, z) = a$. The following proposition involves a little more analysis than that presented in most multivariable calculus courses. However, it shows where the terminology "regular surface" comes from in light of the notion of a regular value of multivariable functions presented in Definition 5.2.9. (Most problems in this textbook involve explicit parametrizations, but the proof of this proposition illustrates the value of the Implicit Function Theorem, usually introduced in a course on mathematical analysis.)

Proposition 5.3.7 *Let $f : U \subset \mathbb{R}^3 \to \mathbb{R}$ be a differentiable function, and let $a \in \mathbb{R}$ be a regular value of f, i.e., a real number such that for all $p \in f^{-1}(a)$, df_p is not $\vec{0}$. Then the surface defined by $S = \{(x, y, z) \in \mathbb{R}^3 \mid f(x, y, z) = a\}$ is a regular surface. Furthermore, the gradient $df_p = \vec{\nabla} f(p)$ is normal to S at p.*

Proof: Let $p \in S$. Since $df_p \neq \vec{0}$, after perhaps relabeling the axes, we have $f_z(p) \neq 0$. We consider the function $G : \mathbb{R}^3 \to \mathbb{R}^3$ defined by $G(x, y, z) = (x, y, f(x, y, z))$ and notice that

$$dG_p = \begin{pmatrix} 1 & 0 & 0 \\ 0 & 1 & 0 \\ f_x & f_y & f_z \end{pmatrix}.$$

Consequently, $\det(dG_p) = f_z(p) \neq 0$, so dG_p is invertible. The Implicit Function Theorem allows us to conclude that there exists a neighborhood U' of p in U such that G is one-to-one on U', $V = G(U')$ is open, and the inverse $G^{-1} : V \to U'$ is differentiable. However, since $(u, v, w) = G(x, y, z) = (x, y, f(x, y, z))$, we will be able to write the inverse function as $(x, y, z) = G^{-1}(u, v, w) = (u, v, g(u, v, w))$ for some differentiable function $g : V \to \mathbb{R}$.

Furthermore, if V' is the projection of V onto the uv-plane, then $h : V' \to \mathbb{R}$ defined by $h_a(u, v) = g(u, v, a)$ is differentiable and

$$G(f^{-1}(a) \cap U') = V \cap \{(u, v, w) \in \mathbb{R}^3 \mid w = a\}$$

which means that

$$f^{-1}(a) \cap U' = \{(u, v, h_a(u, v)) \mid (u, v) \in V'\}.$$

Thus, $f^{-1}(a) \cap U'$ is the graph of h_a, and by Proposition 5.3.6, $f^{-1}(a) \cap U'$ is a coordinate neighborhood of p. Thus, every point $p \in S$ has a coordinate neighborhood, and so $f^{-1}(a)$ is a regular surface in $\mathbb{R}^3$.

To prove that df_p is normal to S at p, suppose that $\vec{X}(u, v) = (x(u, v), y(u, v), z(u, v))$ parametrizes regularly a neighborhood of S around p. Then, differentiating the defining equation $f(x, y, z) = a$ with respect to u and v one obtains

$$\begin{cases} \dfrac{\partial f}{\partial x}\dfrac{\partial x}{\partial u} + \dfrac{\partial f}{\partial y}\dfrac{\partial y}{\partial u} + \dfrac{\partial f}{\partial z}\dfrac{\partial z}{\partial u} = 0, \\ \dfrac{\partial f}{\partial x}\dfrac{\partial x}{\partial v} + \dfrac{\partial f}{\partial y}\dfrac{\partial y}{\partial v} + \dfrac{\partial f}{\partial z}\dfrac{\partial v}{\partial u} = 0. \end{cases}$$

Thus, $df_p = (f_x(p), f_y(p), f_z(p))$ is perpendicular to $\vec{X}_u$ and $\vec{X}_v$ at p. $\square$

(Note: See Section 6.4 in [24] for a statement of the Implicit Function Theorem and some examples.)

We point out that Definition 5.3.1 can be easily modified to provide a definition of regularity for a surface S in $\mathbb{R}^n$. The only required changes are that each parametrization of a neighborhood of S be a

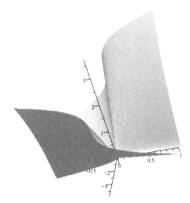

Figure 5.12: Parametrization of Problem 5.3.3.

continuous function $\vec{X} : \mathbb{R}^2 \to \mathbb{R}^n$ and that at each point $p = \vec{X}(u, v)$, the differential $d\vec{X}_{(u,v)}$ be an injective linear transformation $\mathbb{R}^2$ to $\mathbb{R}^n$.

PROBLEMS

1. Show that the function $f : (-1, 1) \to \mathbb{R}$ defined by $f(x) = \tan(\pi x / 2)$ is a homeomorphism.

2. (**CAS**) Let U be the open rectangle $U = (-1.3, 1.3) \times (-1, 1)$. Consider the function $\vec{X} : U \to \mathbb{R}^3$ defined by $f(t, u) = (t^3 - t, t^2, t + u)$, then let $S = \vec{X}(U)$ and restrict the codomain to $\vec{X}$ to S. Show that $\vec{X}$ is continuous and has an inverse. However, show that the inverse of $\vec{X}^{-1} : S \to U$ is not continuous, thereby showing that $\vec{X}$ is not a homeomorphism.

3. Show that the parametrized surface $\vec{X}(u, v) = (u^3 + v + u, u^2 + uv, v^3)$ satisfies the first and second conditions of Definition 5.3.1 but does not satisfy the third at $p = (0, 0, 0)$. (See Figure 5.12.)

4. Consider the hyperbolic paraboloid defined by $S = \{(x, y, z) \in \mathbb{R}^3 \mid z = x^2 - y^2\}$. Check that the following provide systems of coordinates for parts of S.
   (a) $\vec{X}(u, v) = (u + v, u - v, 4uv)$ for $(u, v) \in \mathbb{R}^2$.
   (b) $\vec{X}(u, v) = (u \cosh v, u \sinh v, u^2)$ for $(u, v) \in \mathbb{R}^2$.

5. Consider the cylinder $\mathcal{C}$ that satisfies $x^2 + y^2 = 1$ and $-1 < z < 1$. Find explicit parametrizations (a system of coordinate charts) that exhibit $\mathcal{C}$ to be a regular surface. [Hint: use γ from (5.5).]

6. Explicitly find a set of parametrizations that show the hyperboloid of two sheets defined by $x^2 + y^2 - z^2 = 1$ to be a regular surface.

7. Consider the torus S defined as the image of $\vec{X} : [0, 2\pi] \to \mathbb{R}^3$ with

$$\vec{X}(u, v) = ((2 + \cos v) \cos u, (2 + \cos v) \sin u, \sin v).$$

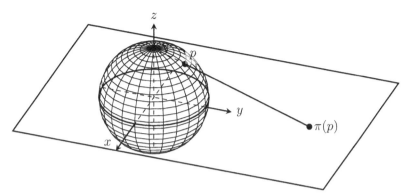

Figure 5.13: Stereographic projection.

This parametrization surjects onto S but does not satisfy the three conditions for a coordinate patch of a regular surface. Find a collection of coordinate patches that explicitly define S as a regular surface. [Hint: try to do it with only three coordinate patches.]

8. Suppose that a coordinate neighborhood of a regular surface can be parametrized by
$$\vec{X}(u, v) = \vec{\alpha}(u) + \vec{\beta}(v),$$
where $\vec{\alpha}$ and $\vec{\beta}$ are regular parametrized curves defined over intervals I and J, respectively. Prove that along coordinate lines (either $u = u_0$ or $v = v_0$) all the tangent planes to the surface are parallel to a fixed line.

9. Let $\vec{X} : U \to \mathbb{R}^3$ be a parametrization of an open set of a regular surface, and let $\vec{p}$ be a point in $\mathbb{R}^3$ that is not in S. Consider the function $F : U \to \mathbb{R}$ defined by
$$F(u, v) = \|\vec{X}(u, v) - \vec{p}\|.$$
Prove that if $\vec{q} = (u_0, v_0)$ is a critical point of F, then the vector $\vec{X}(\vec{q}) - \vec{p}$ is normal to the surface S at $\vec{X}(\vec{q})$.

10. *Stereographic Projection.* One way to define coordinates on the surface of the sphere $\mathbb{S}^2$ given by $x^2 + y^2 + z^2 = 1$ is to use the stereographic projection of $\pi_N : \mathbb{S}^2 - \{N\} \to \mathbb{R}^2$, where $N = (0, 0, 1)$, defined as follows. Given any point $p \in \mathbb{S}^2$, the line (pN) intersects the xy-plane at exactly one point, which is the image of the function $\pi(p)$. If (x, y, z) are the coordinates for p in $\mathbb{S}^2$, let us write $\pi_N(x, y, z) = (u, v)$ (see Figure 5.13).

   (a) Prove that $\pi_N(x, y, z) = \left(\frac{x}{1-z}, \frac{y}{1-z}\right)$.
   (b) Prove that
$$\pi_N^{-1}(u, v) = \left(\frac{2u}{u^2 + v^2 + 1}, \frac{2v}{u^2 + v^2 + 1}, \frac{u^2 + v^2 - 1}{u^2 + v^2 + 1}\right).$$

(c) Define a similar stereographic projection π_S from the south pole $S = (0, 0, -1)$ and then show that the collection $\{\pi_N^{-1}, \pi_S^{-1}\}$ provide coordinate charts that satisfy the three conditions of Definition 5.3.1. [This shows yet once more how $\mathbb{S}^2$ is a regular surface.]

11. Let $\vec{\alpha} : I \to \mathbb{R}^2$ be a plane curve and let $\vec{\beta} : I \to \mathbb{R}^3$ be the pullback of $\vec{\alpha}$ onto the unit sphere via the stereographic projection as described in Problem 5.3.10. In other words, $\vec{\beta}(t) = \pi_N^{-1}(\vec{\alpha}(t))$. Prove that points on the locus of $\vec{\beta}$ where the torsion τ is zero correspond to vertices in $\vec{\alpha}$, that is points where the plane curvature κ_g of $\vec{\alpha}$ satisfies $\kappa_g'(t) = 0$.

12. *Tubes.* Let $\vec{\alpha} : I \to \mathbb{R}^3$ be a regular parametrized curve, where I is a real interval. Let r be a positive real constant. Define the *tube* $\mathcal{T}$ of radius r around $\vec{\alpha}$ as the image of parametrized surface

$$\vec{X}(t, u) = \vec{\alpha}(t) + r(\cos u)\vec{P}(t) + r(\sin u)\vec{B}(t), \text{ with } (t, u) \in I \times [0, 2\pi].$$

(a) Prove that a necessary (though not sufficient) condition for $\vec{X}$ to be a regular surface is that

$$r < 1/(\max_{t \in I} \kappa(t)).$$

(b) Show that when $\vec{X}$ satisfies the regularity condition, the unit normal vector is

$$\vec{N}(t, u) = -\cos u \vec{P}(t) - \sin u \vec{B}(t).$$

(c) Now assume that I is an open interval. Cover $\mathcal{T}$ with a collection of two coordinate charts both satisfying the condition in part (a) that exhibit $\mathcal{T}$ as a regular surface.

13. Let $f(x, y, z) = (x + 2y + 3z - 4)^2$.

(a) Locate the critical points and the critical values of f.

(b) For what values of c is the set $f(x, y, z) = c$ a regular surface?

(c) Repeat (a) and (b) for the function $f(x, y, z) = xy^2z^3$.

14. *Cones.* Let C be a regular planar curve parametrized by $\vec{\alpha} : (a, b) \to \mathbb{R}^3$ lying in a plane $\mathcal{P}$ that does not contain the origin O. The cone Σ over C is the union of all the lines passing through O and p for all points p on C.

(a) Find a parametrization $\vec{X}$ of the cone Σ over C.

(b) Find the points where Σ is not regular.

15. Let S be a regular surface in $\mathbb{R}^3$. Prove that any open subset of S is again a regular surface.

16. Let S_1 and S_2 be regular surfaces in $\mathbb{R}^3$. Prove that $S_1 \cup S_2$ is a regular surface if and only if $S_1 \cap S_2 = \emptyset$.

5.4 Change of Coordinates; Orientability

A particular patch of a surface can be parametrized in a variety of ways. In particular, we could use a different coordinate system in $\mathbb{R}^2$ to describe the domain U of $\vec{X}$. Furthermore, built into the definition of a regular surface is the assumption that we may need to use two different parametrizations that partially cover the same region on the surface.

In this section, we analyze the consequences of using overlapping coordinate charts.

5.4.1 Change of Coordinates

We begin with an example drawn from the multiple ways we illustrated the unit sphere to be a regular surface.

Example 5.4.1 Consider the following two overlapping parametrizations of part of $\mathbb{S}^2$ in $\mathbb{R}^3$:

1. $\vec{X}(\theta, \varphi) = (\cos\theta \sin\varphi, \sin\theta \sin\varphi, \cos\varphi)$, with $(\theta, \varphi) \in U_1 = (0, 2\pi) \times (0, \pi)$;

2. $\vec{Y}(u, v) = (u, v, \sqrt{1 - u^2 - v^2})$, with $(u, v) \in U_2$ where U_2 is the open unit disk, i.e., $U_2 = \{(u, v) \in \mathbb{R}^2 \,|\, u^2 + v^2 < 1\}$.

The intersection of the codomains of these coordinate charts is

$$V \cap \mathbb{S}^2 = \{(x, y, z) \in \mathbb{S}^2 \,|\, z > 0\} - \{(x, 0, z) \in \mathbb{R}^3 \,|\, x \geq 0\},$$

and $V \cap \mathbb{S}^2 = \vec{X}(U_1') = \vec{Y}(U_2')$, where

$$U_1' = (0, 2\pi) \times (0, \pi/2) \quad \text{and}$$
$$U_2' = U_1 - \{(u, 0) \in \mathbb{R}^2 \,|\, v \geq 0\}.$$

Restricting $\vec{X}$ and $\vec{Y}$ respectively to U_1' and U_2', it is relatively easy to see that $\vec{X} = \vec{Y} \circ F$, where F is the function

$$F: \quad U_1' \quad \longrightarrow U_2'$$
$$(\theta, \varphi) \longmapsto (\cos\theta \sin\varphi, \sin\theta \sin\varphi).$$

See Figure 5.14.

The function $F : U_1' \to U_2'$ is a bijection and the inverse function $(\theta, \varphi) = F^{-1}(u, v)$ is given by

$$F^{-1}(u, v) = (\gamma^{-1}(u/\sqrt{u^2 + v^2}, v/\sqrt{u^2 + v^2}), \sin^{-1}(\sqrt{u^2 + v^2})),$$

where γ^{-1} is given in (5.6).

It is tedious but not hard to verify that both F and F^{-1} are not just continuous, making F into a homeomorphism, but differentiable.

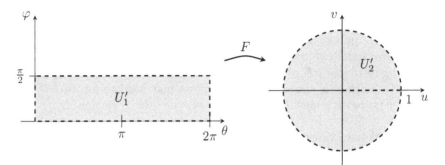

Figure 5.14: Coordinate change by F.

We now consider a general change of coordinates F in $\mathbb{R}^2$. Consider two parametrized surfaces defined by $\vec{X} : U_1 \to \mathbb{R}^3$ and $\vec{Y} : U_2 \to \mathbb{R}^3$, such that $\vec{X} = \vec{Y} \circ F$, where $F : U_1 \to U_2$ is a function between subsets U_1 and U_2 in $\mathbb{R}^2$. The above example illustrates three progressively stringent conditions on the change-of-coordinates function F. First, a necessary condition for $\vec{X} = \vec{Y} \circ F$ and $\vec{X}$ to have the same image in $\mathbb{R}^3$ is for F to be surjective. Second, if one wishes for the coordinates on a surface to be unique or if we are concerned with $\vec{Y}$ being a homeomorphism, we might wish to require that F be a bijection, in which case, if F is already surjective, one merely must restrict the domain of F to make it a bijection. Finally, requiring that both F and F^{-1} be differentiable adds a "smoothness" condition to how the coordinates transform. This latter condition has a name.

Definition 5.4.2 Let U and V be subsets of $\mathbb{R}^n$. We call a function $F : U \to V$ a *diffeomorphism* if it is a bijection, and both F and F^{-1} are differentiable.

Proposition 5.4.3 *Let U and V be subsets of $\mathbb{R}^n$, and let $F : U \to V$ be a diffeomorphism. Suppose that $q = F(q')$. Then the linear transformation $dF_{q'}$ is invertible, and*

$$dF_{q'}^{-1} = d(F^{-1})_q.$$

Proof: Suppose that $F : \mathbb{R}^m \to \mathbb{R}^n$ and $G : \mathbb{R}^n \to \mathbb{R}^s$ are differentiable, multivariable, vector-valued functions. (For what follows, one can also assume that these functions have small domains, just as long as the range of F is a subset of the domain of G.) Using the notion of the differential defined in Equation (5.3), it is an easy exercise to show that the formula for the chain rule on $G \circ F$ in a multivariable context can be written as

$$d(G \circ F)_a = dG_{F(a)} dF_a,$$

where we mean matrix multiplication.

Applying this to the situation of this proposition, where $F^{-1} \circ F = \mathrm{id}_U$, we get

$$d(F^{-1})_{F(q')} \circ dF_{q'} = I_n,$$

where I_n is the $n \times n$ identity matrix. Furthermore, the same reasoning holds with $F \circ F^{-1}$, and we conclude that $dF_{q'}$ is invertible and that $(dF_{q'})^{-1} = d(F^{-1})_q$. $\square$

5.4.2 Change of Coordinates and the Tangent Plane

Suppose that S is a regular surface and $p \in S$. Suppose also that an open set in S around p is parametrized in two ways: by $\vec{X} : U \to \mathbb{R}^3$ and $\vec{Y} : U_2' \to \mathbb{R}^3$, where $\vec{X} = \vec{Y} \circ F$ for a diffeomorphism $F : U \to U'$ between open subsets U and U' in $\mathbb{R}^2$. Let $q \in U$ with $p = \vec{X}(q)$ and define $q' = F(q)$. Let us write for the change of coordinates F,

$$(u, v) = F(s, t) = (f(s, t), g(s, t)).$$

We know that $\mathcal{B} = \{\vec{X}_s(q), \vec{X}_t(q)\}$ and $\mathcal{B}' = \{\vec{Y}_u(q'), \vec{Y}_v(q')\}$ are both bases of the tangent space $T_p S$.

Just as in the proof of Proposition 5.4.3, the chain rule in multiple variables gives

$$d\vec{X}_q = d(\vec{Y} \circ F)_q = d\vec{Y}_{q'} \circ dF_q.$$

The matrix of the differential of this coordinate change is

$$[dF] = \begin{pmatrix} \dfrac{\partial u}{\partial s} & \dfrac{\partial u}{\partial t} \\ \dfrac{\partial v}{\partial s} & \dfrac{\partial v}{\partial t} \end{pmatrix} = \begin{pmatrix} \dfrac{\partial f}{\partial s} & \dfrac{\partial f}{\partial t} \\ \dfrac{\partial g}{\partial s} & \dfrac{\partial g}{\partial t} \end{pmatrix},$$

so with matrices, the chain rule is written as

$$[d\vec{X}_{(s,t)}] = [d\vec{Y}_{F(s,t)}] \cdot \begin{pmatrix} \dfrac{\partial u}{\partial s} & \dfrac{\partial u}{\partial t} \\ \dfrac{\partial v}{\partial s} & \dfrac{\partial v}{\partial t} \end{pmatrix}. \tag{5.7}$$

Writing components $\vec{X} = (X_1, X_2, X_3)$ and $\vec{Y} = (Y_1, Y_2, Y_3)$ if necessary, we find that (5.7) is equivalent to the relations

$$\begin{aligned} \frac{\partial \vec{X}}{\partial s} &= \frac{\partial \vec{Y}}{\partial u} \frac{\partial u}{\partial s} + \frac{\partial \vec{Y}}{\partial v} \frac{\partial v}{\partial s} = \frac{\partial u}{\partial s} \vec{Y}_u + \frac{\partial v}{\partial s} \vec{Y}_v, \\ \frac{\partial \vec{X}}{\partial t} &= \frac{\partial \vec{Y}}{\partial u} \frac{\partial u}{\partial t} + \frac{\partial \vec{Y}}{\partial v} \frac{\partial v}{\partial t} = \frac{\partial u}{\partial t} \vec{Y}_u + \frac{\partial v}{\partial t} \vec{Y}_v. \end{aligned} \tag{5.8}$$

In particular, in coordinates

$$\left[\frac{\partial \vec{X}}{\partial s}\right]_{\mathcal{B}'} = \begin{pmatrix} \dfrac{\partial u}{\partial s} \\ \dfrac{\partial v}{\partial s} \end{pmatrix} \quad \text{and} \quad \left[\frac{\partial \vec{X}}{\partial t}\right]_{\mathcal{B}'} = \begin{pmatrix} \dfrac{\partial u}{\partial t} \\ \dfrac{\partial v}{\partial t} \end{pmatrix}.$$

This means that the matrix $[dF_q]$ is the change of coordinate matrix from $\mathcal{B} = \{\vec{X}_s(q), \vec{X}_t(q)\}$ to $\mathcal{B}' = \{\vec{Y}_u(q'), \vec{Y}_v(q')\}$ coordinates on T_pS.

We formalize our discussion into a proposition in a slightly different way, and introduce notations that we will adopt later, in particular in Chapter 7. We label the coordinates (s,t) as (x_1, x_2), and we label the coordinates (u, v) as $(\bar{x}_1, \bar{x}_2)$. We write the diffeomorphism that relates the coordinates as

$$(\bar{x}_1, \bar{x}_2) = F(x_1, x_2) \quad \text{and} \quad (x_1, x_2) = F^{-1}(\bar{x}_1, \bar{x}_2).$$

Let $\vec{X} : U \to \mathbb{R}^3$ be a parametrization of a coordinate patch $S \cap V$, with coordinates (x_1, x_2) on a regular surface S. Suppose that $\vec{Y} : U' \to \mathbb{R}^3$ gives another parametrization of the same patch with coordinates $(\bar{x}_1, \bar{x}_2)$ so that $\vec{Y} \circ F = \vec{X}$. $q' = F(q)$ and $p = \vec{X}(q)$. Consider

$$\mathcal{B} = \{\vec{X}_{x_1}(q), \vec{X}_{x_2}(q)\} \quad \text{and} \quad \mathcal{B}' = \{\vec{Y}_{\bar{x}_1}(q'), \vec{Y}_{\bar{x}_2}(q')\}$$

as bases of T_pS.

Proposition 5.4.4 *Suppose that $\vec{a} \in T_pS$ with coordinates*

$$[\vec{a}]_\mathcal{B} = \begin{pmatrix} a_1 \\ a_2 \end{pmatrix} \quad \text{and} \quad [\vec{a}]_{\mathcal{B}'} = \begin{pmatrix} \bar{a}_1 \\ \bar{a}_2 \end{pmatrix},$$

then

$$\begin{pmatrix} \bar{a}_1 \\ \bar{a}_2 \end{pmatrix} = [dF_q] \begin{pmatrix} a_1 \\ a_2 \end{pmatrix} = \begin{pmatrix} \dfrac{\partial \bar{x}_1}{\partial x_1} & \dfrac{\partial \bar{x}_1}{\partial x_2} \\ \dfrac{\partial \bar{x}_2}{\partial x_1} & \dfrac{\partial \bar{x}_2}{\partial x_2} \end{pmatrix} \begin{pmatrix} a_1 \\ a_2 \end{pmatrix}. \qquad (5.9)$$

Proof: By the chain rule in (5.8) and (5.7), and assuming the vector functions are evaluated at q or q' as appropriate, we have

$$\vec{a} = a_1 \vec{X}_{x_1} + a_2 \vec{X}_{x_2}$$

$$= a_1 \left(\frac{\partial \bar{x}_1}{\partial x_1} \vec{Y}_{\bar{x}_1} + \frac{\partial \bar{x}_2}{\partial x_1} \vec{Y}_{\bar{x}_2} \right) + a_2 \left(\frac{\partial \bar{x}_1}{\partial x_2} \vec{Y}_{\bar{x}_1} + \frac{\partial \bar{x}_2}{\partial x_2} \vec{Y}_{\bar{x}_2} \right)$$

$$= \left(a_1 \frac{\partial \bar{x}_1}{\partial x_1} + a_2 \frac{\partial \bar{x}_1}{\partial x_2} \right) \vec{Y}_{\bar{x}_1} + \left(a_1 \frac{\partial \bar{x}_2}{\partial x_1} + a_2 \frac{\partial \bar{x}_2}{\partial x_2} \right) \vec{Y}_{\bar{x}_2}.$$

Since $\mathcal{B}'$ is a basis of T_pS, then

$$
\begin{pmatrix} \bar{a}_1 \\ \bar{a}_2 \end{pmatrix} = \begin{pmatrix} \dfrac{\partial \bar{x}_1}{\partial x_1} & \dfrac{\partial \bar{x}_1}{\partial x_2} \\ \dfrac{\partial \bar{x}_2}{\partial x_1} & \dfrac{\partial \bar{x}_2}{\partial x_2} \end{pmatrix} \begin{pmatrix} a_1 \\ a_2 \end{pmatrix}.
$$

The result follows $\qquad\qquad\square$

The matrix product in (5.9) is often written in summation notation as

$$
\bar{a}_i = \sum_{k=1}^{2} \frac{\partial \bar{x}_i}{\partial x_k} a_k.
$$

5.4.3 Orientability

Intuitively speaking, the concept of an orientable surface encapsulates the notion of being able to define an inside and an outside to the surface. As we saw earlier, at any point p of a regular surface S, there are two possible unit normal vectors to the surface, which are opposites to each other. However, we want the direction considered "outward" to vary continuously over the regular surface. This leads us to the first definition of orientability.

Definition 5.4.5 A regular surface $S \subseteq \mathbb{R}^3$ is called *orientable* if there exists a continuous function $n : S \to \mathbb{S}^2$ such that $n(p)$ is perpendicular to T_pS for all $p \in S$. The function n is called an *orientation* on S.

Definition 5.4.6 If S is an oriented regular surface, a pair of vectors $(\vec{v}, \vec{w})$ of the tangent plane T_pS is called a *positively oriented basis* if the vectors form a basis of T_pS and

$$
(\vec{v} \times \vec{w}) \cdot n(p) > 0.
$$

A parametrization $\vec{X} : U \to \mathbb{R}^3$ of an open set of S is also called *positively oriented* if the ordered pair $(\vec{X}_u, \vec{X}_v)$ forms a positively oriented basis of the tangent plane of S at $\vec{X}(u, v)$ for all $(u, v) \in U$.

Example 5.4.7 Every sphere in $\mathbb{R}^3$ is orientable. Suppose that S is a sphere of center M and radius R. The function $n : S \to \mathbb{S}^2$ defined by

$$
n(p) = \frac{1}{R} \overrightarrow{Mp}
$$

is an orientation of $\mathbb{S}^2$. The vector $n(p)$ is what we typically consider the outward pointing unit normal.

Example 5.4.8 The commonly given example of a nonorientable surface in $\mathbb{R}^3$ is the Möbius strip M. Intuitively, the Möbius strip is a surface obtained by taking a long and narrow strip of paper and gluing the short ends together as though to make a cylinder but putting one twist in the strip before gluing (see Figure 5.15).

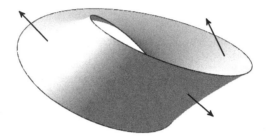

Figure 5.15: Möbius strip.

By looking at Figure 5.15, one can imagine a normal vector pointing outward on the surface of the Möbius strip, but if one follows the direction of this normal vector around on the surface, when it comes back around, it will be pointing inward this time instead of outward.

We now wish to characterize orientability in terms of coordinate patches.

By Definition 5.2.7, if $\vec{X} : U \to \mathbb{R}^3$ with $\vec{X}(s,t)$ is a regular parametrization for a coordinate patch $V \cap S$ around p, then the standard associated unit normal vector function is

$$\vec{N}(s,t) = \frac{\vec{X}_s \times \vec{X}_t}{\|\vec{X}_s \times \vec{X}_t\|}.$$

We point out that this is a continuous vector function $\vec{N} : U \to \mathbb{S}^2$ that maps into the unit sphere. Since $\vec{X} : U \to V \cap S$ is a homeomorphism, for all $q \in U$, the vector function $\vec{N}$ defines a function $n : V \cap S \to \mathbb{S}^2$ by

$$\vec{N} = n \circ \vec{X}.$$

However, so far, this function n is only defined on the particular coordinate patch $V \cap S$ and not necessarily on all of S. To define an orientation on a regular surface, the different coordinate patches must lead to a continuous function n defined over the whole surface.

Proposition 5.4.9 *Let $\vec{X} : U \to \mathbb{R}^3$ be a parametrization of a coordinate patch with coordinates (s,t) on a regular surface S. Let*

$\vec{Y} : U' \to \mathbb{R}^3$ be another parametrization of the same patch with coordinates (u, v) such that $\vec{X} = \vec{Y} \circ F$ for a diffeomorphic change of
coordinates $F : U \to U'$. Then if $q' = F(q)$ and $p = \vec{X}(q)$,

$$\vec{Y}_u(q') \times \vec{Y}_v(q') = \det(dF_q)(\vec{X}_s(q) \times \vec{X}_t(q)).$$

Proof: (Left as an exercise for the reader.) $\square$

We remind the reader that the determinant $\det(dF_q)$ of the differential matrix of the change of coordinate function is called the *Jacobian* of F at q'. (The reader may recall that the Jacobian appears
in the change of variables formula for multiple integrals.)

Definition 5.4.10 With the setup as in Proposition 5.4.9, we say
that a regular reparametrization $\vec{Y}$ such that $\vec{X} = \vec{Y} \circ F$ is a *positively*
(resp. *negatively*) *oriented* reparametrization if $\det(dF_q) > 0$ (resp.
$\det(dF_q) < 0$) for all $q \in U$.

Suppose that, as above, $\vec{X}(s, t)$ and $\vec{Y}(u, v)$ are positively oriented
reparametrizations of the same coordinate patch of a regular surface.
Then the unit normal vector associated to $\vec{Y}$ is

$$\frac{\vec{Y}_u \times \vec{Y}_v}{\|\vec{Y}_u \times \vec{Y}_v\|} = \frac{(\det(dF_{(s,t)}))\vec{X}_u \times \vec{X}_v}{|\det(dF_{(s,t)})| \, \|\vec{X}_u \times \vec{X}_v\|}$$
$$= \frac{\vec{X}_u \times \vec{X}_v}{\|\vec{X}_u \times \vec{X}_v\|},$$

which is equal to the unit normal vector associated to $\vec{X}$. In contrast,
if the reparametrization is negatively oriented, then the corresponding
unit normal vectors are negatives to each other.

This observation leads to the following alternative criterion for
orientability.

Proposition 5.4.11 *A regular surface S is orientable if and only if
it is possible to cover it with a collection of coordinate patches $\vec{X}_{(i)} :
U_i \to V_i \cap S$, with V_i open sets in $\mathbb{R}^3$ and i in some indexing set $\mathcal{I}$,
such that for each $i, j \in \mathcal{I}$ the change of coordinate function*

$$F_{ji} : \vec{X}_{(i)}^{-1}(V_i \cap V_j \cap S) \to \vec{X}_{(j)}^{-1}(V_i \cap V_j \cap S)$$

defined by

$$\vec{X}_{(i)} = \vec{X}_{(j)} \circ F_{ji}$$

satisfies $\det(dF_{ji}) > 0$ *over its domain.*

Proof: The set $V_i \cap V_j \cap S$ is precisely the set on S over which the coordinate patches for $\vec{X}_{(i)}$ and $\vec{X}_{(j)}$ overlap. Hence, the domain and codomain as described above give the correct change of coordinate functions. For each parametrization $\vec{X}_{(i)}$ call $\vec{N}_{(i)}$ the associated unit normal vector function.

We first prove the ($\Longrightarrow$) direction. Suppose that S is orientable and let $n : S \to \mathbb{S}^2$ be an orientation. For each $i \in \mathcal{I}$, we have $\vec{N}_{(i)} = n \circ \vec{X}_{(i)}$ as functions, if $\vec{X}_{(i)}$ is a positively oriented parametrization, or $\vec{N}_{(i)} = -n \circ \vec{X}_{(i)}$ as functions otherwise. If $\vec{X}_{(i)}$ is a negatively oriented parametrization, we replace $\vec{X}_{(i)}$ with the alternate parametrization $\vec{Y}_{(i)}(u, v) = \vec{X}_{(i)}(v, u)$. Since the cross product is anticommutative, the associated unit normal to $\vec{Y}_{(i)}$ is the negative of the associated unit normal to $\vec{X}_{(i)}$, so $\vec{Y}_{(i)}$ is then positively oriented. Then, by Proposition 5.4.9, all change of coordinate functions between all patches are positively oriented.

We now prove the ($\Longleftarrow$) direction. We define the function $n : S \to \mathbb{S}^2$ as follows. For a point $p \in V_i \cap S$, let $q \in U_i$ be the unique point such that $\vec{X}_{(i)}(q) = p$. Define $n(p)$ as $\vec{N}_{(i)}(q)$, the unit normal associated to $\vec{X}_{(i)}$, evaluated at q. Since all change of coordinate functions are positively oriented, by Proposition 5.4.9, the unit normal associated to each parametrization is equal over any overlapping coordinate patches. Hence, this definition of n is well-defined. Furthermore, since each $\vec{N}_{(i)}$ is continuous and each $\vec{X}_{(i)}$ is a homeomorphism, n is also continuous. Hence, S is orientable.   □

Example 5.4.12 We revisit the example of a Möbius strip but now use parametrizations. We can parametrize most of the Möbius strip (all but a line segment) by

$$\vec{X}(u, v) = \left(\left(2 - v \sin \frac{u}{2} \right) \sin u, \left(2 - v \sin \frac{u}{2} \right) \cos u, v \cos \frac{u}{2} \right),$$

with $(u, v) \in (0, 2\pi) \times (-1, 1)$. This coordinate neighborhood omits the boundary of the closed Möbius strip as well as the coordinate line with $u = 0$.

Let p be the point on M given by $\lim_{u \to 0} \vec{X}\left(u, \frac{1}{2} \right) = \left(0, 2, \frac{1}{2} \right)$. To obtain the same point p as a limit with $u \to 2\pi$, we must have a different value for v, namely, $p = \lim_{u \to 2\pi} \vec{X}\left(u, -\frac{1}{2} \right)$. With some calculations, one can find that

$$\begin{aligned}
\vec{X}_u \times \vec{X}_v = {} & \left(-\frac{v}{2} \cos u - \left(2 - v \sin \frac{u}{2} \right) \sin u \cos \frac{u}{2} \right) \vec{i} \\
& + \left(\frac{v}{2} \sin u - \left(2 - v \sin \frac{u}{2} \right) \cos u \cos \frac{u}{2} \right) \vec{j} \\
& - \left(\left(2 - v \sin \frac{u}{2} \right) \sin \frac{u}{2} \right) \vec{k}
\end{aligned}$$

and
$$\|\vec{X}_u \times \vec{X}_v\|^2 = \frac{v^2}{4} + \left(2 - v\sin\frac{u}{2}\right)^2.$$
From this, it is not hard to show that
$$\lim_{u \to 0} \vec{N}\left(u, \frac{1}{2}\right) = \left(-\frac{1}{\sqrt{65}}, -\frac{8}{\sqrt{65}}, 0\right)$$
and $$\lim_{u \to 2\pi} \vec{N}\left(u, -\frac{1}{2}\right) = \left(\frac{1}{\sqrt{65}}, \frac{8}{\sqrt{65}}, 0\right).$$

Consequently, no collection of systems of coordinates of M that includes $\vec{X}$ can satisfy the conditions of orientability. However, the specific system of coordinates $\vec{X}$ is not the problem. Using any collection of coordinate patches to cover M, as soon as one tries to "go all the way around" M, there are two overlapping coordinate patches for which the conditions of Definition 5.4.5 fail.

The Möbius strip has a boundary at $v = \pm 1$ in the above parametrization. Then there are many examples of nonorientable regular surfaces that have a boundary. For example, one can attach a "handle" to the Möbius strip, a tube connecting one part of the strip to another. However, for reasons that can only be explained by theorems in topology, there is not "enough room" in $\mathbb{R}^3$ to fit a closed nonorientable surface that does not have a boundary and does not intersect itself. However, because there is "more room" in higher dimensions, there exist many other closed nonorientable surfaces without boundary in $\mathbb{R}^4$ that do not intersect themselves. See Chapter 9.2 for a few examples.

PROBLEMS

1. Prove the claim in Example 5.4.1 that both F and F^{-1} are differentiable.

2. Consider the change of coordinates from polar to Cartesian described by $(x, y) = F(r, \theta) = (r\cos\theta, r\sin\theta)$. (a) Calculate the matrix of the differential $[dF]$. (b) Determine F^{-1} and calculate $[dF^{-1}]$. (c) Verify Proposition 5.3.3 that $[d(F^{-1})_{(x,y)}] = [dF_{(r,\theta)}]^{-1}$.

3. Consider the coordinate change in two variables
$$F(s, t) = (s^2 - t^2, 2st).$$

   Let $U = \{(s,t) \mid s^2 + t^2 \leq 1 \text{ and } t \geq 0\}$. Show that $F(U)$ is the whole unit disk. Prove that if $\vec{X}(x, y) = (x, y, \sqrt{1 - x^2 - y^2})$, then $\vec{X} \circ F : U \to \mathbb{R}^3$ is a parametrization for the upper half of the unit sphere.

4. Consider the coordinate change in the previous problem but with domain $U' = \{(s,t) \mid t > 0\}$. (a) Determine the set $F(U)$ and show that $F : U \to F(U)$ is a bijection. (b) Determine the function F^{-1} explicitly. (c) Show in what sense $[d(F^{-1})] = [dF]^{-1}$.

5. Consider spherical and cylindrical coordinate systems. Consider the coordinate change function from spherical to cylindrical defined by

$$(r, \theta, z) = F(\rho, \theta, \varphi) = (\rho \sin \varphi, \theta, \rho \cos \varphi).$$

(Note that φ is the angle down from the positive z-axis.) Set

$$U' = (0, \infty) \times (0, 2\pi) \times (0, \pi)$$
$$U = (0, \infty) \times (0, 2\pi) \times \mathbb{R}$$

and consider the coordinate change function $F : U' \to U$.

(a) Determine $F^{-1} : U \to U'$.

(b) Deduce that F is a diffeomorphism.

(c) Verify that for all $[dF_{(r,\theta,z)}]^{-1} = [d(F^{-1})_{(\rho,\theta,\varphi)}]$

6. Repeat the previous exercise with the change of coordinates between spherical and Cartesian coordinates with $(\rho, \theta, \varphi) \in (0, \infty) \times (0, 2\pi) \times (0, \pi)$.

7. Prove that $F(x, y) = (x^3, y^3)$ is a bijection from $\mathbb{R}^2$ to $\mathbb{R}^2$ but not a diffeomorphism. Prove also that $F(x, y) = (x^3 + x, y^3 + y)$ is a diffeomorphism of $\mathbb{R}^2$ onto itself.

8. Prove that the function $F : \mathbb{R}^2 \to \mathbb{R}^2$ defined by

$$F(x, y) = \left(x \cos(x^2 + y^2) - y \sin(x^2 + y^2), x \sin(x^2 + y^2) + y \cos(x^2 + y^2) \right)$$

is a diffeomorphism of $\mathbb{R}^2$ onto itself.

9. Find an example of a diffeomorphism of $\mathbb{R}^2$ onto an open square and show why it is a diffeomorphism.

10. Prove Proposition 5.4.9.

11. Supply the details for Example 5.4.12.

12. (*) Let S be a regular surface in $\mathbb{R}^3$ given as the set of solutions to the equation $F(x, y, z) = a$, where $F : U \subset \mathbb{R}^3 \to \mathbb{R}$ is differentiable and a is a regular value of F. Prove that S is orientable.

CHAPTER 6

First and Second Fundamental Forms

In the previous chapter as we developed the definition for a regular surface, we followed an intuitive principle of wanting to "do geometry" on a surface. By "doing geometry" we mean calculating lengths of curves, angles between intersecting curves, areas of regions bounded by curves, and so on. The first fundamental form, also called the metric tensor, provides a direct product on the tangent space T_pS of a regular surface S that allows for such calculations. As an application, in Section 6.2, we apply the first fundamental form to science of cartography, making maps of portion of the globe.

Sections 6.2 and 6.4 explore how the surface S sits in the ambient Euclidean space. In particular, we introduce the Gauss map and the second fundamental form, each of which describes how the normal vector changes as one moves over the surface. Sections 6.5 through 6.7 introduce the concepts of curvature for a regular surface in $\mathbb{R}^3$, a topic that is far richer than the concept of curvature for a curve.

The Fundamental Theorem of Space Curves affirms that the local geometry of a space curve is determined by two geometric invariants: the curvature and the torsion. In Chapter 7, we will introduce the profound result that a regular surface S in $\mathbb{R}^3$ is fully determined, up to position and orientation in space, by the first and second fundamental forms.

6.1   The First Fundamental Form

Definition 6.1.1 Let S be a regular surface and $p \in S$. The *first fundamental form* $I_p(\cdot, \cdot)$ is the restriction of the usual dot product in $\mathbb{R}^3$ to the tangent plane T_pS. Namely, for $\vec{a}, \vec{b}$ in $T_pS = d\vec{X}_q(\mathbb{R}^2)$, $I_p(\vec{a}, \vec{b}) = \vec{a} \cdot \vec{b}$.

Note that for each point $p \in S$, the first fundamental form $I_p(\cdot, \cdot)$ is defined only on the tangent space. There exists a unique matrix

DOI: 10.1201/9781003295341-6

that represents $I_p(\cdot, \cdot)$ with respect to the standard basis on T_pS, but it is important to remember that this matrix is a matrix of functions with components depending on the point $p \in S$. For any other point $p_2 \in S$, the first fundamental form is still defined the same way, but the corresponding matrix is most likely different.

This definition draws on the ambient space $\mathbb{R}^3$ for an inner product on each tangent plane. However, if around p the surface S can be parametrized by $\vec{X}$, there is a natural basis on T_pS, namely $\vec{X}_u$ and $\vec{X}_v$. We wish to express the inner product $I_p(\cdot, \cdot)$ in terms of this basis on S. Let

$$\vec{a} = a_1 \vec{X}_u(q) + a_2 \vec{X}_v(q) \quad \text{and} \quad \vec{b} = b_1 \vec{X}_u(q) + b_2 \vec{X}_v(q).$$

Then we calculate that

$$\begin{aligned} \vec{a} \cdot \vec{b} = {} & a_1 b_1 \vec{X}_u(q) \cdot \vec{X}_u(q) + a_1 b_2 \vec{X}_u(q) \cdot \vec{X}_v(q) \\ & + a_2 b_1 \vec{X}_u(q) \cdot \vec{X}_v(q) + a_2 b_2 \vec{X}_v(q) \cdot \vec{X}_v(q). \end{aligned}$$

This proves the following proposition.

Proposition 6.1.2 *Let $I_p(\cdot, \cdot) : T_pS^2 \to \mathbb{R}$ be the first fundamental form at a point p on a regular surface S. Given a regular parametrization $\vec{X} : U \to \mathbb{R}^3$ of a neighborhood of p, the matrix associated with the first fundamental form $I_p(\cdot, \cdot)$ with respect to the basis $\{\vec{X}_u, \vec{X}_v\}$ is*

$$g = \begin{pmatrix} g_{11} & g_{12} \\ g_{21} & g_{22} \end{pmatrix},$$

where

$$g_{11} = \vec{X}_u \cdot \vec{X}_u \qquad \text{and} \qquad g_{12} = \vec{X}_u \cdot \vec{X}_v,$$
$$g_{21} = \vec{X}_v \cdot \vec{X}_u \qquad \text{and} \qquad g_{22} = \vec{X}_v \cdot \vec{X}_v,$$

evaluated at (u_0, v_0) when $p = \vec{X}(u_0, v_0)$.

The quantity g is a matrix of real functions from the open domain $U \subset \mathbb{R}^2$. With this notation, if $p = \vec{X}(q)$, then the first fundamental form $I_p(\cdot, \cdot)$ at the point p can be expressed as the bilinear form

$$I_p(\vec{a}, \vec{b}) = \vec{a}^T g(q) \vec{b} = \begin{pmatrix} a_1 & a_2 \end{pmatrix} g(q) \begin{pmatrix} b_1 \\ b_2 \end{pmatrix}$$

for all $\vec{a}, \vec{b} \in T_pS$.

One should remark immediately that for any differentiable function $\vec{X} : U \to \mathbb{R}^3$, one has $\vec{X}_u \cdot \vec{X}_v = \vec{X}_v \cdot \vec{X}_u$. Thus, $g_{12} = g_{21}$ and the matrix g is symmetric.

Example 6.1.3 (The xy-Plane) As the simplest possible example, consider the xy-plane. It is a regular surface parametrized by a single system of coordinates $\vec{X}(u,v) = (u,v,0)$. Obviously, we obtain

$$g = \begin{pmatrix} 1 & 0 \\ 0 & 1 \end{pmatrix}.$$

This should have been obvious from the definition of the first fundamental form. The xy-plane is its own tangent space for all p, and the parametrization $\vec{X}$ induces the basis

$$\{\vec{X}_u, \vec{X}_v\} = \{(1,0,0),(0,1,0)\}.$$

Therefore, for any $\vec{v} = \begin{pmatrix} v_1 \\ v_2 \end{pmatrix}$ and $\vec{w} = \begin{pmatrix} w_1 \\ w_2 \end{pmatrix}$, with coordinates given in terms of the standard basis, we have

$$\vec{v} \cdot \vec{w} = v_1 w_1 + v_2 w_2 = \begin{pmatrix} v_1 & v_2 \end{pmatrix} \begin{pmatrix} 1 & 0 \\ 0 & 1 \end{pmatrix} \begin{pmatrix} w_1 \\ w_2 \end{pmatrix}.$$

Example 6.1.4 (Cylinder) Consider the right circular cylinder parametrized by $\vec{X}(u,v) = (\cos u, \sin u, v)$, with $(u,v) \in (0, 2\pi) \times \mathbb{R}$. We calculate

$$\vec{X}_u = (-\sin u, \cos u, 0) \qquad \text{and} \qquad \vec{X}_v = (0,0,1),$$

and thus,

$$g = \begin{pmatrix} 1 & 0 \\ 0 & 1 \end{pmatrix}.$$

Obviously, a cylinder and the xy-plane are not the same surface. As we will see later, a plane and a cylinder share many properties. However, the second fundamental form, which carries information about how the normal vector behaves on the cylinder, will distinguish between the cylinder and the xy-plane.

Example 6.1.5 (Spheres) Consider the longitude-colatitude parametrization on the sphere

$$\vec{X}(u,v) = (\cos u \sin v, \sin u \sin v, \cos v),$$

with $(u,v) \in (0, 2\pi) \times (0, \pi)$. The first derivatives of $\vec{X}$ are

$$\vec{X}_u = (-\sin u \sin v, \cos u \sin v, 0),$$
$$\vec{X}_v = (\cos u \cos v, \sin u \cos v, -\sin v).$$

We deduce that for this parametrization

$$g = \begin{pmatrix} g_{11} & g_{21} \\ g_{12} & g_{22} \end{pmatrix} = \begin{pmatrix} \sin^2 v & 0 \\ 0 & 1 \end{pmatrix}.$$

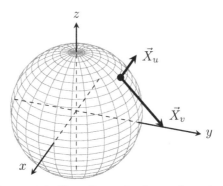

Figure 6.1: Coordinate basis on the sphere.

Figure 6.1 shows the sphere with this parametrization along with the basis $\{\vec{X}_u, \vec{X}_v\}$ at the point $\vec{X}(\pi/2, \pi/4)$.

Some classical differential geometry texts use the letters $E = g_{11}$, $F = g_{12}$, $G = g_{22}$. The classical notation was replaced by the "tensor notation" g_{ij} as the notion of a tensor became more prevalent in differential geometry. Though we do not emphasize the tensor concept in this book, this book provides a brief discussion of tensor notation in Appendix A.1.

As a first application of how the first fundamental form allows one to do geometry on a regular surface, consider the problem of calculating the arc length of a curve on the surface. Let S be a regular surface with a coordinate neighborhood parametrized by $\vec{X} : U \to \mathbb{R}^3$, where $U \subset \mathbb{R}^2$ is open. Consider a curve on S given by $\vec{\gamma}(t) = \vec{X}(u(t), v(t))$, where $(u(t), v(t)) = \vec{\alpha}(t)$ is a differentiable parametrized plane curve in the domain U. The arc length formula over the interval $[t_0, t]$ is

$$s(t) = \int_{t_0}^{t} \|\vec{\gamma}'(\tau)\| \, d\tau.$$

After some reworking, we can rewrite the formula as

$$s(t) = \int_{t_0}^{t} \sqrt{g_{11}(u'(\tau))^2 + 2g_{12}u'(\tau)v'(\tau) + g_{22}(v'(\tau))^2} \, d\tau. \quad (6.1)$$

We need to be careful with this notation; each function $g_{ij}(u, v)$ is evaluated at $(u(t), v(t))$, so the square of the integrand is in fact

$$g_{11}(u(\tau), v(\tau))(u'(\tau))^2 + 2g_{12}(u(\tau), v(\tau))u'(t)v'(\tau)$$
$$+ g_{22}(u(\tau), v(\tau))(v'(\tau))^2.$$

Using the first fundamental form, we can rewrite the arc length

formula as

$$s(t) = \int_{t_0}^{t} \sqrt{I_{\vec{\gamma}(t)}\big(\vec{\alpha}'(\tau), \vec{\alpha}'(\tau)\big)}\, d\tau \, . \qquad (6.2)$$

The reader should note that this is an abuse of notation since, for all $p \in S$, $I_p(\cdot, \cdot)$ is a bilinear form on the tangent space T_pS, but $\vec{\alpha}'(t)$ is a vector function in $\mathbb{R}^2$. However, the justification behind this notation is that for all t, the coordinates of $\vec{\alpha}'(t)$ in the standard basis of $\mathbb{R}^2$ are precisely the coordinates of $\vec{\gamma}'(t)$ in the basis $\{\vec{X}_u, \vec{X}_v\}$ based at the point $\vec{\gamma}(t)$.

In essence, Equation (6.2) relates the geometry in the particular coordinate neighborhood of S to the plane geometry in the tangent plane T_pS. More precisely, while doing geometry in U, by using $I_p(\cdot, \cdot)$ instead of the usual dot product in $\mathbb{R}^3$, one obtains information about what happens on the regular surface S as opposed to simply what happens on the tangent plane T_pS. This approach is useful for more than just calculating arc length.

Consider two plane curves $\vec{\alpha}(t)$ and $\vec{\beta}(t)$ in the domain of $\vec{X}$, with $\vec{\alpha}(t_0) = \vec{\beta}(t_0) = q$. Also let $\vec{\gamma} = \vec{X} \circ \vec{\alpha}$ and $\vec{\delta} = \vec{X} \circ \vec{\beta}$ be the corresponding curves on the regular surface S. Then at the point $p = \vec{X}(q)$, the curves $\vec{\gamma}(t)$ and $\vec{\delta}(t)$ form an angle of θ, with

$$\cos\theta = \frac{\vec{\gamma}'(t_0) \cdot \vec{\delta}'(t_0)}{\|\vec{\gamma}'(t_0)\| \, \|\vec{\delta}'(t_0)\|} = \frac{I_p(\vec{\alpha}'(t_0), \vec{\beta}'(t_0))}{\sqrt{I_p(\vec{\alpha}'(t_0), \vec{\alpha}'(t_0))}\sqrt{I_p(\vec{\beta}'(t_0), \vec{\beta}'(t_0))}}.$$

$$(6.3)$$

(Again, the comment following Equation (6.2) applies here as well.)

In the basis $\{\vec{X}_u, \vec{X}_v\}$, the vectors $\vec{X}_u$ and $\vec{X}_v$ have the coordinates $(1,0)$ and $(0,1)$, respectively. Therefore, Equation (6.3) implies that the angle φ of the coordinate curves of a parametrization $\vec{X}(u,v)$ is given by

$$\cos\varphi = \frac{\vec{X}_u \cdot \vec{X}_v}{\|\vec{X}_u\| \, \|\vec{X}_v\|} = \frac{I_p((1,0),(0,1))}{\sqrt{I_p((1,0),(1,0))}\sqrt{I_p((0,1),(0,1))}}$$

$$= \frac{g_{12}}{\sqrt{g_{11}g_{22}}}.$$

Thus, all the coordinate curves of a parametrization are orthogonal if and only if $g_{12}(u,v) = 0$ for all (u,v) in the domain of $\vec{X}$. If $\vec{X}$ satisfies this property, we call $\vec{X}$ an *orthogonal parametrization*.

Example 6.1.6 (Loxodromes) Consider the longitude-colatitude parametrization of the sphere, given in Example 6.1.5. Recall that a meridian of the sphere is any curve on the sphere with u fixed and that the coefficients for the first fundamental form are

$$g_{11} = \sin^2 v, \quad g_{12} = g_{21} = 0, \quad g_{22} = 1.$$

A loxodrome on the sphere is a curve that makes a constant angle β with every meridian. Let $\vec{\alpha}(t) = (u(t), v(t))$ be a curve in the domain $(0, 2\pi) \times (0, \pi)$ and $\vec{\gamma}(t) = \vec{X}(\vec{\alpha}(t))$. If $\vec{\gamma}(t)$ makes the same angle β with all the meridians, then by Equation (6.3),

$$\cos\beta = \frac{\begin{pmatrix} u' & v' \end{pmatrix} \begin{pmatrix} \sin^2 v & 0 \\ 0 & 1 \end{pmatrix} \begin{pmatrix} 0 \\ 1 \end{pmatrix}}{\sqrt{(u')^2 \sin^2 v + (v')^2} \sqrt{1}} = \frac{v'}{\sqrt{(u')^2 \sin^2 v + (v')^2}}.$$

Then

$$(u')^2 \sin^2 v \cos^2 \beta + (v')^2 \cos^2 \beta = (v')^2$$
$$\Longleftrightarrow (v')^2 \sin^2 \beta = (u')^2 \sin^2 v \cos^2 \beta$$
$$\Longleftrightarrow (u')^2 \sin^2 v = (v')^2 \tan^2 \beta$$
$$\Longrightarrow \pm u' \cot \beta = \frac{v'}{\sin v},$$

where the $\pm$ makes sense in that a loxodrome can travel either toward the north pole or toward the south pole. Now integrating both sides of the last equation with respect to t and then performing a substitution, one obtains

$$\int \pm u' \cot \beta \, dt = \int \frac{v'}{\sin v} \, dt \Longleftrightarrow \int \pm \cot \beta \, du = \int \frac{1}{\sin v} \, dv$$
$$\Longleftrightarrow (\pm \cot \beta) u + C = \ln \tan\left(\frac{v}{2}\right),$$

where we have chosen a form of antiderivative of $\frac{1}{\sin v}$ that suits our calculations best. This establishes an equation between u and v that any loxodrome must satisfy. From this, one can deduce the following parametrization for the loxodrome:

$$(u(t), v(t)) = \left((\pm \tan \beta)(\ln t - C), 2 \tan^{-1} t\right).$$

(Figure 6.2 was plotted using $\tan \beta = 4$ and $C = 0$.)

Not only does the first fundamental form allow us to talk about lengths of curves and angles between curves on a regular surface, but it also provides a way to calculate the area of a region on the regular surface.

 Proposition 6.1.7 *Let* $\vec{X} : U \subset \mathbb{R}^2 \to S$ *be the parametrization for a coordinate neighborhood of a regular surface. Then if Q is a compact subset of U and $\mathcal{R} = \vec{X}(Q)$ is a region of S, then the area of $\mathcal{R}$ is given by*

$$A(\mathcal{R}) = \iint_Q \sqrt{\det(g)} \, du \, dv. \tag{6.4}$$

Figure 6.2: Loxodrome on the sphere.

Proof: Recall the formula for surface area introduced in multivariable calculus that gives the surface area of the region $\mathcal{R}$ on S as

$$A(\mathcal{R}) = \iint_{\mathcal{R}} dA = \iint_{Q} \|\vec{X}_u \times \vec{X}_v\| \, du \, dv. \qquad (6.5)$$

(See [26, Section 7.4] or [31, Section 17.7] for an explanation of Equation (6.5).)

By Problem 3.1.6, for any two vectors $\vec{v}$ and $\vec{w}$ in $\mathbb{R}^3$, the following identity holds:

$$(\vec{v} \times \vec{w}) \cdot (\vec{v} \times \vec{w}) = (\vec{v} \cdot \vec{v})(\vec{w} \cdot \vec{w}) - (\vec{v} \cdot \vec{w})^2.$$

Therefore,

$$\|\vec{X}_u \times \vec{X}_v\|^2 = (\vec{X}_u \cdot \vec{X}_u)(\vec{X}_v \cdot \vec{X}_v) - (\vec{X}_u \cdot \vec{X}_v)^2 = g_{11}g_{22} - g_{12}^2, \quad (6.6)$$

and the proposition follows because $g_{12} = g_{21}$. $\qquad \square$

Proposition 6.1.7 is interesting in itself as it leads to another property of the first fundamental form. It is obvious from the definition of the first fundamental form that $g_{11}(u, v) \geq 0$ and $g_{22}(u, v) \geq 0$ for all $(u, v) \in U$. However, since $\|\vec{X}_u \times \vec{X}_v\|^2 = \det(g)$ and since $\vec{X}_u \times \vec{X}_v$ is never $\vec{0}$ on a regular surface, we deduce that $\det(g) = g_{11}g_{22} - g_{12} > 0$ for all $(u, v) \in U$. In other words, the matrix of functions g is always a *positive definite* matrix.

Example 6.1.8 Consider the regular surface of revolution S parametrized by $\vec{X}(u, v) = (v \cos u, v \sin u, \ln v)$, with $(u, v) \in [0, 2\pi] \times (0, \infty) = U$. (From a practical standpoint, it is not so important that U is not an open subset of $\mathbb{R}^2$, but it should be understood that S may require two coordinate neighborhoods to satisfy the criteria

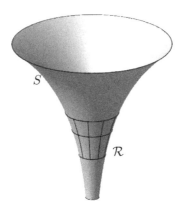

Figure 6.3: Surface area patch for Example 6.1.8.

for a regular surface. See Figure 6.3.) Let $Q = [0, 2\pi] \times [1, 2]$, and let's calculate the area of $\mathcal{R} = \vec{X}(Q)$.

It's not hard to show that

$$(g_{ij}) = \begin{pmatrix} v^2 & 0 \\ 0 & 1 + \frac{1}{v^2} \end{pmatrix}.$$

Then Proposition 6.1.7 gives

$$A(\mathcal{R}) = \int_0^{2\pi} \int_1^2 \sqrt{v^2 + 1}\, dv\, du = 2\pi \left[\frac{1}{2} v \sqrt{v^2 + 1} + \frac{1}{2} \sinh^{-1} v \right]_1^2$$

$$= \pi \left(2\sqrt{5} - \sqrt{2} + \ln \left(\frac{2 + \sqrt{5}}{1 + \sqrt{2}} \right) \right).$$

As we have done for many quantities on curves and surfaces, we wish to see how the functions in the first fundamental form change under a coordinate transformation. In order to avoid confusion with various coordinate systems, we introduce some notations that will become more common in future topics. Suppose that an open subset V of a regular surface S can be parametrized by two different sets of coordinates (x_1, x_2) and $(\bar{x}_1, \bar{x}_2)$. If we write $\vec{X}(x_1, x_2)$ and $\vec{Y}(\bar{x}_1, \bar{x}_2)$ for the specific parametrizations, then the function $F = \vec{Y}^{-1} \circ \vec{X}$ is a diffeomorphism between two open subsets of $\mathbb{R}^2$ and

$$(\bar{x}_1, \bar{x}_2) = F(x_1, x_2).$$

For convenience, we will often simply write

$$\begin{cases} \bar{x}_1 = \bar{x}_1(x_1, x_2), \\ \bar{x}_2 = \bar{x}_2(x_1, x_2) \end{cases}$$

to indicate the dependency of one set of variables on the other. Vice versa, we can write

$$(x_1, x_2) = F^{-1}(\bar{x}_1, \bar{x}_2) \quad \text{or simply} \quad \begin{cases} x_1 = x_1(\bar{x}_1, \bar{x}_2), \\ x_2 = x_2(\bar{x}_1, \bar{x}_2). \end{cases}$$

Now $\vec{Y}(\bar{x}_1, \bar{x}_2) = \vec{X}(x_1(\bar{x}_1, \bar{x}_2), x_2(\bar{x}_1, \bar{x}_2))$ for the respective parametrizations. Let g_{ij} be the coefficient functions in the first fundamental form for S parametrized by x_1, x_2, and let $\bar{g}_{ij}$ be the coefficient functions for S parametrized by $\bar{x}_1, \bar{x}_2$. Then by the chain rule,

$$\bar{g}_{ij} = \vec{Y}_{\bar{x}_i} \cdot \vec{Y}_{\bar{x}_j} = \left(\vec{X}_{x_1} \frac{\partial x_1}{\partial \bar{x}_i} + \vec{X}_{x_2} \frac{\partial x_2}{\partial \bar{x}_i} \right) \cdot \left(\vec{X}_{x_1} \frac{\partial x_1}{\partial \bar{x}_j} + \vec{X}_{x_2} \frac{\partial x_2}{\partial \bar{x}_j} \right).$$

After some reorganization, we find

$$\bar{g}_{ij} = \sum_{k=1}^{2} \sum_{l=1}^{2} \frac{\partial x_k}{\partial \bar{x}_i} \frac{\partial x_l}{\partial \bar{x}_j} g_{kl}. \tag{6.7}$$

This type of transformation of coordinates is our first (nontrivial) encounter with tensors. Because the g_{ij} functions satisfy this particular identity under a change of coordinates, we call the matrix of functions $g = (g_{ij})$ the components of a tensor of type $(0, 2)$. The collection of the quantities g_{ij} defined at each point of $p \in S$ is referred to as the components of *metric tensor* of S.

The value of the first fundamental form is that, given the metric tensor, one can use the appropriate formulas for arc length, for angles between curves, and for areas of regions without knowing the specific parametrization of the surface. An additional benefit of the first fundamental form and the metric tensor is that one can still use them to study the geometry of surfaces in higher dimensions, as we will see in Section 9.2.

Because of Equation (6.2), a few mathematicians and most physicists use an alternative notation for the metric tensor. These authors describe the metric tensor by saying that the "line element" in a coordinate patch is given by

$$ds^2 = g_{11}(u, v) \, du^2 + 2g_{12}(u, v) \, du \, dv + g_{22}(u, v) \, dv^2.$$

Older texts in differential geometry that used the E, F, and G letters for g_{11}, g_{12}, and g_{22} write this as

$$ds^2 = E(u, v) \, du^2 + 2F(u, v) \, du \, dv + G(u, v) \, dv^2.$$

This notation has the advantage that one can write it concisely on a single line instead of in matrix format. However, one should not forget that it not only leads to Equation (6.2) for arc length of a curve on a surface but that it also leads to Equation (6.3) for angles between curves on the surface and the formula in Proposition 9.2.2 that gives the area of regions on the surface.

Example 6.1.9 (Intrinsic Geometry) If we are given the metric coefficients $g_{ij}(u,v)$ of a surface as functions of u and v, we do not know the actual equations of the surface in space in order to compute quantities like the length of a curve or the area of a coordinate patch. Such computations are an example of "intrinsic geometry of a surface," namely the study of geometric computations that only depend on the metric coefficients.

An important example of intrinsic geometry is given by the metric coefficient functions $g_{11}(u,v) = 1/v^2$, $g_{12}(u,v) = 0 = g_{21}(u,v)$, and $g_{22}(u,v) = 1/v^2$ defined on the domain consisting of all points in the (u,v) plane with $v > 0$, the "upper half-plane."

For example, we can compute the length of the curve given by $(u(t), v(t)) = (t, c)$ where $a \le t \le b$ by calculating the integral

$$\int_a^b \sqrt{\frac{1}{c^2}u'(t)^2 + 2(0)u'(t)v'(t) + \frac{1}{c^2}v'(t)^2}\, dt$$

$$= \int_a^b \sqrt{\frac{1}{c^2} + 0 + 0}\, dt = \int_a^b (1/c)\, dt = \frac{1}{c}(b - a).$$

We can also calculate the length of the curve $(u(t), v(t)) = (a, t)$ where $0 < c \le t \le d$ by

$$\int_c^d \sqrt{\frac{1}{t^2}0 + 0 + \frac{1}{t^2}(1)}\, dt = \int_c^d \frac{dt}{t} = [\ln(t)]_c^d = \ln(d) - \ln(c).$$

Note that the limit of this length as d goes to infinity is infinity and the same is true as c goes to 0. It follows that the perimeter of the region $\mathcal{R}$ defined by $a \le u \le b$, and $c \le v \le d$ is

$$(b - a)(1/c) - (b - a)(1/d) + 2(\ln(d) - \ln(c)).$$

Since the area depends only on the metric coefficient $g_{ij}(u,v)$, we can compute the area of the region $\mathcal{R}$ in this metric by integrating $\sqrt{g_{11}(u,v)g_{22}(u,v)} = \sqrt{(1/v^2)(1/v^2)} = 1/v^2$ to get

$$\int_a^b \int_c^d \frac{1}{v^2}\, dv\, du = (b - a)\left(\frac{1}{c} - \frac{1}{d}\right).$$

Note that the limit of the area of this domain using these metric coefficients goes to $(b - a)(1/c)$ as d goes to infinity, whereas the area becomes infinitely large as c goes to 0.

The upper half-plane with this alternative metric tensor is called the *Poincaré Upper Half-Plane*.

PROBLEMS

1. Suppose that a regular surface S has a metric tensor with coefficient functions given by

$$g_{11}(u, v) = u^2 + v^2, \ g_{12} = uv, \ g_{22} = v^2.$$

(a) Let p be a point with coordinates $(u, v) = (2, 3)$. Let $\vec{a}$ be a vector with coordinates $\binom{3}{-2}$ in T_pS. Calculate the length of $\vec{a}$.

(b) Let $\vec{b}$ be another vector in T_pS with coordinates $\binom{-1}{1}$. Calculate the angle between $\vec{a}$ and $\vec{b}$.

(c) Let p' be another point on S with coordinates $(u', v') = (-1, 2)$. Let $\vec{c} \in T_{p'}S$ with coordinates with respect to the standard basis $\binom{5}{2}$. What is the length of $\vec{c}$ with respect to the first fundamental form?

2. Suppose that a regular surface S has a metric tensor with coefficient functions given by

$$g_{11}(u, v) = e^v, \ g_{12} = 1, \ g_{22} = e^u.$$

(a) Determine the domain for (u, v) in which $g_{11}g_{22} - g_{12}^2 > 0$ holds.

(b) Let p be a point with coordinates $(u, v) = (1, 2)$. Let $\vec{a}$ be a vector with coordinates $\binom{3}{-1}$ in T_pS. Calculate the length of $\vec{a}$.

(c) (**CAS**) Write down the definite integral that gives the arclength of the curve $(u, v) = (1 + t, 2t)$ for $0 \le t \le 1$. Determine this length up to 4 digits of accuracy.

For Problems 6.1.3 through 6.1.10, calculate the coefficients of the first fundamental form (metric tensor).

3. $\vec{X}(u, v) = \vec{a} + u\vec{b} + v\vec{c}$, where $\vec{a}, \vec{b}, \vec{c} \in \mathbb{R}^3$ are constant vectors.

4. $\vec{X}(u, v) = (u^2 - v, uv, u)$.

5. The helicoid: $\vec{X}(u, v) = (v \cos u, v \sin u, au)$, where $a \in \mathbb{R}$ is constant.

6. The ellipsoid: $\vec{X}(u, v) = (a \cos u \sin v, b \sin u \sin v, c \cos v)$.

7. The elliptic paraboloid: $\vec{X}(u, v) = (av \cos u, bv \sin u, v^2)$.

8. The hyperbolic paraboloid: $\vec{X}(u, v) = (au \cosh v, bu \sinh v, u^2)$.

9. $\vec{X}(u, v) = (a \cosh v \cos u, b \cosh v \sin u, c \sinh v)$, a hyperboloid of one sheet.

10. $\vec{X}(u, v) = (a \cos u \sinh v, b \sin u \sinh v, c \cosh v)$, one component of a hyperboloid of two sheets.

11. Let $0 < r < R$ be reals and consider the torus parametrized by

$$\vec{X}(u, v) = ((R + r \cos v) \cos u, (R + r \cos v) \sin u, r \sin v).$$

(a) Show that if the domain of $\vec{X}$ is $U = (0, 2\pi) \times (0, 2\pi)$, then the parametrization is regular and that if $U = [0, 2\pi] \times [0, 2\pi]$ the parametrization is surjective.

(b) Calculate the metric tensor of this torus.

(c) Use the metric tensor to calculate the area of the torus.

12. Let U be an open subset of $\mathbb{R}^2$, and let $f : U \to \mathbb{R}$ be a two-variable function. Explicitly calculate the metric tensor for the graph of f, which can be parametrized by $\vec{X} : U \to \mathbb{R}^3$, with $\vec{X}(u, v) = (u, v, f(u, v))$. Prove that the coordinate lines are orthogonal if and only if $f_u f_v = 0$.

13. Let γ be a loxodrome on the sphere as described in Example 6.1.6. Prove that the arc length of the loxodrome between the north and south poles is $\pi \sec \beta$ (regardless of the constant of integration C).

14. Consider the surface of revolution parametrized by

$$\vec{X}(u, v) = (f(v) \cos u, f(v) \sin u, g(v)),$$

with f and g chosen so that the surface is regular (see Problem 5.2.14). Calculate the metric tensor.

15. (**CAS**) Consider the portion of a hyperboloid of two sheets given by

$$\vec{X}(u, v) = (\cos u \sinh v, \sin u \sinh v, \cosh v).$$

Find the surface area of this shape with $(u, v) \in [0, 2\pi] \times [0, r]$.

16. Let $\vec{\alpha}(t)$ be a regular space curve and consider the tangential surface S parametrized by $\vec{X}(t, u) = \vec{\alpha}(t) + u\vec{T}(t)$. Calculate the metric tensor for $\vec{X}$.

17. *Tubes.* Let $\vec{\alpha}(t)$ be a regular space curve. We call the *tube* of radius r around $\vec{\alpha}$ the surface that is parametrized by $\vec{X}(t, u) = \vec{\alpha}(t) + (r \cos u)\vec{P}(t) + (r \sin u)\vec{B}(t)$. Calculate the metric tensor for $\vec{X}$. Supposing that the tube is regular, prove that the area of the tube is $2\pi r$ times the length of $\vec{\alpha}$.

18. Consider the cone parametrized by $\vec{X}(u, v) = (v \cos u, v \sin u, v)$.

(a) Let $\mathcal{C}$ be a curve in $\mathbb{R}^2$ such that the tangent vector makes a constant angle of β with the position vector. Use polar coordinates to prove. that $\mathcal{C}$ can be parametrized by $\vec{\alpha}(t) = (f(t) \cos t, f(t) \sin t)$ where $f(t) = Re^{(\cot \beta)t}$ and $R > 0$.

(b) Consider the spiral curve $\vec{\gamma}(t) = \vec{X}(t, f(t))$ on the cone. Calculate the arc length of $\vec{\gamma}$ for $0 \le t \le 2\pi$. (It will depend on R and β.)

19. Consider the parametrization $\vec{Y} : \mathbb{R}^2 \to \mathbb{R}^3$ of the unit sphere by stereographic projection (see Problem 5.3.10 and use the parametrization $\vec{Y}(u, v) = \pi^{-1}(u, v)$).

(a) Calculate the corresponding metric tensor functions g_{ij}.

(b) Consider the usual parametrization $\vec{X}$ of the sphere (as presented in Example 6.1.5.) Give appropriate domains for and explicitly determine the formula for the change of coordinate system given by $F^{-1} = \vec{Y}^{-1} \circ \vec{X} = \pi \circ \vec{X}$.

(c) Explicitly verify Equation (6.7) in this context.

20. Consider the parametrization $\vec{X} : \mathbb{R}^2 - \{(0,0)\} \to \mathbb{R}^3$ defined by $\vec{X}(x,y) = (x, y, \frac{1}{2}\ln(x^2 + y^2))$.

   (a) Show that $\vec{X}$ is another parametrization for the Log Trumpet discussed in Example 6.1.8.

   (b) Calculate the components of the metric tensor associate to this parametrization.

   (c) Use this and the parametrization given in Example 6.1.8 to illustrate how your result exemplifies the coordinate change properties of the metric tensor components as described in Equation 6.7.

21. Suppose that the first fundamental form has a matrix of the form

$$g = \begin{pmatrix} 1 & 0 \\ 0 & f(u,v) \end{pmatrix}.$$

Prove that all the v-coordinate lines have equal arc length over any interval $u \in [u_1, u_2]$. In this case, the v-coordinate lines are called parallel.

22. Let (g_{ij}) be the metric tensor of some surface S parametrized by the coordinates $(x_1, x_2) \in U$, where U is some open subset in $\mathbb{R}^2$. Suppose that we reparametrize the surface with coordinates $(\bar{x}_1, \bar{x}_2)$ in such a way that

$$\begin{cases} x_1 = f(\bar{x}_1, \bar{x}_2), \\ x_2 = \bar{x}_2, \end{cases}$$

where $f : V \to U$ is a function with $V \subset \mathbb{R}$. Let $(\bar{g}_{kl})$ be the metric tensor to S under this parametrization.

   (a) Prove that $(\bar{g}_{kl})$ is a diagonal matrix if and only if the function f satisfies the differential equation

$$\frac{\partial f}{\partial \bar{x}_2} = -\frac{g_{12}\big(f(\bar{x}_1, \bar{x}_2), \bar{x}_2\big)}{g_{11}\big(f(\bar{x}_1, \bar{x}_2), \bar{x}_2\big)}.$$

   (b) (**ODE**) Use the above result and the existence theorem for differential equations to prove that every regular surface admits an orthogonal parametrization.

23. Let C be a regular value of the function $F(x, y, z)$ and consider the surface S defined by $F(x, y, z) = C$. Without loss of generality, suppose that the variables x and y can be used to parametrize a neighborhood U of a point $p \in S$. Use implicit differentiation to calculate the metric tensor functions g_{ij} in terms of derivatives of F.

24. Consider the parametrization $\vec{X}(u, v) = (\cos u \sin v, \sin u \sin v, \cos v)$ of the unit sphere. It is easy to show that the unit normal vector $\vec{N}$ to the sphere at (u, v) is again the vector $\vec{N}(u, v) = \vec{X}(u, v)$. Let $f(u, v)$ be a nonnegative real function in two variables such that $f(u, 0)$ is constant, $f(u, \pi)$ is constant, and for all fixed v_0, $f(u, v_0)$ is periodic 2π. A *normal variation* to a given surface S is a surface created by

going out a distance of $f(u, v)$ along the normal vector $\vec{N}$ of S, given by

$$\vec{Y}(u, v) = \vec{X}(u, v) + f(u, v)\vec{N}(u, v).$$

A normal variation of the unit sphere is therefore given by

$$\vec{Y}(u, v) = (1 + f(u, v))\vec{X}(u, v).$$

(a) Calculate the metric tensor for the parametrization $\vec{Y}$.

(b) Use the explicit function $f(u, v) = \cos^2(2u)\cos^2(2v - \pi/2)$. Calculate the metric tensor for $\vec{Y}$.

25. Consider the metric coefficients on the upper half-plane defined in Example 6.1.9.

(a) Compute the length of the curve $(u, v) = (\sqrt{2}\cos t, \sqrt{2}\sin t)$ with $\pi/4 \leq t \leq 3\pi/4$ from $(1, 1)$ to $(-1, 1)$ using these metric coefficients.

(b) Show that the length of the curve is less than the length of the segment $(u(t), v(t)) = (t, 1)$ with $-1 \leq t \leq 1$ using these metric coefficients.

(c) Show that the length of the curve $(u(t), v(t)) = (R\cos t, R\sin t)$ with $\pi/4 \leq t \leq 3\pi/4$ is independent of R.

26. Consider the metric defined in the previous problem on the upper half-plane.

(a) Compute the area above the curve $(u(t), v(t)) = (R\cos t + m, R\sin t)$ with $\pi/4 \leq t \leq 3\pi/4$ and below the line $v = d$, for some $d > R$. Observe that the area is independent of the constant m. What happens to the area of this region as d approaches infinity?

(b) Repeat the same question with $(u, v) = (R\cos t + m, R\sin t)$ but with $0 \leq t \leq \pi$. [Hint: This will involve an improper integral.]

27. Consider the general metric coefficients $g_{11}(u, v) = g_{22}(u, v) = f(v)$ and $g_{12}(u, v) = g_{21}(u, v) = 0$ defined on the upper half-plane. Find an expression for the length of the curve $(u(t), v(t))$ for $a \leq t \leq b$.

6.2   Map Projections (Optional)

6.2.1   Metric Properties of Maps of the Earth

In order to illustrate properties and uses of the first fundamental form, we propose to briefly discuss maps of the Earth. Throughout history, maps of portions of the Earth have helped political leaders understand the geography of regions over which they exert influence. Starting particularly in the age of exploration, traders and navigators needed maps that describe large regions of the Earth. For navigation purposes, it would be ideal if a map accurately represented distances between points, angles between directions, and areas of regions. Distances, angles, and areas are the metric properties of a map. However, because a map is typically drawn on a flat piece of paper whereas the Earth is (very close to) spherical, no map can accurately reflect all metric properties at the same time.

Points on the Earth are usually located using coordinates of latitude and longitude. In other places in this text (Example 6.1.5), we introduced the mathematicians' customary way of parametrizing the unit sphere: the polar coordinates angle θ, which is essentially the longitude, and then the colatitude φ, which is the angle down from the positive z-axis. To cover the sphere, we use $0 \leq \theta < 2\pi$ and $0 \leq \varphi \leq \pi$. There is no absolute consensus on which coordinate is listed first. If we denote geographer's latitude by φ_g, then $\varphi + \varphi_g = \pi/2$. Hence, the North Pole has $\varphi_g = \pi/2$, the equator is at $\varphi_g = 0$, and the South Pole has $\varphi_g = -\pi/2$. (Of course, latitude and longitude are typically described with degrees.)

For the purposes of our discussion, we will use the parametrization of the sphere

$$\vec{X}(\varphi, \theta) = (R \sin \varphi \cos \theta, R \sin \varphi \sin \theta, R \cos \varphi) \qquad (6.8)$$

with colatitude $0 \leq \varphi \leq \pi$ and longitude $-\pi < \theta \leq \pi$ and where R is the radius of the Earth. Coordinate lines with θ constant are called *meridians*. Note that colatitude lines are precisely latitude lines.

Points on a map are given by a pair of coordinates, say x and y, which are standard Cartesian coordinates of the plane. A map projection corresponds to a function from (φ, θ)- to (x, y)-coordinates,

$$M(\varphi, \theta) = (x(\varphi, \theta), y(\varphi, \theta)). \qquad (6.9)$$

Maps generally do not show the x and y grid but rather the curves in the xy-plane that correspond to constant θ and φ coordinates.

Using the parametrization of the Earth in (6.8), the metric properties of the Earth are encapsulated by the first fundamental form, namely

$$g_{\text{Earth}} = \begin{pmatrix} \vec{X}_\varphi \cdot \vec{X}_\varphi & \vec{X}_\varphi \cdot \vec{X}_\theta \\ \vec{X}_\theta \cdot \vec{X}_\varphi & \vec{X}_\theta \cdot \vec{X}_\theta \end{pmatrix} = \begin{pmatrix} R^2 & 0 \\ 0 & R^2 \sin^2 \varphi \end{pmatrix}.$$

We deduce that the line element ds satisfies

$$ds^2 = R^2 (d\varphi^2 + \sin^2 \varphi d\theta^2),$$

which means in other words

$$ds_{\text{Earth}} = R \sqrt{(\varphi')^2 + \sin^2 \varphi (\theta')^2} \, dt.$$

The surface area element is

$$dS_{\text{Earth}} = \sqrt{\det g}\, d\varphi\, d\theta = R^2 \sin\varphi\, d\varphi\, d\theta.$$

For the purposes of calculations, some authors set $R = 1$ so that the radius of the Earth becomes the unit length in length and area calculations. By keeping R explicit, we can give it any necessary value in any given system of units.

Now the map function M in (6.9) is a function $M : [0, \pi] \times [-\pi, \pi] \to \mathbb{R}^2$. This can be understood as an alternate parametrization of a portion of the plane using the coordinates φ and θ. It has a first fundamental form defined by

$$g_{\text{Map}} = \begin{pmatrix} M_\varphi \cdot M_\varphi & M_\varphi \cdot M_\theta \\ M_\theta \cdot M_\varphi & M_\theta \cdot M_\theta \end{pmatrix},$$

where the dot products require us to view $M(\varphi, \theta)$ and its derivatives as a vector. Note that at each point p on a surface, the first fundamental form is an inner product $I_p(\ ,\ )$ on the tangent plane T_pS to the surface at p. However, in the case of a map, the surface is a subset of $\mathbb{R}^2$ and the tangent plane is $\mathbb{R}^2$ itself.

Definition 6.2.1 A *conformal map* is a map in which the angles of intersection of paths on the map are the same as the angles of intersection of paths on the Earth.

Conformal maps are particularly useful for navigation since they can tell a pilot that at least he or she is heading in a desired direction.

Proposition 6.2.2 *A map is conformal if there is a positive function* $w(\varphi, \theta)$ *such that*

$$g_{Map} = w(\varphi, \theta) g_{Earth}.$$

Proof: Let $\vec{a}$ and $\vec{b}$ be vectors in the tangent plane to the Earth at a point $p = (\varphi, \theta)$, with components given with respect to the $(\vec{X}_\varphi, \vec{X}_\theta)$ ordered basis of the tangent plane. The angle α between them has a cosine of

$$\cos\alpha = \frac{I_p(\vec{a}, \vec{b})}{\sqrt{I_p(\vec{a}, \vec{a})}\sqrt{I_p(\vec{b}, \vec{b})}}.$$

If there is a positive function $w(\varphi, \theta)$ as described in the proposition, then the first fundamental form of the map is

$$I_p^{\text{Map}}(\ ,\ ) = w(p) I_p(\ ,\ ).$$

Then cosine of the angle between directions in the map is

$$\frac{I_p^{\mathrm{Map}}(\vec{a},\vec{b})}{\sqrt{I_p^{\mathrm{Map}}(\vec{a},\vec{a})}\sqrt{I_p^{\mathrm{Map}}(\vec{b},\vec{b})}} = \frac{w(p)I_p(\vec{a},\vec{b})}{\sqrt{w(p)I_p(\vec{a},\vec{a})}\sqrt{w(p)I_p(\vec{b},\vec{b})}}$$

$$= \frac{I_p(\vec{a},\vec{b})}{\sqrt{I_p(\vec{a},\vec{a})}\sqrt{I_p(\vec{b},\vec{b})}} = \cos\alpha.$$

The proposition follows.   □

Maps that *preserve area* are maps such that there exists a constant D so that the area of any region measured in the map is a factor D times that surface area of the corresponding region on the Earth. Hence, for all regions $\mathcal{R}$ in the $\varphi\theta$-plane, we have

$$\iint_{\mathcal{R}} dS_{\mathrm{Map}} = D \iint_{\mathcal{R}} dS_{\mathrm{Earth}}$$

$$\Longleftrightarrow \iint_{\mathcal{R}} \sqrt{\det g_{\mathrm{Map}}}\, d\varphi\, d\theta = D \iint_{\mathcal{R}} \sqrt{\det g_{\mathrm{Earth}}}\, d\varphi\, d\theta.$$

The Mean Value Theorem for integrals gives the following.

Proposition 6.2.3 *A map of the Earth is area-preserving if and only if there is a positive constant D such that*

$$\det(g_{Map}) = D^2 \det(g_{Earth}).$$

The arc length element in a map is $ds = \sqrt{(x'(t))^2 + (y'(t))^2}\, dt$, which, in particular, does not depend on the position in the plane but only on the components of the velocity vector. The explicit dependence of ds_{Earth} on φ shows that no map can calculate arc lengths of all paths correctly. In other words, there is no map with a given proportionality factor such that the arc length of the path in the map is that factor times the arc length of the corresponding path on the Earth. However, it is possible that certain sets of paths in a given map have the same relative arc lengths as on the Earth.

6.2.2 Azimuthal Projections

A first category of map projection is the azimuthal projections. One point P on the Earth is chosen as the center of the map and serves as the origin in the xy-coordinate system. If P is the North Pole, then the longitude lines through P appear on the map as equally spaced rays radiating from the origin, and the latitude circles appear on the map as circles centered on the map with radii determined by the particular projection. (If P is the South Pole, longitude and latitude lines have a similar representation on the map. However, if P is not one of the poles, then the radial lines in the map and circles at given radii around the origin have more complicated interpretations in latitude and longitude on the Earth.)

Choosing P as the North Pole, any azimuthal projection maps the (φ, θ) coordinates to polar coordinates $(r, \theta) = (r(\varphi), \theta)$ on the map. Hence, the map function has the form

$$(x, y) = M(\varphi, \theta) = (r(\varphi) \cos \theta, r(\varphi) \sin \theta),$$

where r is a nonnegative increasing function with $r(0) = 0$. The first fundamental form of any azimuthal projection is

$$g_{\text{Map}} = \begin{pmatrix} r'(\varphi)^2 & 0 \\ 0 & r(\varphi)^2 \end{pmatrix}.$$

Different choices of the function $r(\varphi)$ correspond to different azimuthal projections. Even if we always pick the center of the map to be the North Pole, there are a variety of classical choices. Some give a map for the whole Earth (except for the South Pole), while others only map the northern hemisphere.

Example 6.2.4 (Orthographic Projection) One of the geometrically simplest azimuthal projections is the orthographic projection. It maps only the northern hemisphere. We imagine the map as the tangent plane T to the North Pole. A point Q in the northern hemisphere of the Earth is mapped to the point Q' in the tangent via orthogonal projection onto the tangent plane T. The resulting map fills a circular disc in the plane. This disc is then scaled uniformly by a factor c to any reasonable radius for a map. (If the disc were not scaled, it would have a radius equal to the radius of the Earth, which would make for a very large map!)

In this case, the function $r(\varphi)$ is simply $r(\varphi) = cR \sin \varphi$ for $0 \le \varphi \le \pi/2$.

Example 6.2.5 (Gnomic Projection) Another simple azimuthal projection is the gnomic projection. It also maps only the northern hemisphere and this time excludes the equator. We imagine the map as the tangent plane T to the North Pole. Let O be the center of the Earth. A point Q in the northern hemisphere of the Earth is mapped to the point Q' in the tangent plane where Q' is the intersection of the ray $\overrightarrow{OQ}$ with the tangent plane T. The resulting map fills the whole plane. As usual, the tangent plane is then scaled uniformly by a reasonable factor c. In practice, one only depicts a certain bounded portion of the map.

In this case, the function $r(\varphi)$ is simply $r(\varphi) = cR\tan\varphi$ for $0 \leq \varphi < \pi/2$.

One interesting property about the gnomic projection is that every arc of a great circle on the Earth is mapped to a straight line segment. Indeed, a great circle on a sphere corresponds to the intersection of the sphere and a plane $\mathcal{P}$ through the origin. Via the gnomic projection, the arc of such a great circle maps to the intersection of $\mathcal{P} \cap T$, which is a line.

Example 6.2.6 (Stereographic Projection) Problem 5.3.10 introduced stereographic projection as an alternate parametrization of the sphere. However, as presented in that problem, it defines an azimuthal projection with the South Pole as the center. In this example, we make a few adjustments to the presentation in Problem 5.3.10 to conform to the presentation here.

Like the gnomic projection, the stereographic projection assumes the map is the tangent plane T to the North Pole. (This a variation from the presentation in Problem 5.3.10, where the map was assumed to be the plane through the equator.) A point Q on the Earth (except for the South Pole) is mapped to the point Q', which is the intersection of T and the ray out of the South Pole through Q. As usual, the tangent plane is then scaled uniformly by a reasonable factor c. The stereographic projection maps the whole Earth except for the South Pole and covers the whole plane.

We leave it as an exercise to show that in this case, the function $r(\varphi)$ is

$$r(\varphi) = \frac{2cR\sin\varphi}{1 + \cos\varphi} \qquad \text{for } 0 \leq \varphi < \pi.$$

Example 6.2.7 (Area-Preserving Azimuthal) Suppose that we want to devise an area-preserving azimuthal projection. Then by Proposition 6.2.3, the radial function $r(\varphi)$ must be such that

$$\det(g_{\text{Map}}) = r(\varphi)^2(r'(\varphi))^2 = D^2 R^4 \sin^2\varphi$$

for some positive constant D. Taking a square root of both sides, we get $r(\varphi)r'(\varphi) = DR^2\sin\varphi$. Integrating both sides with respect to φ and using substitution, we find that

$$-DR^2\cos\varphi = \int r(\varphi)r'(\varphi)\,d\varphi = \int r\,dr = \frac{1}{2}r^2 + C,$$

for some constant of integration. Since we want $r(0)$ with $\varphi = 0$, we have $C = -DR^2$ and thus

$$r(\varphi) = R\sqrt{2D(1 - \cos\varphi)} = 2\sqrt{D}R\sin\frac{\varphi}{2}$$

Figure 6.4: Cylindrical projection.

for $0 \leq \varphi < \pi$. Hence, there exists an area-preserving azimuthal projection that covers the whole Earth except for the South Pole and its map is a disc of radius $2\sqrt{D}R$.

6.2.3 Cylindrical Projections

A cylindrical projection map is one based off of the intuition of taking a globe, wrapping a sheet of paper as a cylindrical tube around the equator of the globe, projecting the surface of the globe onto the sheet of paper, and then unwrapping the cylinder of paper to lay it flat. See Figure 6.4 for a visual.

More precisely, one point P on the Earth is chosen as a pole. If P is the North Pole, then longitude lines through P appear on the map as equally spaced vertical parallel lines and latitude circles appear on the map as horizontal lines, with spacing determined by the particular projection. The choice of point P corresponds to a choice of an axis for the cylinder because we assume the cylinder has axis $\overleftrightarrow{OP}$, where O is the center of the sphere.

We always scale the map horizontally so that a unit in the x-direction corresponds to a unit in θ. With P as the North Pole, then (φ, θ) on the Earth maps to $(x, y) = M(\varphi, \theta) = (\theta, h(\varphi))$, where h is a decreasing function with $h(\pi/2) = 0$. That h is decreasing corresponds to setting the North Pole $\varphi = 0$ as up on the map. The first fundamental form of such maps is

$$g_{\text{Map}} = \begin{pmatrix} (h'(\varphi))^2 & 0 \\ 0 & 1 \end{pmatrix}.$$

Different choices of the function $h(\varphi)$ correspond to different cylindrical projections. Also $h(\varphi)$ may be scaled linearly simply to adjust the vertical size of the map. As with the azimuthal projections, even setting P as the North Pole, there are a number of classical choices.

Example 6.2.8 (Radial Cylindrical) Let O represent the center of the Earth. The radial cylindrical projection is a geometrically simple projection in which each point Q on the Earth, except for the North Pole and the South Pole, is mapped to the point Q' on the cylinder that is the intersection with the ray $\overrightarrow{OQ}$. The cylinder is then unrolled and, as usual, scaled by a reasonable factor c. The map then consists of the strip $[0, 2\pi] \times \mathbb{R}$.

In this projection, we have $h(\varphi) = c \cot \varphi$, where c is a constant.

Example 6.2.9 (Mercator) The well-known Mercator projection is a cylindrical projection that is conformal. By Proposition 6.2.2, the Mercator projection must be such that there is a function $w(\varphi, \theta)$ such that

$$\begin{pmatrix} (h'(\varphi))^2 & 0 \\ 0 & 1 \end{pmatrix} = w(\varphi, \theta) \begin{pmatrix} R^2 & 0 \\ 0 & R^2 \sin^2 \varphi \end{pmatrix}.$$

Obviously, we must have $w(\varphi, \theta) = 1/(R^2 \sin^2 \varphi)$ and then we also need

$$h'(\varphi) = -\frac{1}{\sin \varphi} \qquad \text{with} \qquad h\left(\frac{\pi}{2}\right) = 0.$$

The choice of the negative sign for $h'(\varphi)$ comes from the requirement that $h(\varphi)$ is decreasing. Integrating, we find that

$$h(\varphi) = -\int_{\pi/2}^{\varphi} \frac{du}{\sin u} = -\ln(\csc \varphi + \cot \varphi),$$

for $0 < \varphi \leq \pi/2$. Using trigonometric identities, we can rewrite this function as $h(\varphi) = \ln(\tan(\varphi/2))$. Hence, the map for the Mercator projection is $M(\varphi, \theta) = \left(\theta, \ln\left(\tan \frac{\pi}{2}\right)\right)$.

Many other classes of map projections exist. For example, pseudo-cylindrical projections attempt to fix the distortions that invariably occur near the poles in cylindrical projections. Indeed, on a sphere, latitude circles decrease in radius the closer one gets to the poles. However, under any cylindrical projection, every latitude circle becomes a horizontal line segment of the same length. A pseudocylindrical projection has for its map

$$(x, y) = M(\varphi, \theta) = \left(\frac{w(\varphi)\theta}{2\pi}, h(\varphi)\right)$$

where

- as for cylindrical projections, $h(\varphi)$ is a decreasing function with $h(\pi/2) = 0$;

- $w(\varphi)$ is a nonnegative function that gives the width of the map at a given latitude φ.

For many maps, the function $w(\varphi)$ satisfies $w(0) = w(\pi) = 0$ and $w(\pi/2) = 1$, so the width is 0 at the North and South Pole and 1 at the equator.

6.2.4   Coordinate Changes on the Sphere

With various maps for the Earth (sphere) at our disposal, we take the opportunity to illustrate the coordinate change transformation property of the metric tensor as given in Equation (6.7). To locate points on the sphere (Earth), we will consider the standard latitude and longitude (φ, θ) coordinates and the (x, y) coordinates as given by the stereographic projection described in Example 6.2.6.

For simplicity, let us use $R = c = 1$. Then the stereographic map is

$$(x, y) = M(\varphi, \theta) = \left(\frac{2 \sin \varphi \cos \theta}{1 + \cos \varphi}, \frac{2 \sin \varphi \sin \theta}{1 + \cos \varphi} \right).$$

Using similar geometry as required by Problem 5.3.10, we can show that the (x, y) coordinates parametrize the unit sphere by

$$\vec{Y}(x, y) = \left(\frac{4x}{4 + x^2 + y^2}, \frac{4y}{4 + x^2 + y^2}, \frac{4 - x^2 - y^2}{4 + x^2 + y^2} \right).$$

In order to verify Equation (6.7), let us call the (x_1, x_2) coordinates the (x, y) coordinates and we call the $(\bar{x}_1, \bar{x}_2)$ coordinates the (φ, θ) coordinates. In these labels, we already know that

$$(\bar{g}_{ij}) = \begin{pmatrix} 1 & 0 \\ 0 & \sin^2 \varphi \end{pmatrix}.$$

Using the parametrization $\vec{Y}(x, y)$, we find that

$$(g_{ij}) = \begin{pmatrix} \vec{Y}_x \cdot \vec{Y}_x & \vec{Y}_x \cdot \vec{Y}_y \\ \vec{Y}_y \cdot \vec{Y}_x & \vec{Y}_y \cdot \vec{Y}_y \end{pmatrix} = \begin{pmatrix} \dfrac{16}{(4 + x^2 + y^2)^2} & 0 \\ 0 & \dfrac{16}{(4 + x^2 + y^2)^2} \end{pmatrix}.$$

(6.10)

Furthermore, the differential $[dM_{(\varphi,\theta)}]$, also known as the Jacobian matrix of the coordinate change for (φ, θ) coordinates to (x, y) coordinates, is

$$[dM_{(\varphi,\theta)}] = \begin{pmatrix} \dfrac{\partial x}{\partial \varphi} & \dfrac{\partial x}{\partial \theta} \\ \dfrac{\partial y}{\partial \varphi} & \dfrac{\partial y}{\partial \theta} \end{pmatrix} = \begin{pmatrix} \dfrac{2\cos\theta}{1+\cos\varphi} & -\dfrac{2\sin\varphi\sin\theta}{1+\cos\varphi} \\ \dfrac{2\sin\theta}{1+\cos\varphi} & \dfrac{2\sin\varphi\cos\theta}{1+\cos\varphi} \end{pmatrix}.$$

Recall that Equation (6.7) states that

$$\bar{g}_{ij} = \sum_{k=1}^{2}\sum_{l=1}^{2} \frac{\partial x_k}{\partial \bar{x}_i}\frac{\partial x_l}{\partial \bar{x}_j} g_{kl}.$$

This can be rewritten more explicitly as

$$\bar{g}_{ij} = \frac{\partial x_1}{\partial \bar{x}_i}\frac{\partial x_1}{\partial \bar{x}_j}g_{11} + \frac{\partial x_1}{\partial \bar{x}_i}\frac{\partial x_2}{\partial \bar{x}_j}g_{12} + \frac{\partial x_2}{\partial \bar{x}_i}\frac{\partial x_1}{\partial \bar{x}_j}g_{21} + \frac{\partial x_2}{\partial \bar{x}_i}\frac{\partial x_2}{\partial \bar{x}_j}g_{22}.$$

We consider our specific example. However, note that as a simplification, we have $g_{12} = g_{21} = 0$. Also, we point out ahead of time that

$$x^2 + y^2 + 4 = \left(\frac{2\sin\varphi\cos\theta}{1+\cos\varphi}\right)^2 + \left(\frac{2\sin\varphi\sin\theta}{1+\cos\varphi}\right)^2 + 4 = \frac{8}{1+\cos\varphi}.$$

Thus

$$\frac{16}{(x^2+y^2+4)^2} = \frac{(1+\cos\varphi)^2}{4}.$$

So we get the following confirmation:

$$\begin{aligned}
\bar{g}_{11} &= \frac{\partial x_1}{\partial \bar{x}_1}\frac{\partial x_1}{\partial \bar{x}_1}g_{11} + \frac{\partial x_2}{\partial \bar{x}_1}\frac{\partial x_2}{\partial \bar{x}_1}g_{22} \\
&= \left(\frac{2\cos\theta}{1+\cos\varphi}\right)^2 \frac{16}{(4+x^2+y^2)^2} + \left(\frac{2\sin\theta}{1+\cos\varphi}\right)^2 \frac{16}{(4+x^2+y^2)^2} \\
&= \frac{4\cos^2\theta}{(1+\cos\varphi)^2}\frac{(1+\cos\varphi)^2}{4} + \frac{4\sin^2\theta}{(1+\cos\varphi)^2}\frac{(1+\cos\varphi)^2}{4} \\
&= 1;
\end{aligned}$$

$$\begin{aligned}
\bar{g}_{12} &= \frac{\partial x_1}{\partial \bar{x}_1}\frac{\partial x_1}{\partial \bar{x}_2}g_{11} + \frac{\partial x_2}{\partial \bar{x}_1}\frac{\partial x_2}{\partial \bar{x}_2}g_{22} \\
&= \left(\frac{2\cos\theta}{1+\cos\varphi}\right)\left(-\frac{2\sin\varphi\sin\theta}{1+\cos\varphi}\right)\frac{16}{(4+x^2+y^2)^2} \\
&\quad + \left(\frac{2\sin\theta}{1+\cos\varphi}\right)\left(\frac{2\sin\varphi\cos\theta}{1+\cos\varphi}\right)\frac{16}{(4+x^2+y^2)^2} = 0;
\end{aligned}$$

we can verify directly that $\bar{g}_{21} = \bar{g}_{12} = 0$; and finally

$$
\begin{aligned}
\bar{g}_{22} &= \frac{\partial x_1}{\partial \bar{x}_2}\frac{\partial x_1}{\partial \bar{x}_2}g_{11} + \frac{\partial x_2}{\partial \bar{x}_2}\frac{\partial x_2}{\partial \bar{x}_2}g_{22} \\
&= \frac{16}{(4 + x^2 + y^2)^2}\left(\left(-\frac{2\sin\varphi\sin\theta}{1 + \cos\varphi}\right)^2 + \left(\frac{2\sin\varphi\cos\theta}{1 + \cos\varphi}\right)^2\right) \\
&= \frac{4\sin^2\varphi}{(1 + \cos\varphi)^2}\frac{(1 + \cos\varphi)^2}{4} \\
&= \sin^2\varphi.
\end{aligned}
$$

These calculations illustrate the coordinate change transformation described by Equation (6.7)

PROBLEMS

1. Find the possible maps for cylindrical projections that are area-preserving.

2. Verify the calculations for the metric tensor given in Equation (6.10).

3. Show that the metric tensor of the map for the general pseudocylindrical projection is

$$
g_{\text{Map}} = \frac{1}{4\pi^2}\begin{pmatrix} 4\pi^2 h'(\varphi) + \theta^2 w'(\varphi) & \theta w(\varphi)w'(\varphi) \\ \theta w(\varphi)w'(\varphi) & w(\varphi)^2 \end{pmatrix}.
$$

4. The Sinusoidal Projection is a pseudocylindrical map projection that uses $w(\varphi) = \sin\varphi$ and $h(\varphi) = c\left(\frac{\pi}{2} - \varphi\right)$, where c is a positive constant.

   (a) Prove that the lengths of the latitude lines on the map are a fixed multiple of the lengths of the latitude lines on the Earth.

   (b) Prove that this map also preserves areas.

   (c) Decide if the map is conformal.

   (d) Describe the shape of the image of the map.

6.3 The Gauss Map

In Section 2.2, given any closed, simple, regular curve $\gamma : I \to \mathbb{R}^2$ in the plane, we defined the tangential indicatrix to be the curve given by $\vec{T} : I \to \mathbb{R}^2$, the unit tangent vector of $\gamma(t)$. The image of $\vec{T}$ lies on the unit circle, and though $\vec{T}$ is a closed curve, it need not be regular as its locus may stop and double back. Regardless of the parametrization, the tangential indicatrix is well defined up to a change in sign.

On any regular surface S in $\mathbb{R}^3$, the tangent plane T_pS at a point $p \in S$ is a two-dimensional subspace of $\mathbb{R}^3$, and hence the vectors that are normal to S at p form a one-dimensional subspace. Thus, there exist exactly two possible choices for a unit normal vector. By definition, if the surface is orientable, we can specify its orientation by a continuous function $n : S \to \mathbb{S}^2$, where $\mathbb{S}^2$ is the unit sphere in $\mathbb{R}^3$.

Definition 6.3.1 Let S be an oriented regular surface in $\mathbb{R}^3$ with an orientation $n : S \to \mathbb{S}^2$. In the classical theory of surfaces, the function n is also called the *Gauss map*.

It is important to remain aware of the distinction between the functions n and $\vec{N}$. Let S be an oriented surface with orientation $n : S \to \mathbb{S}^2$. Suppose that $\vec{X} : U \to \mathbb{R}^3$, where U is an open subset of $\mathbb{R}^2$, parametrizes a neighborhood of S, then the function $\vec{N} : U \to \mathbb{S}^2$ is defined in reference to $\vec{X}$ by Equation (5.2). The functions n and $\vec{N}$ are related by the fact that if $\vec{X}$ is a positively oriented parametrization, then

$$\vec{N} = n \circ \vec{X}.$$

Example 6.3.2 Consider as an example a sphere S in $\mathbb{R}^3$ equipped with the orientation n, with the unit vectors normal to S pointing away from the center $\vec{c}$ of the sphere. The sphere can be given as the solution to the equation

$$\|\vec{x} - \vec{c}\|^2 = R^2. \tag{6.11}$$

Recall that $\|\vec{v}\|^2 = \vec{v} \cdot \vec{v}$. If $\vec{x}(t)$ is any curve on the sphere, then by differentiating the relationship in Equation (6.11), one obtains

$$\vec{x}\,'(t) \cdot (\vec{x}(t) - \vec{c}) = 0.$$

The tangent plane to the sphere S at a point $\vec{p}$ consists of all possible vectors $\vec{x}'(t_0)$, where $\vec{x}(t)$ is a curve on S, with $\vec{x}(t_0) = \vec{p}$. Therefore, at any point $\vec{p}$ on the sphere, there are two options for unit normal vectors:

$$\frac{\vec{p} - \vec{c}}{\|\vec{p} - \vec{c}\|} \quad \text{or} \quad -\frac{\vec{p} - \vec{c}}{\|\vec{p} - \vec{c}\|}.$$

However, it is the former that provides the outward-pointing orientation n. Thus the Gauss map for the sphere of radius R and center $\vec{c}$ is explicitly

$$n(\vec{p}) = \frac{\vec{p} - \vec{c}}{R}.$$

Furthermore, if S is itself the unit sphere centered at the origin, then the Gauss map is the identity function.

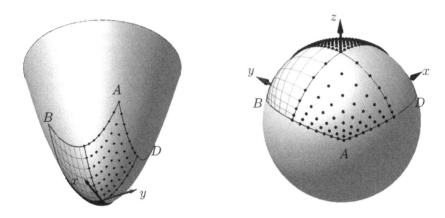

Figure 6.5: Gauss map of the elliptic paraboloid.

Example 6.3.3 Consider the elliptic paraboloid S defined by the equation $z = x^2 + y^2$. This is an orientable surface, and suppose it is oriented with the unit normal always pointing in a positive z-direction. (Figure 6.5 shows this is possible.) Using the function $F(x, y, z) = z - x^2 - y^2$, the elliptic paraboloid is given by the equation $F(x, y, z) = 0$. By Proposition 5.3.7, for all $(x, y, z) \in S$, the gradient $\vec{\nabla}F(x, y, z)$ is a normal vector, so (since the z-direction is positive) the Gauss map is

$$n(x, y, z) = \frac{(-2x, -2y, 1)}{\sqrt{4x^2 + 4y^2 + 1}}.$$

Figure 6.5 shows how the Gauss map acts on the patch

$$\{(x, y, z) \in S \mid -1 \le x \le 1 \text{ and } -1 \le y \le 1\}.$$

It is not hard to show that the image of the Gauss map on the unit sphere is the upper hemisphere. Indeed, using cylindrical coordinates, the Gauss map is

$$n(r, \theta, z) = \frac{(-2r \cos \theta, -2r \sin \theta, 1)}{\sqrt{4r^2 + 1}}.$$

The function $f(r) = 1/\sqrt{4r^2 + 1}$ is a bijection from $[0, +\infty)$ to $(0, 1]$ and that it is decreasing. Thus, for any $z_0 \in (0, 1]$, there exists a unique $r_0 \in [0, +\infty)$ with $f(r_0) = z_0$. Then, with $\theta \in [0, 2\pi]$, the image of $n(r_0, \theta, z_0)$ is a circle on the unit sphere at height z_0. Thus, the image of n is

$$n(S) = \{(x, y, z) \in \mathbb{S}^2 \mid z > 0\}.$$

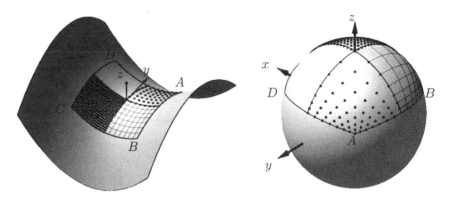

Figure 6.6: Gauss map on the hyperbolic paraboloid.

Example 6.3.4 In contrast to Example 6.3.3, consider the hyperbolic paraboloid given by the parametrization $\vec{X} : \mathbb{R}^2 \to \mathbb{R}^3$, with

$$\vec{X}(u,v) = (u, v, u^2 - v^2).$$

We calculate

$$\vec{X}_u = (1, 0, 2u) \qquad \text{and} \qquad \vec{X}_v = (0, 1, -2v),$$

and thus,

$$\vec{N} = \frac{\vec{X}_u \times \vec{X}_v}{\|\vec{X}_u \times \vec{X}_v\|} = \frac{(-2u, 2v, 1)}{\sqrt{4u^2 + 4v^2 + 1}}.$$

Like Figure 6.5, Figure 6.6 shows the Gauss map for a "square" on the hyperbolic paraboloid. These two examples manifest a central behavior of the Gauss map. On the elliptic paraboloid, the Gauss map preserves the orientation of the square in the sense that if one travels along the boundary in a clockwise sense on the surface, one also travels along the boundary of the image of the square mapped by n in a clockwise sense. On the other hand, with the hyperbolic paraboloid, the Gauss map reverses the orientation of the square.

Proposition 6.3.5 *Let S be a regular surface in $\mathbb{R}^3$, and let $\vec{X} : U \to \mathbb{R}^3$ be a parametrization of a coordinate patch $V = \vec{X}(U)$, where U is an open set in $\mathbb{R}^2$. Define the vector function $\vec{N} : U \to \mathbb{R}^3$ by*

$$\vec{N} = \frac{\vec{X}_u \times \vec{X}_v}{\|\vec{X}_u \times \vec{X}_v\|}.$$

If $\vec{X}$ is of class C^r, then $\vec{N}$ is of class C^{r-1}.

Proof: By Problem 3.1.14, we deduce that if $\vec{F} : U \to \mathbb{R}^3$, $\vec{G} : U \to \mathbb{R}^3$, and $h : U \to \mathbb{R}$ are of class C^1, then the three functions $\vec{F} \cdot \vec{G} : U \to \mathbb{R}$, $\vec{F} \times \vec{G} : U \to \mathbb{R}^3$, and $h\vec{F} : U \to \mathbb{R}^3$ are also of class C^1, and their partial derivatives follow appropriate product rules. By definition,

$$\vec{N} = \frac{\vec{X}_u \times \vec{X}_v}{\sqrt{(\vec{X}_u \times \vec{X}_v) \cdot (\vec{X}_u \times \vec{X}_v)}}.$$

One can see that any partial derivative of any order of $\vec{N}$ will involve the partial derivative of a combination of the above three types of products and the derivative of an expression of the form

$$f(u, v) = \left((\vec{X}_u \times \vec{X}_v) \cdot (\vec{X}_u \times \vec{X}_v)\right)^{-k/2},$$

where k is an odd positive integer. Since S is regular, $\|\vec{X}_u \times \vec{X}_v\|$ is never 0, so a particular higher derivative of $\vec{N}$ will exist if that higher derivative exists for $\vec{X}_u$ and for $\vec{X}_v$. The proposition follows. $\square$

One would be correct to think that the different behaviors in Figures 6.5 and 6.6 can be illustrated and quantified by a function between tangent spaces to S and $\mathbb{S}^2$, respectively. This perspective is encapsulated in the notion of the differential of functions between regular surfaces. We develop the more general theory of differentials of functions between manifolds in Chapter 11 of [24]. However, the differential of the Gauss map, though only a particular case of the differentials of functions between surfaces, serves a central role in the rest of this chapter.

Let S be a regular oriented surface with orientation n, and let $\vec{X} : U \to \mathbb{R}^3$ be a regular positively oriented parametrization of a coordinate patch $\vec{X}(U)$ of S. Suppose also that S is of class C^2, which, according to Proposition 6.3.5, implies that $\vec{N}$ is of class C^1. Since $\vec{N} \cdot \vec{N} = 1$ for all $(u, v) \in U$, we have

$$\vec{N} \cdot \vec{N}_u = 0 \qquad \text{and} \qquad \vec{N} \cdot \vec{N}_v = 0 \qquad (6.12)$$

for all $(u, v) \in U$. Therefore, both derivatives $\vec{N}_u$ and $\vec{N}_v$ are vector functions such that $\vec{N}_u(q)$ and $\vec{N}_v(q)$ are in T_pS when $\vec{X}(q) = p$.

The vector function $\vec{N} : U \to \mathbb{R}^3$ is itself a parametrized surface, with its image lying in the unit sphere, though not necessarily giving a regular parametrization of $\mathbb{S}^2$. The simple fact in Equation (6.12) indicates that if $\vec{N}$ admits a tangent plane at $q \in U$, then

$$T_pS = T_{\vec{N}(q)}(\mathbb{S}^2) \qquad (6.13)$$

as subspaces of $\mathbb{R}^3$. In other words, the sets of vectors $\{\vec{X}_u, \vec{X}_v\}$ and $\{\vec{N}_u, \vec{N}_v\}$ span the same subspace T_pS.

The differential of the Gauss map at a point $p \in S$ is a linear transformation $dn_p : T_pS \rightarrow T_{n(p)}(\mathbb{S}^2)$, but by virtue of Equation 6.13, one identifies it as a linear transformation $dn_p : T_pS \rightarrow T_pS$. For all $\vec{v} \in T_pS$, one defines $dn_p(\vec{v}) = \vec{w}$ if there exists a curve $\vec{\alpha} : I \rightarrow U$ such that

$$\begin{cases} \vec{\alpha}(0) = q, \text{ with } \vec{X}(q) = p, \\ \dfrac{d}{dt}\left(\vec{X} \circ \vec{\alpha}\right)(t)\big|_{t=0} = \vec{v}, \text{ and} \\ \dfrac{d}{dt}\left(\vec{N} \circ \vec{\alpha}\right)(t)\big|_{t=0} = \vec{w}. \end{cases} \quad (6.14)$$

In other words, by writing $\vec{X}(t) = \vec{X} \circ \vec{\alpha}(t)$ and $\vec{N}(t) = \vec{N} \circ \vec{\alpha}(t)$, the differential of the Gauss map satisfies $dn_p(\vec{X}'(t)) = \vec{N}'(t)$.

Using coordinate lines $\vec{X}(t, v_0)$ or $\vec{X}(u_0, t)$ for the curve $\vec{\alpha}(t)$ in the above definition, it is easy to see that

$$\begin{aligned} dn_p(\vec{X}_u) &= \vec{N}_u, \\ dn_p(\vec{X}_v) &= \vec{N}_v. \end{aligned} \quad (6.15)$$

Though the Gauss map $n : S \rightarrow \mathbb{S}^2$ is independent of any co-ordinate systems on S, the differential dn_p requires reference to a regular positively oriented parametrization of a neighborhood of p. However, though different parametrizations of a neighborhood of p induce different coordinate bases on T_pS, the definition in Equation 6.14 remains unchanged and consequently, as a linear transformation of T_pS to itself, is independent of the parametrization.

PROBLEMS

1. Describe the region of the unit sphere covered by the image of the Gauss map for the following surfaces:
   (a) The hyperbolic paraboloid given by $z = x^2 - y^2$.
   (b) The hyperboloid of one sheet given by $x^2 + y^2 - z^2 = 1$.

2. Describe the region of the unit sphere covered by the image of the Gauss map for the following surfaces:
   (a) The cone with opening angle α given by $z^2 \tan^2 \alpha = x^2 + y^2$.
   (b) The right circular cylinder $x^2 + y^2 = R^2$, with R a constant.

3. What regular surface S has a single point on $\mathbb{S}^2$ as the image of the Gauss map?

4. Determine whether there is a connected regular surface whose Gauss map consists of n distinct points.

5. Show that the graph of a differentiable function $z = f(x, y)$ has a Gauss map that lies inside either the upper hemisphere or the lower hemisphere.

6. Prove that the image of the Gauss map of a regular surface S is an arc of a great circle if and only if S is a generalized cylinder. (See Example 6.7.2.)

7. Consider the surface S obtained as a one-sheeted cone over a regular plane curve C (see Problem 5.3.14). Prove that the image of the Gauss map for S is a curve on the unit sphere. Find a parametrization for the image of the Gauss map in terms of a parametrization of C. [Hint: Without loss of generality, suppose that C lies in the plane $z = a$.]

8. Consider the torus S as described in Exercise 5.2.13.

   (a) Prove that the Gauss map $n : S \to \mathbb{S}^2$ is a double cover, that is to say that for each point $q \in \mathbb{S}^2$, there exist exactly 2 points $p \in S$ such that $n(p) = q$.

   (b) Using the parametrization $\vec{X}$ from Exercise 5.2.13, let $p = \vec{X}(0, \pi/4)$. Let $\vec{a}$ be the vector in T_pS with components in $\binom{1}{1}$ with respect to the standard basis on T_pS (i.e., determined by $\vec{X}_u$ and $\vec{X}_v$). Calculate directly $dn_p(\vec{a})$. [Hint: Use (6.14) and use a suitable choice of curve $\vec{\alpha} : I \to \mathbb{R}^2$ in the domain.]

9. Consider the unit hyperboloid of one sheet S, satisfying $x^2 + y^2 - z^2 = 1$. Recall that we can parametrize this surface by

$$\vec{X}(u, v) = (\cosh v \cos u, \cosh v \sin u, \sinh v)$$

with $(u, v) \in [0, 2\pi) \times \mathbb{R}$. Let $p = (5/4, 0, 3/4) \in S$ and let $\vec{a}$ be the vector in $\mathbb{R}^3$ with coordinates $(3, 4, 5)$.

   (a) Find the vectors $\vec{X}_u$ and $\vec{X}_v$ at p; show that $\vec{a} \in T_pS$; and give the coordinates of $\vec{a}$ with respect to the basis $\{\vec{X}_u, \vec{X}_v\}$ of T_S.

   (b) Use (6.14) and use a suitable choice of curve $\vec{\alpha} : I \to \mathbb{R}^2$ in the domain of (u, v) to directly determine $dn_p(\vec{a})$.

6.4   The Second Fundamental Form

The examples in the previous section illustrate how the differential of the Gauss map dn_p qualifies how much the surface S is curving at the point p. Despite the abstract definition for dn_p, it is not difficult to calculate the matrix for $dn_p : T_pS \to T_pS$. However, in order for the differential dn_p to exist at all points $p \in S$, we will need to assume that S is a surface of class C^2.

Let S be a regular oriented surface of class C^2 with orientation n, and let $\vec{X} : U \to \mathbb{R}^3$ be a positively oriented parametrization of a neighborhood V of a point p on S. Since $\vec{N}$ satisfies

$$\vec{N} \cdot \vec{X}_i = 0 \qquad \text{for } i = 1, 2,$$

then by differentiating with respect to another variable, the product rule gives

$$\vec{N}_j \cdot \vec{X}_i + \vec{N} \cdot \vec{X}_{ij} = 0 \qquad \text{for } i, j = 1, 2.$$

Note that since $\vec{X}_{ij} = \vec{X}_{ji}$, by interchanging i and j, one deduces that

$$\vec{N}_i \cdot \vec{X}_j = -\vec{N} \cdot \vec{X}_{ij} = \vec{N}_j \cdot \vec{X}_i. \tag{6.16}$$

(Recall that the notation $\vec{f}_i$ in Equation (6.16) and in what follows refers to taking the derivative of the multivariable vector function $\vec{f}$ with respect to the ith variable.) This leads to the following proposition.

Proposition 6.4.1 *Using the above setup, the linear map dn_p is a self-adjoint operator with respect to the first fundamental form $I_p(\cdot, \cdot)$.*

Proof: Let $\vec{v}, \vec{w} \in T_p S$ and write $\vec{v} = v_1 \vec{X}_u + v_2 \vec{X}_v$ and $\vec{w} = w_1 \vec{X}_u + w_2 \vec{X}_v$. Recall that the first fundamental form $I_p(\vec{v}, \vec{w})$ is the dot product $\vec{v} \cdot \vec{w}$ when viewing $T_p S$ as a subset of $\mathbb{R}^3$. Also recall from Equation (6.15) that $dn_p(\vec{X}_u) = \vec{N}_u$ and similarly for v. Hence,

$$
\begin{aligned}
I_p(dn_p(\vec{v}), \vec{w}) &= I_p(v_1 \vec{N}_u + v_2 \vec{N}_v, \vec{w}) \\
&= (v_1 \vec{N}_u + v_2 \vec{N}_v) \cdot (w_1 \vec{X}_u + w_2 \vec{X}_v) \\
&= \sum_{i,j=1}^{2} v_i w_j \vec{N}_i \cdot \vec{X}_j \\
&= \sum_{i,j=1}^{2} v_i w_j \vec{N}_j \cdot \vec{X}_i \qquad \text{(by Equation (6.16))} \\
&= (v_1 \vec{X}_u + v_2 \vec{X}_v) \cdot (w_1 \vec{N}_u + w_2 \vec{N}_v) \\
&= I_p(\vec{v}, dn_p(\vec{w})).
\end{aligned}
$$

$\square$

Definition 6.4.2 Let S be an oriented regular surface of class C^2 with orientation n, and let p be a point of S. We define the *second fundamental form* as the quadratic form on $T_p S$ defined by

$$II_p(\vec{v}) = -dn_p(\vec{v}) \cdot \vec{v} = -I_p(dn_p(\vec{v}), \vec{v}).$$

The first fundamental form allows one to measure lengths, angles, and area of regions on a parametrized surface. The second fundamental form provides a measure for how much the normal vector changes if one travels away from p in a particular direction $\vec{v}$, with $\vec{v} \in T_p S$.

Recall from linear algebra that every quadratic form Q on a vector space V of dimension n is of the form

$$Q(\vec{v}) = \vec{v}^t M v$$

for some $n \times n$ matrix. Define the functions $L_{ij} : U \to \mathbb{R}$ by

$$II_p(\vec{v}) = \vec{v}^T \begin{pmatrix} L_{11}(q) & L_{12}(q) \\ L_{21}(q) & L_{22}(q) \end{pmatrix} \vec{v},$$

where $p = \vec{X}(q)$, for all $\vec{v} \in T_pS$. Since

$$II_p(\vec{v}) = -dn_p(\vec{v}) \cdot \vec{v} = -\vec{v}^T dn_p(\vec{v}),$$

by writing $\vec{v} = \begin{pmatrix} a \\ b \end{pmatrix}$ in the coordinate basis $\{\vec{X}_u, \vec{X}_v\}$, one obtains

$$II_p \begin{pmatrix} a \\ b \end{pmatrix} = -(a\vec{N}_1 + b\vec{N}_2) \cdot (a\vec{X}_1 + b\vec{X}_2)$$

$$= -(a^2 \vec{N}_1 \cdot \vec{X}_1 + ab\vec{N}_1 \cdot \vec{X}_2 + ab\vec{N}_2 \cdot \vec{X}_1 + b^2 \vec{N}_2 \cdot \vec{X}_2).$$

Therefore,

$$L_{ij} = -\vec{N}_j \cdot \vec{X}_i \qquad \text{for all } 1 \le i, j \le 2, \tag{6.17}$$

and by Equation (6.16),

$$L_{ij} = \vec{N} \cdot \vec{X}_{ij}. \tag{6.18}$$

This provides a convenient way to calculate the L_{ij} functions and, hence, the second fundamental form. By Proposition 6.4.1, the matrix (L_{ij}) is a symmetric matrix, so $L_{12} = L_{21}$. In classical differential geometry texts, authors often refer to the coefficients of the second fundamental form using the letters e, f, and g, as follows:

$$e = L_{11}, \qquad f = L_{12} = L_{21}, \qquad g = L_{22}.$$

 Example 6.4.3 (Spheres) Let S be the sphere of radius R and consider the coordinate patch parametrized by

$$\vec{X}(u, v) = (R \cos u \sin v, R \sin u \sin v, R \cos v).$$

The unit normal vector associated to this parametrization is

$$\vec{N}(u, v) = (-\cos u \sin v, -\sin u \sin v, -\cos v).$$

(This unit normal vector is inward pointing. Recall that to get an outward pointing normal vector, we could simply interchange the role of u and v.) The second derivatives are

$$\vec{X}_{11} = (-R \cos u \sin v, -R \sin u \sin v, 0),$$

$$\vec{X}_{12} = \vec{X}_{21} = (-R \sin u \cos v, R \cos u \cos v, 0),$$

$$\vec{X}_{22} = (-R \cos u \sin v, -R \sin u \sin v, -R \cos v).$$

Thus, the matrix for the second fundamental form is

$$(L_{ij}) = \begin{pmatrix} R \sin^2 v & 0 \\ 0 & R \end{pmatrix}.$$

(a) Elliptic point.   (b) Hyperbolic point.   (c) Parabolic point.

Figure 6.7: Elliptic, hyperbolic, and parabolic points.

At a point $p \in S$ in a coordinate neighborhood parametrized positively by $\vec{X}(u,v)$, if $\vec{X}(q) = p$, the height function

$$h(s,t) = \frac{1}{2}II_p\begin{pmatrix} s \\ t \end{pmatrix} = \frac{1}{2}(L_{11}(q)s^2 + 2L_{12}(q)st + L_{22}(q)t^2)$$

is called the *osculating paraboloid*. This paraboloid provides the second-order approximation of the surface near p in reference to the normal vector $\vec{N}(q)$ in the following sense. Setting $q = (u_0, v_0)$, the second-order Taylor approximation of $\vec{X}$ is

$$\vec{X}(u,v) \approx \vec{X}(u_0, v_0) + \vec{X}_u(u_0, v_0)(u - u_0) + \vec{X}_v(u_0, v_0)(v - v_0)$$
$$+ \frac{1}{2}\vec{X}_{uu}(u_0, v_0)(u - u_0)^2 + \vec{X}_{uv}(u_0, v_0)(u - u_0)(v - v_0)$$
$$+ \frac{1}{2}\vec{X}_{vv}(u_0, v_0)(v - v_0)^2. \quad (6.19)$$

Since $\vec{N}$ is perpendicular to $\vec{X}_u$ and $\vec{X}_v$, setting $s = u - u_0$ and $t = v - v_0$,

$$h(s,t) \approx \left(\vec{X}(s + u_0, t + v_0) - \vec{X}(u_0, v_0)\right) \cdot \vec{N}(u_0, v_0),$$

with the approximation accurate up through the second order. Furthermore, from the second derivative test in multivariable calculus, we know that the point $(s,t) = (0,0)$ is a local extremum if $L_{11}L_{22} - L_{12}^2 > 0$ and is a saddle point if $L_{11}L_{22} - L_{12}^2 < 0$. This leads to the following definition.

Definition 6.4.4 Let S be a regular orientable surface of class C^2. A point p on S is called:
1) *elliptic* if $\det(L_{ij}) > 0$;
2) *hyperbolic* if $\det(L_{ij}) < 0$;

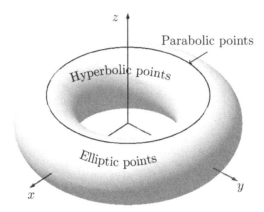

Figure 6.8: Torus.

3) *parabolic* if $\det(L_{ij}) = 0$ but not all $L_{ij} = 0$; and

4) *planar* if $L_{ij} = 0$ for all i, j.

It is not hard to check (Problem 6.4.8) that if (x_1, x_2) and $(\bar{x}_1, \bar{x}_2)$ are two coordinate systems for the same open set of an oriented surface S, then

$$\det(\bar{L}_{kl}) = \left(\frac{\partial(x_1, x_2)}{\partial(\bar{x}_1, \bar{x}_2)}\right)^2 \det(L_{ij}). \qquad (6.20)$$

Furthermore, the change of coordinates between (x_1, x_2) and $(\bar{x}_1, \bar{x}_2)$ is a diffeomorphism between two open sets in $\mathbb{R}^2$. Therefore, the Jacobian in Equation (6.20) is never 0 and Definition 6.4.4 is independent of any particular parametrization of S.

 Example 6.4.5 Consider the torus parametrized by

$$\vec{X}(u, v) = \big((2 + \cos v) \cos u, (2 + \cos v) \sin u, \sin v\big).$$

The unit normal vector and the second derivatives of the vector function $\vec{X}$ are

$$\vec{N} = (-\cos u \cos v, -\sin u \cos v, -\sin v),$$
$$\vec{X}_{11} = (-(2 + \cos v) \cos u, -(2 + \cos v) \sin u, 0),$$
$$\vec{X}_{12} = (\sin u \sin v, -\cos u \sin v, 0),$$
$$\vec{X}_{22} = (-\cos v \cos u, -\cos v \sin u, -\sin v).$$

Thus,

$$L_{11} = (2 + \cos v) \cos v, \qquad L_{12} = 0, \qquad L_{22} = 1.$$

Thus, in this example, it is easy to see that $\det(L_{ij}) = (2+\cos v)\cos v$. Since $-1 \leq \cos v \leq 1$, we know $2 + \cos v \geq 1$, so the sign of $\det(L_{ij})$ is the sign of $\cos v$. Hence, the parabolic points on the torus are where $v = \pm\frac{\pi}{2}$, the elliptic points are where $-\frac{\pi}{2} < v < \frac{\pi}{2}$, and the hyperbolic points are where $\frac{\pi}{2} < v < \frac{3\pi}{2}$ (see Figure 6.8).

As Figure 6.7 implies, the only quadratic surface possessing at least one planar point is a plane. However, similar to the "undecided" case in the second derivative test from multivariable calculus, on a general surface S, the existence of a planar point does not imply that S is a plane; it merely implies that third-order behavior, as opposed to second-order, governs the local geometry of the surface with respect to the normal vector. In this case, a variety of possibilities can occur.

Example 6.4.6 (Monkey Saddle) The simplest surface that illustrates third-order behavior is the *monkey saddle* parametrized by $\vec{X}(u, v) = (u, v, u^3 - 3uv^2)$. We calculate that

$$\vec{X}_{uu} = (0, 0, 6u),$$
$$\vec{X}_{uv} = (0, 0, -6v),$$
$$\vec{X}_{vv} = (0, 0, -6u),$$

and hence, even without calculating the unit normal vector function, we deduce that $L_{ij}(0,0) = 0$ for all i, j. Consequently, $(0,0,0)$ is a planar point (see Figure 6.9 for a picture). Near $(0,0,0)$, the best approximating quadratic to the surface is in fact the plane $z = 0$, but obviously such an approximation describes the surface poorly. (The terminology "monkey saddle" comes from the shape of the surface having room for two legs and a tail.)

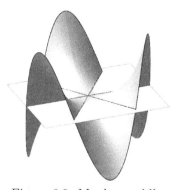

Figure 6.9: Monkey saddle.

Example 6.4.5 illustrates a general fact about the local shape of a surface encapsulated in the following proposition.

Proposition 6.4.7 *Let S be a regular surface of class C^2, and let V be a coordinate neighborhood on S.*

1) If $p \in V$ is an elliptic point, then there exists a neighborhood V' of p such that T_pS does not intersect $V' - \{p\}$.

2) If $p \in V$ is a hyperbolic point, then T_pS intersects every deleted neighborhood V' of p.

Proof: Let $\vec{X} : U \to \mathbb{R}^3$ be a parametrization of the coordinate neighborhood V and suppose that $\vec{X}(0,0) = p$. Consider the real-valued function on U

$$h(u,v) = (\vec{X}(u,v) - \vec{X}(0,0)) \cdot \vec{N}(0,0).$$

Since $\vec{N}(0,0)$ is a unit vector perpendicular to T_pS, then $h(u,v)$ is the height function of signed distance between the surface S at $p' = \vec{X}(u,v)$ and the tangent plane T_pS.

The second-order Taylor series of $\vec{X}$ near $(0,0)$ is given by Equation (6.19), with a remainder function $\vec{R}(u,v)$ such that

$$\lim_{(u,v) \to (0,0)} \frac{\vec{R}(u,v)}{u^2 + v^2} = \vec{0}. \tag{6.21}$$

Thus

$$h(u,v) = \frac{1}{2}\vec{X}_{uu}(0,0) \cdot \vec{N}(0,0)u^2 + \vec{X}_{uv}(0,0) \cdot \vec{N}(0,0)uv$$

$$+ \frac{1}{2}\vec{X}_{vv}(0,0) \cdot \vec{N}(0,0)v^2) + \vec{R}(u,v) \cdot \vec{N}(0,0)$$

$$= \frac{1}{2}(L_{11}(0,0)u^2 + 2L_{12}(0,0)uv + L_{22}(0,0)v^2)$$

$$+ \vec{R}(u,v) \cdot \vec{N}(0,0)$$

$$= \frac{1}{2}II_p\left(\begin{pmatrix} u \\ v \end{pmatrix}\right) + \vec{R}(u,v) \cdot n(p).$$

Solving the quadratic equation

$$L_{11}u^2 + 2L_{12}uv + L_{22}v^2 = 0 \tag{6.22}$$

for u in terms of v leads to

$$L_{11}u = \left(-L_{12} \pm \sqrt{(L_{12})^2 - L_{11}L_{22}}\right)v.$$

If we assume that $(u,v) \neq (0,0)$, Equation (6.22) has no solutions if $\det(L_{ij}) > 0$ and has solutions if $\det(L_{ij}) < 0$. Hence, if p is hyperbolic, $II_p(\binom{u}{v})$ changes sign in a neighborhood of $(0,0)$, and if p is elliptic, it does not.

Define functions $\bar{R}_1$ and $\bar{R}_2$ by

$$\bar{R}_i(u,v) = \frac{R_i(u,v)}{u^2 + v^2} \qquad \text{for } i = 1,2.$$

Then solving $h(u,v) = 0$ amounts to solving

$$(L_{11} + \bar{R}_1(u,v))u^2 + 2L_{12}uv + (L_{22} + \bar{R}_2(u,v))v^2 = 0.$$

Since both $\bar{R}_1$ and $\bar{R}_2$ have a limit of 0 as (u,v) approaches $(0,0)$,

$$\lim_{(u,v)\to(0,0)} (L_{11}(0,0) + \bar{R}_1(u,v))(L_{22}(0,0) + \bar{R}_2(u,v)) - L_{12}(0,0)^2$$
$$= L_{11}(0,0)L_{22}(0,0) - L_{12}(0,0)^2$$

Thus, if p is an elliptic point, there is a neighborhood of $(0,0)$ in which $h(u,v) = 0$ does not have solutions except for $(u,v) = (0,0)$, and if p is hyperbolic, every neighborhood of $(0,0)$ has points through which $h(u,v)$ changes sign. $\qquad\square$

Let us return now to considering the differential of the Gauss map. With Equation (6.18), it is now possible to explicitly calculate the matrix for $dn_p : T_pS \to T_pS$ in terms of the oriented basis $\mathcal{B} = (\vec{X}_u, \vec{X}_v)$, where $\vec{X} : U \to \mathbb{R}^3$ is a positively oriented parametrization of a neighborhood around p. Since $\vec{N}_u$ and $\vec{N}_v$ lie in T_pS, there exist functions $a_j^i(u,v)$ defined on U such that

$$\begin{aligned} \vec{N}_u &= a_1^1 \vec{X}_u + a_1^2 \vec{X}_v, \\ \vec{N}_v &= a_2^1 \vec{X}_u + a_2^2 \vec{X}_v. \end{aligned} \qquad (6.23)$$

Using numerical indices to represent corresponding derivatives, Equation (6.23) can be written as

$$\vec{N}_j = \sum_{i=1}^{2} a_j^i \vec{X}_i, \qquad (6.24)$$

where we denote $\vec{X}_1 = \vec{X}_u$ and $\vec{X}_2 = \vec{X}_v$. (It is important to remember that the superscripts in a_j^i also correspond to indices and not to powers. This notation, though perhaps awkward at first, is the standard notation for components of a tensor. It is useful to remember that the superscript represents a row index while the subscript is a column index.)

From a standard theorem in linear algebra, the matrix of dn_p with respect to the basis $\mathcal{B}$ is

$$[dn_p]_\mathcal{B}^\mathcal{B} = \left([\vec{N}_u]_\mathcal{B} \quad [\vec{N}_v]_\mathcal{B}\right) = \begin{pmatrix} a_1^1 & a_2^1 \\ a_1^2 & a_2^2 \end{pmatrix}.$$

However, from Equations (6.16), (6.18), and (6.24),

$$-L_{ij} = \vec{N}_i \cdot \vec{X}_j = \left(\sum_{k=1}^{2} a_i^k \vec{X}_k \right) \cdot \vec{X}_j,$$

and so

$$-L_{ij} = \sum_{k=1}^{2} a_i^k g_{kj} = \sum_{k=1}^{2} g_{jk} a_i^k. \tag{6.25}$$

In matrix notation, Equation (6.25) means that

$$-\begin{pmatrix} L_{11} & L_{12} \\ L_{21} & L_{22} \end{pmatrix} = \begin{pmatrix} g_{11} & g_{12} \\ g_{21} & g_{22} \end{pmatrix} \begin{pmatrix} a_1^1 & a_2^1 \\ a_1^2 & a_2^2 \end{pmatrix}. \tag{6.26}$$

Since the metric tensor is a positive definite matrix at any point on a regular surface, we can multiply both sides of (6.26) by the inverse of (g_{ij}), and conclude that the matrix for dn_p given in terms of the ordered basis $\mathcal{B} = (\vec{X}_u, \vec{X}_v)$ is

$$\begin{pmatrix} a_1^1 & a_2^1 \\ a_1^2 & a_2^2 \end{pmatrix} = -\begin{pmatrix} g_{11} & g_{12} \\ g_{21} & g_{22} \end{pmatrix}^{-1} \begin{pmatrix} L_{11} & L_{12} \\ L_{21} & L_{22} \end{pmatrix}. \tag{6.27}$$

This matrix formula is more common in modern texts but classical differential geometry texts refer to the individual component equations implicit in the above formula as the *Weingarten equations*.

Note that the (g_{ij}) and (L_{ij}) matrices are symmetric but that the (a_j^i) matrix need not be symmetric.

Since all the above matrices are 2×2, then $\det(L_{ij}) = \det(-L_{ij})$ and since $\det(g_{ij}) > 0$, the determinant $\det(L_{ij})$ has the same sign as $\det(a_j^i)$. Therefore, one deduces the following reformulation of Definition 6.4.4.

Proposition 6.4.8 *Let S be an oriented regular surface of class C^2 with orientation n. Then, a point $p \in S$ is called*

1) *elliptic if $\det(dn_p) > 0$;*

2) *hyperbolic if $\det(dn_p) < 0$;*

3) *parabolic if $\det(dn_p) = 0$ but $dn_p \neq 0$;*

4) *planar if $dn_p = 0$.*

PROBLEMS

1. Calculate the second fundamental form (i.e., the matrix of functions (L_{ij})) for the ellipsoid $\vec{X}(u,v) = (a \cos u \sin v, b \sin u \sin v, c \cos v)$.

2. Calculate the second fundamental form for the parabolic hyperboloid $\vec{X}(u,v) = (au, bv, uv)$.

3. Calculate the second fundamental form for the catenoid $\vec{X}(u,v) =$ $(\cosh v \cos u, \cosh v \sin u, v)$.

4. Calculate the second fundamental form (i.e., the matrix of functions (L_{ij})) using the following parametrizations of the right cylinder:
   (a) $\vec{X}(u,v) = (\cos u, \sin u, v)$.
   (b) $\vec{Y}(u,v) = (\cos(u+v), \sin(u+v), v)$.
   Using the parametrization $\vec{Y}$, find all vectors $\vec{v} \in T_p S$ such that $II_p(\vec{v}) = 0$. Show that these correspond to the straight lines on the cylinder. ∗

5. Calculate the L_{ij} and a_j^i matrices of functions for the inverse of stereographic projection π^{-1} described in Exercise 5.3.10.

6. Consider Enneper's surface parametrized by

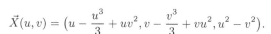

$$\vec{X}(u,v) = \left(u - \frac{u^3}{3} + uv^2, v - \frac{v^3}{3} + vu^2, u^2 - v^2\right).$$

   Show that
   (a) the coefficients of the first fundamental form are

   $$g_{11} = g_{22} = (1 + u^2 + v^2)^2, \qquad g_{12} = 0;$$

   (b) the coefficients of the second fundamental form are

   $$L_{11} = 2, \qquad L_{12} = 0, \qquad L_{22} = -2.$$

7. Suppose that an open set V of a regular oriented surface has two systems of coordinates (x_1, x_2) and $(\bar{x}_1, \bar{x}_2)$ and suppose that V is parametrized by $\vec{X}(x_1, x_2)$ in terms of $(x_1, x_2) \in U$ and by $\vec{Y}(\bar{x}_1, \bar{x}_2)$ in terms of $(\bar{x}_1, \bar{x}_2)$. Call L_{ij} the terms of the second fundamental form in terms of the (x_1, x_2) coordinates and call $\bar{L}_{ij}$ the terms of the second fundamental form in terms of the $(\bar{x}_1, \bar{x}_2)$ coordinates. Prove that

$$\bar{L}_{kl} = \sum_{i,j=1}^{2} \frac{\partial x_i}{\partial \bar{x}_k} \frac{\partial x_j}{\partial \bar{x}_l} L_{ij}.$$

8. Use the previous exercise to show that under the same conditions

$$\det(\bar{L}_{kl}) = \left(\frac{\partial(x_1, x_2)}{\partial(\bar{x}_1, \bar{x}_2)}\right)^2 \det(L_{ij}).$$

9. Suppose we are under the same conditions as in Problem 6.4.7. Call $[dF]$ the 2×2 matrix of the differential of the coordinate change, that is

$$[dF] = \left(\frac{\partial x_i}{\partial \bar{x}_j}\right)_{i,j=1,2}.$$

   Call $\bar{a}_i^j$ the coefficients of dn_p in terms of the coordinate system $(\bar{x}_1, \bar{x}_2)$. Prove that, as matrices,

$$(\bar{a}_i^j) = (dF)\,(a_k^l)\,(dF)^{-1}.$$

10. Calculate the second fundamental form (L_{ij}) and the matrix for dn_p for function graphs $\vec{X}(u, v) = (u, v, f(u, v))$. Determine which points are elliptic, hyperbolic, or parabolic. (This exercise coupled with Proposition 6.4.8 is a proof of the second derivative test from multivariable calculus.)

11. Prove that all points on the tangential surface of a regular space curve are parabolic.

12. Give an example of a surface with an isolated parabolic point.

13. Consider a regular surface S given by the equation $F(x, y, z) = 0$. Use implicit differentiation to provide a criterion for determining whether points are elliptic, hyperbolic, parabolic, or planar. [Hint: Assume that z can be expressed as a function of x and y. Then by implicit differentiation,

$$\frac{\partial z}{\partial x} = -\frac{F_x}{F_z} \quad \text{and} \quad \frac{\partial z}{\partial y} = -\frac{F_y}{F_z}.$$

Then using the parametrization of S by $\vec{X}(x, y) = (x, y, z(x, y))$ and these implicit derivatives, it is possible to calculate $\vec{N}$ and then also the L_{ij} matrix.]

6.5   Normal and Principal Curvatures

One way to analyze the shape of a regular surface S near a point p is to consider curves on S through p and analyze the normal component of their principal curvature vector. In order to use the techniques of differential geometry, in this section, we will always assume that the surface S is of class C^2 and that the curves on S are regular curves also of class C^2.

 Definition 6.5.1 Let S be a regular surface of class C^2, and let $\vec{X} : U \to \mathbb{R}^3$ be a parametrization of a coordinate neighborhood V of S. Let $\vec{\gamma} : I \to \mathbb{R}^3$ be a parametrization of class C^2 for a curve C that lies on S in V. The *normal curvature* of S along C is the function

$$\kappa_n(t) = \frac{1}{s'}\vec{T}' \cdot \vec{N} = \kappa(\vec{P} \cdot \vec{N}) = \kappa \cos \theta,$$

where θ is the angle between the principal normal vector $\vec{P}$ of the curve and the normal vector $\vec{N}$ of the surface.

Interestingly enough, though the curvature of a space curve in general depends on the second derivative of the curve, once one has the second fundamental form for a surface, the normal curvature at a point depends only on the direction.

Proposition 6.5.2 *Let p be a point on S in the neighborhood V and suppose that $\vec{\gamma}$ is the parametrization of a curve C on S such that $\vec{\gamma}(0) = p$ and $\vec{T}(0) = \vec{w} \in T_pS$. Then at p, we write*

$$\kappa_n(0) = II_p(\vec{w}).$$

Proof: Suppose that $\vec{\gamma} = \vec{X} \circ \vec{\alpha}$, where $\vec{\alpha}(t) = (u(t), v(t))$ is a curve in the domain U. The normal vector of S along the curve is given by the function $\vec{N}(t) = \vec{N}(\vec{\alpha}(t))$. Since $\vec{T}(t) \cdot \vec{N}(t) = 0$ for all $t \in I$, then

$$\vec{T}' \cdot \vec{N} = -\vec{T} \cdot \vec{N}',$$

and, hence,

$$\kappa_n(t) = -\frac{1}{s'(t)} \vec{T} \cdot \vec{N}'.$$

Since $\vec{N}(t) = \vec{N}(u(t), v(t))$, then $\vec{N}'(t) = \vec{N}_u u'(t) + \vec{N}_v v'(t)$, and therefore, $\vec{N}'(0) = dn_p \left(\begin{smallmatrix} u'(0) \\ v'(0) \end{smallmatrix}\right) = dn_p(s'(0)\vec{w})$. Thus, at the point p, the normal curvature of S along C is

$$\kappa_n = -\frac{1}{s'(0)} \vec{T}(0) \cdot dn_p(s'(0)\vec{w}) = -\vec{w} \cdot dn_p(\vec{w}) = II_p(\vec{w}).$$

$\square$

Corollary 6.5.3 *All curves C on S passing through the point p with direction $\vec{w} \in T_pS$ have the same normal curvature at p.*

Because of Corollary 6.5.3, given a point p on a regular surface S, one knows everything about the local change in the Gauss map by knowing the values of the second fundamental form on the unit circle in T_pS around p. In particular, most interesting are the optimal values of $II_p(\vec{w})$ for $\|\vec{w}\| = 1$ in T_pS. To find these optimal values, set $\vec{w} = \left(\begin{smallmatrix} a \\ b \end{smallmatrix}\right)$ in the coordinate basis $\{\vec{X}_u, \vec{X}_v\}$ and optimize

$$II_p(\vec{w}) = L_{11}a^2 + 2L_{12}ab + L_{22}b^2,$$

with variables a, b subject to the constraint

$$I_p(\vec{w}, \vec{w}) = g_{11}a^2 + 2g_{12}ab + g_{22}b^2 = 1.$$

Using Lagrange multipliers, one finds that there exists λ such that the optimization problem is solved when

$$\begin{cases} L_{11}a + L_{12}b = \lambda(g_{11}a + g_{12}b), \\ L_{21}a + L_{22}b = \lambda(g_{21}a + g_{22}b), \end{cases}$$

or, in matrix form,

$$\begin{pmatrix} L_{11} & L_{12} \\ L_{21} & L_{22} \end{pmatrix} \begin{pmatrix} a \\ b \end{pmatrix} = \lambda \begin{pmatrix} g_{11} & g_{12} \\ g_{21} & g_{22} \end{pmatrix} \begin{pmatrix} a \\ b \end{pmatrix}. \tag{6.28}$$

This remark leads to the following fundamental proposition.

Proposition 6.5.4 *The maximum and minimum values κ_1 and κ_2 of $II_p(\vec{w})$ restricted to the unit circle are the negatives of the eigenvalues of dn_p. Furthermore, there exists an orthonormal basis $\{\vec{e}_1, \vec{e}_2\}$ of T_pS such that $dn_p(\vec{e}_1) = -\kappa_1 \vec{e}_1$ and $dn_p(\vec{e}_2) = -\kappa_2 \vec{e}_2$.*

Proof: By multiplying on the left by the inverse g^{-1} of the metric tensor, Equation (6.28) becomes

$$g^{-1}L\begin{pmatrix} a \\ b \end{pmatrix} = \lambda \begin{pmatrix} a \\ b \end{pmatrix}.$$

Since the matrix of the differential of the Gauss map is $[dn_p] = -g^{-1}L$, the first part of the proposition follows.

   Since the first fundamental form $I_p(\cdot, \cdot)$ is positive definite and since, by Proposition 6.4.1, dn_p is self-adjoint with respect to this form, then by the Spectral Theorem, dn_p is diagonalizable. (Many linear algebra textbooks discuss the Spectral Theorem exclusively using the dot product as the positive definite form. See [21, Section 7, Chapter XV] for a more general presentation of the Spectral Theorem, which we use here.)

   Now, since dn_p is self-adjoint with respect to the first fundamental form,

$$I_p(dn_p(\vec{e}_1), \vec{e}_2) = I_p(-\kappa_1 \vec{e}_1, \vec{e}_2) = -\kappa_1 I_p(\vec{e}_1, \vec{e}_2),$$
$$\|$$
$$I_p(\vec{e}_1, dn_p(\vec{e}_2)) = I_p(\vec{e}_1, -\kappa_2 \vec{e}_2) = -\kappa_2 I_p(\vec{e}_1, \vec{e}_2),$$

and thus,

$$(\kappa_1 - \kappa_2)I_p(\vec{e}_1, \vec{e}_2) = 0.$$

Hence, if $\kappa_1 \neq \kappa_2$, then $I_p(\vec{e}_1, \vec{e}_2) = \vec{e}_1 \cdot \vec{e}_2 = 0$, where in the latter dot product we view $\vec{e}_1$ and $\vec{e}_2$ as vectors in $\mathbb{R}^3$, and if $\kappa_1 = \kappa_2$, then by the Spectral Theorem, any orthonormal basis satisfies the claim of the proposition. $\square$

Definition 6.5.5 Let S be a regular surface, and let p be a point on S. The maximum and minimum normal curvatures κ_1 and κ_2 at p are called the *principal curvatures* of S at p. The corresponding directions, i.e., unit eigenvectors $\vec{e}_1$ and $\vec{e}_2$ with $dn_p(\vec{e}_i) = -\kappa_i \vec{e}_i$, are called *principal directions* at p.

In a plane, the second fundamental form is identically 0, all normal curvatures including the principal curvatures are 0, and hence, all directions at all points are principal directions. Similarly, it is not hard to show that at every point of the sphere the normal curvature in every direction is the same, and hence, all directions are principal.

Example 6.5.6 (Ellipsoids) As a contrast to the plane or sphere, consider the ellipsoid parametrized by

$$\vec{X}(u,v) = (a\cos u \sin v, b\sin u \sin v, c\cos v).$$

It turns out that the formulas for κ_1 and κ_2 as functions of (u,v) are quite long so we shall calculate the principal curvatures at the point corresponding to $(u,v) = \left(\frac{\pi}{3}, \frac{\pi}{6}\right)$. It is not too hard to calculate the coefficients of the first and second fundamental forms

$$g_{11} = a^2 \sin^2 u \sin^2 v + b^2 \cos^2 u \sin^2 v, \qquad L_{11} = \frac{abc \sin^2 v}{\sqrt{\det g}},$$

$$g_{12} = (b^2 - a^2)\sin u \cos u \sin v \cos v, \qquad L_{12} = 0,$$

$$g_{22} = a^2 \cos^2 u \cos^2 v + b^2 \sin^2 u \cos^2 v + c^2 \sin^2 v, \qquad L_{22} = \frac{abc}{\sqrt{\det g}}.$$

At $(u,v) = \left(\frac{\pi}{3}, \frac{\pi}{6}\right)$, this leads to

$$(a_i^j) = \frac{8abc}{(12a^2b^2 + 3a^2c^2 + b^2c^2)^{3/2}} \begin{pmatrix} -3a^2 - 9b^2 - 4c^2 & -12a^2 + 12b^2 \\ -3a^2 + 3b^2 & -12a^2 - 4b^2 \end{pmatrix},$$

and, after some algebra simplifications, the eigenvalues of this matrix are

$$\lambda = \frac{4abc\left(-15a^2 - 13b^2 - 4c^2 \pm \sqrt{(9a^2 - 5b^2 - 4c^2)^2 + 36(a^2 - b^2)^2}\right)}{(12a^2b^2 + 3a^2c^2 + b^2c^2)^{3/2}}.$$

From this we see that at this particular point for (u,v), the eigenvalues are equal if and only if $a^2 - b^2 = 0$ and $9a^2 - 5b^2 - 4c^2 = 0$, which is equivalent to $a^2 = b^2 = c^2$.

If a curve on the surface given by $\gamma(t) = \vec{X}(u(t), v(t))$ is such that at $\gamma(t_0)$ its direction $\gamma'(t_0)$ is a principal direction with principal curvature κ_i, then by Equation (6.28), with $\lambda = -\kappa_i$, we have

$$\begin{cases} L_{11}u' + L_{12}v' = -\kappa_i(g_{11}u' + g_{12}v'), \\ L_{21}u' + L_{22}v' = -\kappa_i(g_{21}u' + g_{22}v'). \end{cases}$$

Eliminating κ_i from these two equations leads to the relationship

$$(L_{11}g_{21} - L_{21}g_{11})(u')^2 + (L_{11}g_{22} - L_{22}g_{11})u'v'$$
$$+(L_{12}g_{22} - L_{22}g_{12})(v')^2 = 0, \tag{6.29}$$

or equivalently,

$$\begin{vmatrix} (v')^2 & -u'v' & (u')^2 \\ g_{11} & g_{12} & g_{22} \\ L_{11} & L_{12} & L_{22} \end{vmatrix} = 0.$$

Also, in light of the Weingarten equations in Equation (6.27), we can summarize the above two formulas by

$$\begin{pmatrix} v' \\ -u' \end{pmatrix} \cdot dn_p \begin{pmatrix} u' \\ v' \end{pmatrix} = 0.$$

 Definition 6.5.7 A regular curve on a regular oriented surface given by $\gamma(t) = \vec{X}(u(t), v(t))$ in a coordinate neighborhood parametrized by $\vec{X} : U \to \mathbb{R}^3$ that satisfies Equation (6.29) for all t is called a *line of curvature*.

The lines of curvature on a surface form an orthogonal family of curves on the surface, that is, two sets of curves intersecting at right angles. With an appropriate change of variables and in an interval of t where $u'(t) \neq 0$, using the chain rule $\frac{dv}{du} = \frac{v'(t)}{u'(t)}$, (6.29) can be rewritten as

$$(L_{12}g_{22} - L_{22}g_{12})\left(\frac{dv}{du}\right)^2 + (L_{11}g_{22} - L_{22}g_{11})\frac{dv}{du} + (L_{11}g_{21} - L_{21}g_{11}) = 0.$$
$$(6.30)$$

If in the neighborhood we are studying $u'(t) = 0$ for some t, then since we assume the curve is regular, $v'(t)$ cannot also be 0, and hence, we can rewrite Equation (6.29) with u as a function of v. In general, solving the above differential equation is an intractable problem. Of course, as long as the coefficients of the first and second fundamental form are not proportional to each other, one can easily solve algebraically for $\frac{dv}{du}$ in Equation (6.30) and obtain two distinct solutions of $\frac{dv}{du}$ as a function of u and v. Then according to the theory of differential equations (see [2, Section 9.2] for a reference), given any point (u_0, v_0) and for each of the two solutions of $\frac{dv}{du}$ as a function of u and v, there exists a unique function $v = f(u)$ solving Equation (6.30).

Consequently, a regular oriented surface of class C^2 can be covered by lines of curvature wherever the coefficients of the first and second fundamental form are continuous and are not proportional to each other. Points where this cannot be done have a special name.

Definition 6.5.8 Let S be a regular oriented surface of class C^2. An *umbilical point* is a point p on S such that, given a parametrization $\vec{X}$ of a neighborhood of p, the corresponding first and second fundamental forms have coefficients that are proportional, namely,

$$L_{11}g_{12} - L_{12}g_{11} = 0, \ L_{11}g_{22} - L_{22}g_{11} = 0, \ \text{and} \ L_{22}g_{12} - L_{12}g_{22} = 0.$$

Proposition 6.5.9 *Let S be a regular oriented surface. A point p on S is an umbilical point if and only if the eigenvalues of dn_p are equal.*

Proof: Let $\vec{X}$ be a parametrization of a neighborhood of p. With respect to this parametrization and the associated basis $\{\vec{X}_u, \vec{X}_v\}$ on T_pS, the matrix of dn_p is $[dn_p] = (a_i^j)$. Proposition 6.5.4 implies that (a_i^j) is diagonalizable.

The matrix (a_i^j) has equal eigenvalues if and only if there exists an invertible matrix B such that

$$(a_i^j) = B \begin{pmatrix} \lambda & 0 \\ 0 & \lambda \end{pmatrix} B^{-1},$$

and then

$$(a_i^j) = B(\lambda I)B^{-1} = \lambda BB^{-1} = \lambda I.$$

Consequently (a_i^j) has equal eigenvalues if and only if it is already diagonal, with elements on the diagonal being equal. Then $\lambda I = (a_i^j) = -g^{-1}L$, and therefore, $L = -\lambda g$, which is tantamount to saying that the coefficients of the first and second fundamental forms are proportional. The proposition follows. $\qquad\square$

The proof of Proposition 6.5.4 shows that if $\kappa_1 \neq \kappa_2$, then the orthonormal basis of principal directions is unique up to signs, while if $\kappa_1 = \kappa_2$, any orthogonal basis is principal. Therefore, in light of Proposition 6.5.9, at umbilical points, there is no preferred basis of principal directions, and thus it makes sense that the only solutions to Equation (6.29) at umbilical points have $u'(t) = v'(t) = 0$, that is, $(u(t), v(t)) = (u_0, v_0)$. Note that at an umbilical point, Equation (6.29) degenerates to the trivial equation $0 = 0$.

Solving Equation (6.29) gives the lines of curvature on a surface. At every point p on the surface that is not umbilical, there are two lines of curvature through p, and they intersect at right angles. This simple remark leads to the following nice characterization.

Proposition 6.5.10 *The coordinate lines of a parametrization $\vec{X}$ of a surface are curvature lines if and only if $g_{12} = L_{12} = 0$.*

Proof: Suppose that coordinate lines are curvature lines. Since any two lines of curvature intersect at a point at a right angle, we know that $g_{12} = \vec{X}_u \cdot \vec{X}_v = 0$. Furthermore, coordinate lines are given by $\vec{\gamma}(t) = \vec{X}(t, v_0)$ or $\vec{\gamma}(t) = \vec{X}(u_0, t)$. If $\vec{\gamma}(t) = \vec{X}(t, v_0)$, then $u'(t) = 1$ and $v'(t) = 0$, and if $\vec{\gamma}(t) = \vec{X}(u_0, t)$, then $u'(t) = 0$ and $v'(t) = 1$. Then Equation (6.29) implies that both of the following hold:

$$L_{11}g_{21} - L_{21}g_{11} = 0 \qquad \text{and} \qquad L_{12}g_{22} - L_{22}g_{12} = 0.$$

Since $g_{12} = 0$, we have $L_{21}g_{11} = 0$ and $L_{12}g_{22} = 0$. However, since $\det(g) = g_{11}g_{22} - g_{12}^2 > 0$, then at all points on S the functions g_{11} and g_{22} cannot both be 0 at the same time. Since $L_{12} = L_{21}$, we deduce that $L_{12} = 0$.

Conversely, if $g_{12} = L_{12} = 0$, then (g_{ij}) and (L_{ij}) are both diagonal matrices, making (a_j^i) a diagonal matrix, and hence, $\vec{X}_u$ and $\vec{X}_v$ are eigenvectors of dn_p. Hence, coordinate lines are curvature lines.

$\square$

As a linear transformation from T_pS to itself, dn_p is independent of a parametrization of a neighborhood of p. Therefore, if p is not an umbilical point, the principal directions $\{\vec{e}_1, \vec{e}_2\}$ as eigenvectors of dn_p provide a basis of T_pS that possesses more geometric meaning than the coordinate basis of any particular parametrization of a neighborhood of p.

In the orthonormal basis $\{\vec{e}_1, \vec{e}_2\}$, a unit vector $\vec{w} \in T_pS$ is written as $\vec{w} = \cos\theta\vec{e}_1 + \sin\theta\vec{e}_2$ for some angle θ. Using these coordinates, the normal curvature of S at p in the direction of $\vec{w}$ is

$$
\begin{aligned}
II_p(\vec{w}) &= -\vec{w} \cdot dn_p(\vec{w}) \\
&= -(\cos\theta\vec{e}_1 + \sin\theta\vec{e}_2) \cdot dn_p(\cos\theta\vec{e}_1 + \sin\theta\vec{e}_2) \\
&= -(\cos\theta\vec{e}_1 + \sin\theta\vec{e}_2) \cdot (\cos\theta dn_p(\vec{e}_1) + \sin\theta dn_p(\vec{e}_2)) \\
&= -(\cos\theta\vec{e}_1 + \sin\theta\vec{e}_2) \cdot (-\cos\theta\kappa_1\vec{e}_1 - \sin\theta\kappa_2\vec{e}_2) \\
II_p(\vec{w}) &= (\cos^2\theta)\kappa_1 + (\sin^2\theta)\kappa_2. \tag{6.31}
\end{aligned}
$$

Equation (6.31) is called *Euler's curvature formula*.

Another useful geometric characterization of the behavior of S near p is called the *Dupin indicatrix*. The Dupin indicatrix consists of all vectors $\vec{w} \in T_pS$ such that $II_p(\vec{w}) = \pm 1$. If $\vec{w} = (w_1, w_2) = (\rho\cos\theta, \rho\sin\theta)$ are expressions of $\vec{w}$ in Cartesian and polar coordinates referenced in terms of the orthonormal frame $\{\vec{e}_1, \vec{e}_2\}$, then Euler's curvature formula gives

$$\pm 1 = II_p(\vec{w}) = \rho^2 II_p(\vec{w}) = \kappa_1\rho^2\cos^2\theta + \kappa_2\rho^2\sin^2\theta.$$

Thus, the Dupin indicatrix satisfies the equation

$$\kappa_1 w_1^2 + \kappa_2 w_2^2 = \pm 1. \tag{6.32}$$

Since the principal curvatures at a point p are the negatives of the eigenvalues of dn_p, Proposition 6.4.8 provides a characterization of whether a point is elliptic, hyperbolic, parabolic, or planar in terms of the principal curvatures. More precisely, a point $p \in S$ is

1) elliptic if κ_1 and κ_2 have the same sign;

2) hyperbolic if κ_1 and κ_2 have opposite signs;

3) parabolic if exactly one of κ_1 and κ_2 is 0;

4) planar if $\kappa_1 = \kappa_2 = 0$.

The Dupin indicatrix justifies this terminology because p is elliptic or hyperbolic if and only if Equation (6.32) is the equation for a single ellipse or two hyperbolas, respectively. Furthermore, the half-axes for the corresponding ellipse or hyperbola are

$$\sqrt{\frac{1}{|\kappa_1|}} \quad \text{and} \quad \sqrt{\frac{1}{|\kappa_2|}}.$$

Figure 6.10 illustrates the Dupin indicatrix in the elliptic and hyperbolic cases. In this figure, the we have $|\kappa_1| > |\kappa_2|$.

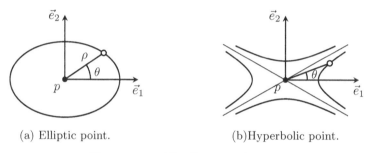

(a) Elliptic point.   (b)Hyperbolic point.

Figure 6.10: The Dupin indicatrix.

If a point is parabolic, then the Dupin indicatrix is simply a pair of parallel lines equidistant from one of the principal direction lines, and if a point is planar, the Dupin indicatrix is the empty set.

When a point is hyperbolic, it is possible to find the asymptotes of the Dupin indicatrix without referring directly to the principal curvatures.

Definition 6.5.11 Let p be a point on a regular oriented surface S of class C^2. An *asymptotic direction* of S at p is a unit vector $\vec{w}$ in T_pS such that the normal curvature is 0. An *asymptotic curve* on S is a regular curve C such that at every point $p \in C$, the unit tangent vector at p is an asymptotic direction.

Since the principal curvatures at a point p are the maximum and minimum normal curvatures, at an elliptic point, there exist no asymptotic directions, which one can see from the Dupin indicatrix. If, on the other hand, p is hyperbolic, then the principal curvatures have opposite signs. If a unit vector $\vec{w}$ is an asymptotic direction at p and makes an angle θ with $\vec{e}_1$, then by Euler's formula we deduce that

$$\cos^2 \theta = -\frac{\kappa_2}{\kappa_1 - \kappa_2}.$$

The right-hand side is always positive and less than 1, so this equation leads to four solutions for $\vec{w}$, i.e., two mutually negative pairs, each representing one of the asymptotes of the Dupin indicatrix hyperbola. As we see in Problem 6.5.13, the asymptotic curves in fact provide a more subtle description of the behavior of a surface near a point than the Dupin indicatrix does since it describes more than second-order phenomena.

As with the lines of curvature, it is not difficult to find a differential equation that characterizes asymptotic curves on a surface. Suppose that $\vec{X} : U \to \mathbb{R}^3$ parametrizes a neighborhood of $p \in S$ and that $\vec{\gamma} : I \to S$ is a curve on S defined by $\vec{\gamma} = \vec{X} \circ \vec{\alpha}$, with $\vec{\alpha} : I \to U$ and $\vec{\alpha}(t) = (u(t), v(t))$. By Proposition 6.5.2, if we call $\kappa_n(t)$ the normal curvature of S at $\vec{\gamma}(t)$ in the direction of $\vec{\gamma}'(t)$, then

$$\kappa_n(t) = II_{\vec{\gamma}(t)}(\vec{T}(t)) = \frac{1}{(s'(t))^2} II_{\vec{\gamma}(t)}\left(\begin{pmatrix} u' \\ v' \end{pmatrix} \right).$$

Thus, since an asymptotic curve must satisfy $\kappa_n(t) = 0$ for all t, then any asymptotic curve must satisfy the differential equation

$$L_{11}(u')^2 + 2L_{12}u'v' + L_{22}(v')^2 = 0.$$

Example 6.5.12 (Surfaces of Revolution) Consider a surface of revolution parametrized by

$$\vec{X}(u,v) = (f(v) \cos u, f(v) \sin u, h(v)),$$

with $f(v) > 0$, $u \in (0, 2\pi)$, and $v \in (a, b)$. It is not hard to show (see Problem 6.6.13) that the coefficients of the first fundamental form are

$$g_{11} = f(v)^2, \quad g_{12} = g_{21} = 0, \quad g_{22} = (f'(v))^2 + (h'(v))^2,$$

and the coefficients of the second fundamental form are

$$L_{11} = -\frac{fh'}{\sqrt{(f')^2 + (h')^2}}, \quad L_{12} = L_{21} = 0, \quad L_{22} = \frac{f''h' - f'h''}{\sqrt{(f')^2 + (h')^2}}.$$

By Problem 6.5.7, we deduce that the meridians (where $u = \text{const.}$) and the parallels (where $v = \text{const.}$) are lines of curvature. We can also conclude that the principal curvatures are

$$\kappa_1(u,v) = -\frac{h'(v)}{f(v)\sqrt{(f'(v))^2 + (h'(v))^2}},$$

$$\kappa_2(u,v) = \frac{f''(v)h'(v) - f'(v)h''(v)}{(f'(v))^2 + (h'(v))^{3/2}},$$

$$(6.33)$$

though, as written, we make no assumption that $\kappa_1 > \kappa_2$. Clearly, the first and second fundamental forms depend only on the coordinate v, and hence, all properties of points, such as whether they are elliptic, hyperbolic, parabolic, planar, or umbilical, depend only on v. Setting $\kappa_1 = \kappa_2$ in Equation (6.33) produces an equation that determines for what v the points on the surface are umbilical points.

PROBLEMS

1. Find the principal directions and the principal curvatures of the quadric surface $z = ax^2 + 2bxy + cy^2$ at $(0,0,0)$ in terms of the constants a, b, c.

2. Find the principal directions and the principal curvatures of the surface $z = 4x^2 - x^4 - y^2$ at its critical points (as a function in two variables).

3. Provide the details for Example 6.5.6.

4. Consider the ellipsoid with half-axes a, b, and c.

   (a) Prove the ellipsoid has four umbilical points when a, b, and c are distinct.

   (b) Calculate the coordinates of the umbilical points when all half-axes have different length.

   (c) What happens when two of the half-axes are equal?

5. Determine the asymptotic curves of the hyperboloid of one sheet parametrized by

$$\vec{X}(u, v) = (\cosh v \cos u, \cosh v \sin u, \sinh v).$$

   [Hint: Use Example 6.5.12.]

6. (ODE) Determine the asymptotic curves and the lines of curvature of the helicoid

$$\vec{X}(u, v) = (v \cos u, v \sin u, cu).$$

7. Let $\vec{X}$ be the parametrization for a neighborhood of S. Prove that if (g_{ij}) and (L_{ij}) are diagonal matrices, then the lines of curvature are the coordinate curves (curves on S where $u =$const. or $v =$const.).

8. Let $\vec{X}$ be the parametrization for a neighborhood of S. Prove that the coordinate curves are asymptotic curves if and only if $L_{11} = L_{22} = 0$.

9. (ODE) Determine the asymptotic curves of the catenoid

$$\vec{X}(u, v) = (\cosh v \cos u, \cosh v \sin u, v).$$

10. Consider Enneper's surface. Use the results of Problem 6.4.6 to show the following:

   (a) The principal curvatures are

$$\kappa_1 = \frac{2}{(1 + u^2 + v^2)^2}, \qquad \kappa_2 = -\frac{2}{(1 + u^2 + v^2)^2}.$$

(b) The lines of curvature are the coordinate curves.

(c) The asymptotic curves are $u + v =$const. and $u - v =$const.

11. Find equations for the lines of curvature when the surface is given by $z = f(x, y)$.

12. Find equations for the lines of curvature when the surface is given by $z = f(r, \theta)$ using polar coordinates.

13. Consider the monkey saddle given by the graph of the function $z = x^3 - 3xy^2$. Prove that the set of asymptotic curves that possesses $(0, 0, 0)$ as a limit point consists of three straight lines through $(0, 0, 0)$ with equal angles between them.

14. Let S_1 and S_2 be two regular surfaces that intersect along a regular curve C. Let p be a point on C, and call λ_1 and λ_2 the normal curvatures of S at p in the direction of C. Prove that the curvature κ of C at p satisfies

$$\kappa^2 \sin^2 \theta = \lambda_1^2 + \lambda_2^2 - 2\lambda_1 \lambda_2 \cos^2 \theta,$$

where θ is the angle between S_1 and S_2 at p (calculated using the normals to S_1 and S_2 at p).

15. Consider a sphere with a bump at the north pole, parametrized according to a surface of revolution as in Example 6.5.12 with

$$f(v) = R(v) \sin v \qquad \text{and} \qquad h(v) = R(v) \cos v,$$

where $R(v)$ is a function that parametrizes the bump.

(a) Find an equation in v that determines where the umbilical points on this bumped sphere occur.

(b) Suppose that $R(v) = R_b e^{-cv^2}$, where c is such that $R(\pi) = R_0$. Show that the equation in the previous part becomes $\tan v = v + 4c^2 v^3$.

(c) If $R_0 = 1$ and $R_2 = 2$, numerically find the value of v where the bumped sphere has umbilical points.

16. Let S be a regular oriented surface and p a point on S. Two nonzero vectors $\vec{u}_1, \vec{u}_2 \in T_p S$ are called *conjugate* if

$$I_p(dn_p(\vec{u}_1), \vec{u}_2) = I_p(\vec{u}_1, dn_p(\vec{u}_2)) = 0. \qquad (6.34)$$

Prove the following:

(a) A curve C on S parametrized by $\vec{\gamma} : I \to S$ is a line of curvature if and only if the unit tangent vector $\vec{T}$ and any normal vector to $\vec{T}$ in $T_{\vec{\gamma}(t)}(S)$ are conjugate to each other.

(b) Let $\vec{\gamma}_1(t)$ and $\vec{\gamma}_2(t)$ be regular space curves, and define a surface S by the parametrization $\vec{X}(u, v) = \vec{\gamma}_1(u) + \vec{\gamma}_2(v)$. Surfaces constructed in this manner are called *translation surfaces*. Show that the coordinate lines of S are conjugate lines.

6.6   Gaussian and Mean Curvatures

We now arrive at two fundamental geometric invariants that encapsulate a considerable amount of useful information about the local shape of a surface. Again, in this section, we must assume that S is a regular surface of class C^2.

Definition 6.6.1 Let κ_1 and κ_2 be the principal curvatures of a regular oriented surface S at a point p. Define

1) the *Gaussian curvature* of S at p as the product $K = \kappa_1 \kappa_2$;

2) the *mean curvature* of S at p as the average $H = \frac{\kappa_1 + \kappa_2}{2}$ of the principal curvatures.

By Problem 6.4.9, the matrix for the Gauss map in terms of any particular coordinate system on a neighborhood of p on S is conjugate to the corresponding matrix for a different coordinate system via the change of basis matrix $\left(\frac{\partial x_i}{\partial \bar{x}_j}\right)$. Therefore, since $\det(BAB^{-1}) = \det(A)$ and $\mathrm{Tr}(BAB^{-1}) = \mathrm{Tr}(A)$, which is a standard result in linear algebra, the eigenvalues of the Gauss map, the principal curvatures, the Gaussian curvature, and the mean curvature are invariant under coordinate changes. Furthermore, since $\det(-A) = \det(A)$ when A is a square matrix with an even number of rows, then

$$K = \kappa_1 \kappa_2 = \det(a_j^i) = \det(dn_p),$$

$$H = \frac{\kappa_1 + \kappa_2}{2} = -\frac{1}{2}\,\mathrm{Tr}(a_j^i) = -\frac{1}{2}\,\mathrm{Tr}(dn_p).$$

Consequently, as claimed above, the Gaussian curvature and the mean curvature (up to a sign) of S at p are geometric invariants in that they do not depend on any particular coordinate system on S in a neighborhood of the point p. We leave it as an exercise to the reader to also show that the Gaussian curvature and the mean curvature do not depend on the orientation or position of S in space.

In order to calculate the mean curvature $H(u, v)$, one has no choice but to calculate the matrix of the differential of the Gauss map $(a_j^i) = [dn_p]$. However, from a computational perspective, Equation (6.27) leads to a much simpler formula for the Gaussian curvature function $K(u, v)$. Recall that $\det(AB) = \det(A)\det(B)$ for all square matrices of the same size and also that $\det(A^{-1}) = 1/\det(A)$. Then Equation (6.27) implies that

$$K = \frac{\det(L_{ij})}{\det(q_{ij})} = \frac{L_{11}L_{22} - L_{12}^2}{g_{11}g_{22} - g_{12}^2}. \tag{6.35}$$

This equation lends itself readily to calculations since one does not need to fully compute the matrix of the Gauss map, let alone find its eigenvalues. However, we can also obtain an alternate characterization of the Gaussian curvature.

Proposition 6.6.2 *Let S be a regular oriented surface of class C^2, and let V be a neighborhood parametrized by $\vec{X} : U \subset \mathbb{R}^2 \to \mathbb{R}^3$. Define the unit normal vector $\vec{N}$ as*

$$\vec{N} = \frac{\vec{X}_u \times \vec{X}_v}{\|\vec{X}_u \times \vec{X}_v\|}.$$

Then over the domain U, the Gaussian curvature is the unique function $K(u,v)$ satisfying

$$\vec{N}_u \times \vec{N}_v = K(u,v)\vec{X}_u \times \vec{X}_v.$$

Proof: From the definition of the (a^i_j) functions in Equation (6.23), we have

$$\vec{N}_u \times \vec{N}_v = (a^1_1\vec{X}_u + a^2_1\vec{X}_v) \times (a^1_2\vec{X}_u + a^2_2\vec{X}_v).$$

It is then easy to see that

$$\vec{N}_u \times \vec{N}_v = \det(a^i_j)\vec{X}_u \times \vec{X}_v.$$

The result follows since $K = \det(a^i_j)$. $\square$

Corollary 6.6.3 *If V is a region of a regular oriented surface S of class C^2, then*

$$\iint_V K\,dS = \iint_{n(V)} dS,$$

where the latter integral is the signed area on the unit sphere of the image of V under the Gauss map.

Example 6.6.4 (Spheres) Consider the sphere parametrized by the vector function

$$\vec{X}(u,v) = (R\cos u \sin v, R\sin u \sin v, R\cos v),$$

with $(u,v) \in (0, 2\pi) \times (0, \pi)$. Example 6.1.5 and Example 6.4.3 gave us

$$(g_{ij}) = \begin{pmatrix} R^2\sin^2 v & 0 \\ 0 & R^2 \end{pmatrix} \quad \text{and} \quad (L_{ij}) = \begin{pmatrix} R\sin^2 v & 0 \\ 0 & R \end{pmatrix}.$$

Using Equation (6.35), it is easy to compute that the Gaussian curvature function on the sphere is

$$K(u,v) = \frac{R^2\sin^2 v}{R^4\sin^2 v} = \frac{1}{R^2},$$

which is a constant function. (Note that we assumed that $v \neq 0, \pi$, which correspond to the north and south poles of the parametrization. However, another parametrization that includes the north and south poles would show that at these points as well we have $K = 1/R^2$.)

The matrix of the Gauss map is then

$$[dn_p] = -g^{-1}L = \begin{pmatrix} -\frac{1}{R} & 0 \\ 0 & -\frac{1}{R} \end{pmatrix},$$

from which we immediately deduce that the principal curvatures are $\kappa_1(u,v) = \kappa_2(u,v) = \frac{1}{R}$. This shows that all points on the sphere are umbilical points. In addition, the mean curvature is also a constant function $H(u,v) = \frac{1}{R}$.

Example 6.6.5 (Function Graphs) Consider the graph of a function $f : U \subset \mathbb{R}^2 \to \mathbb{R}$. The graph can be parametrized by $\vec{X} : U \to \mathbb{R}^3$, with

$$\vec{X}(u,v) = (u, v, f(u,v)).$$

Problem 6.4.10 asks the reader to calculate the matrix (L_{ij}). It is not hard to show that

$$(g_{ij}) = \begin{pmatrix} 1 + (f_u)^2 & f_u f_v \\ f_v f_u & 1 + (f_v)^2 \end{pmatrix} \quad \text{and}$$

$$(L_{ij}) = \frac{1}{\sqrt{1 + f_u^2 + f_v^2}} \begin{pmatrix} f_{uu} & f_{uv} \\ f_{vu} & f_{vv} \end{pmatrix},$$

where f_u is the typical shorthand to mean $\frac{\partial f}{\partial u}$. But then $\det(g) = 1 + f_u^2 + f_v^2$, and we find that the Gaussian curvature function on a function graph is

$$K(u,v) = \frac{1}{(1 + f_u^2 + f_v^2)^2} \begin{vmatrix} f_{uu} & f_{uv} \\ f_{vu} & f_{vv} \end{vmatrix} = \frac{f_{uu} f_{vv} - f_{uv}^2}{(1 + f_u^2 + f_v^2)^2}.$$

This result allows us to rephrase the second derivative test in the calculus of a function f from $\mathbb{R}^2$ to $\mathbb{R}$ as follows: If f has continuous second partial derivatives and (u_0, v_0) is a critical point (i.e., $f_u(u_0, v_0) = f_v(u_0, v_0) = 0$), then

1) (u_0, v_0) is a local maximum if $K(u_0, v_0) > 0$ and $f_{uu}(u_0, v_0) < 0$;
2) (u_0, v_0) is a local minimum if $K(u_0, v_0) > 0$ and $f_{uu}(u_0, v_0) > 0$;
3) (u_0, v_0) is a saddle point if $K(u_0, v_0) < 0$;
4) the test is inconclusive if $K(u_0, v_0) = 0$.

In the language we have introduced in this chapter, local minima and local maxima of the function $z = f(u,v)$ are elliptic points, saddle points are hyperbolic points, and points where the second derivative test is inconclusive are either parabolic or planar points.

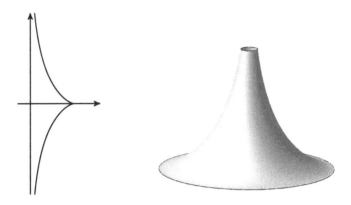

Figure 6.11: Tractrix and pseudosphere.

Example 6.6.6 (Pseudosphere) A *tractrix* is a curve in the plane
with parametric equations

$$\vec{\alpha}(t) = (\operatorname{sech} t, t - \tanh t),$$

and it has the y-axis as an asymptote. The *pseudosphere* is defined
as half of the surface of revolution of a tractrix about its asymptote.
More precisely, we can parametrize the pseudosphere by

$$\vec{X}(u, v) = (\operatorname{sech} v \cos u, \operatorname{sech} v \sin u, v - \tanh v),$$

with $u \in [0, 2\pi)$ and $v \in [0, \infty)$. (See Figure 6.11 for a picture of
the tractrix and the pseudosphere.) We leave it as an exercise for the
reader to prove that

$$(g_{ij}) = \begin{pmatrix} \operatorname{sech}^2 v & 0 \\ 0 & \tanh^2 v \end{pmatrix} \quad \text{and}$$

$$(L_{ij}) = \begin{pmatrix} -\operatorname{sech} v \tanh v & 0 \\ 0 & \operatorname{sech} v \tanh v \end{pmatrix}.$$

The Gaussian curvature for the pseudosphere is $K = -1$, which motivates the name "pseudosphere" since it is analogous to the unit sphere
but with a constant Gaussian curvature of -1 instead of 1.

Proposition 6.6.7 *Let S be a regular oriented surface of class C^2
with Gauss map $n : S \to \mathbb{S}^2$, and let p be a point on S. Call $K(p)$ the
Gaussian curvature of S at p, and suppose that $K(p) \neq 0$. Let B_ε be
the ball of radius ε around p and define $V_\varepsilon = B_\varepsilon \cap S$. Then*

$$|K(p)| = \lim_{\varepsilon \to 0} \frac{Area(n(V_\varepsilon))}{Area(V_\varepsilon)}.$$

In other words, the absolute value of the Gaussian curvature at a point p is the limit around p of the ratio of the surface area on $\mathbb{S}^2$ mapped under the Gauss map to the corresponding surface area on S.

Proof: Let $\vec{X} : U \to \mathbb{R}^3$ be a regular parametrization of a neighborhood of $p = \vec{X}(u_0, v_0)$, where U is an open subset of $\mathbb{R}^2$. Since $\vec{X}(U)$ is an open set, $V_\varepsilon \subset \vec{X}(U)$ for all ε small enough. Furthermore, since we assumed that $K(p) \neq 0$, by the continuity of the Gaussian curvature function $K : U \to \mathbb{R}$, we deduce that K does not change sign for ε chosen small enough. Therefore, from now on, we assume that K does not change sign.

Define $U_\varepsilon = \vec{X}^{-1}(V_\varepsilon)$, the preimage of the neighborhood V_ε under $\vec{X}$. Since $\vec{X}$ is bijective as a regular parametrization, $\{(u_0, v_0)\}$ is the unique point in all U_ε for all $\varepsilon > 0$. Then by the formula for surface area,

$$\text{Area}(V_\varepsilon) = \iint_{V_\varepsilon} dA = \iint_{U_\varepsilon} \|\vec{X}_u \times \vec{X}_v\| \, du \, dv.$$

Similarly, on the unit sphere,

$$\text{Area}(n(V_\varepsilon)) = \iint_{U_\varepsilon} \|\vec{N}_u \times \vec{N}_v\| \, du \, dv.$$

However, by Proposition 6.6.2, we also have

$$\text{Area}(n(V_\varepsilon)) = \iint_{U_\varepsilon} |K(u,v)| \|\vec{X}_u \times \vec{X}_v\| \, du \, dv.$$

By the Mean Value Theorem for double integrals, for every $\varepsilon > 0$, there exist points $(u_\varepsilon, v_\varepsilon)$ and $(u'_\varepsilon, v'_\varepsilon)$ in the open set U_ε such that

$$\iint_{U_\varepsilon} \|\vec{X}_u \times \vec{X}_v\| \, du \, dv = \|\vec{X}_u(u_\varepsilon, v_\varepsilon) \times \vec{X}_v(u_\varepsilon, v_\varepsilon)\|$$

and

$$\iint_{U_\varepsilon} \|\vec{N}_u \times \vec{N}_v\| \, du \, dv = |K(u'_\varepsilon, v'_\varepsilon)| \, \|\vec{X}_u(u'_\varepsilon, v'_\varepsilon) \times \vec{X}_v(u'_\varepsilon, v'_\varepsilon)\|.$$

Thus, since $\lim_{\varepsilon \to 0} (u_\varepsilon, v_\varepsilon) = \lim_{\varepsilon \to 0} (u'_\varepsilon, v'_\varepsilon) = (u_0, v_0)$, we have

$$\lim_{\varepsilon \to 0} \frac{\text{Area}(n(V_\varepsilon))}{\text{Area}(V_\varepsilon)} = \lim_{\varepsilon \to 0} \frac{|K(u'_\varepsilon, v'_\varepsilon)| \, \|\vec{X}_u(u'_\varepsilon, v'_\varepsilon) \times \vec{X}_v(u'_\varepsilon, v'_\varepsilon)\|}{\|\vec{X}_u(u_\varepsilon, v_\varepsilon) \times \vec{X}_v(u_\varepsilon, v_\varepsilon)\|}$$

$$= \frac{|K(u_0, v_0)| \, \|\vec{X}_u(u_0, v_0) \times \vec{X}_v(u_0, v_0)\|}{\|\vec{X}_u(u_0, v_0) \times \vec{X}_v(u_0, v_0)\|}$$

$$= |K(u_0, v_0)| = |K(p)|.$$

$\square$

PROBLEMS

1. In Example 6.6.5 we calculated the Gaussian curvature of function graphs. Calculate the mean curvature.

2. Using an appropriate parametrization, find the Gaussian curvature of the hyperboloid of one sheet, $x^2 + y^2 - z^2 = 1$.

3. Consider the circular paraboloid given by $z = x^2 + y^2$. Show that the Gaussian curvature at any point is $K = 4/(1+4z)^2$ and that the mean curvature is $H = -4z/(1+4z)^{3/2}$. [Hint: Use an appropriate parametrization of the paraboloid and use Exercise 6.5.12.]

4. Calculate the Gaussian and mean curvature functions of the general ellipsoid

$$\vec{X}(u, v) = (a \cos u \sin v, b \sin u \sin v, c \cos v).$$

5. Calculate the Gaussian curvature of the torus parametrized by

$$\vec{X}(u, v) = \big((a + b \cos v) \cos u, (a + b \cos v) \sin u, b \sin v\big)$$

where $a > b$ are constants and $(u, v) \in (0, 2\pi) \times (0, 2\pi)$.

6. Consider a regular space curve $\vec{\gamma} : I \to \mathbb{R}^3$ and the tangential surface defined by

$$\vec{X}(t, u) = \vec{\gamma}(t) + u\vec{\gamma}'(t)$$

for $(t, u) \in I \times \mathbb{R}$. Calculate the mean and Gaussian curvature functions.

7. Let $\vec{\alpha}(t)$ and $\vec{\beta}(t)$ be differentiable vector functions with common domain I. Define the *secant surface* between the two resulting curves by

$$\vec{X}(t, u) = (1 - u)\vec{\alpha}(t) + u\vec{\beta}(t)$$

for $(t, u) \in I \times \mathbb{R}$. Assume that the corresponding surface is regular.

   (a) Prove that $K(u, v) = 0$ for any point with $u = \frac{1}{2}$.

   (b) Prove that $K(u, v) = 0$ if and only if $u = \frac{1}{2}$ or $(\vec{\beta}(t) - \vec{\alpha}(t))$ is in the plane spanned by $\vec{\alpha}'(t)$ and $\vec{\beta}'(t)$.

8. Prove that the metric coefficients, the differential of the Gauss map, and hence, the Gaussian and mean curvatures of a surface at a point do not depend on the position or orientation of the surface in space. In particular, suppose that S is a regular surface of class C^2 and $p \in S$ and that $f : \mathbb{R}^3 \to \mathbb{R}^3$ is an isometry; prove that the differential of the Gauss map to S at p is the same as the differential of the Gauss map to $f(S)$ at $f(p)$.

9. If M is a nonorientable surface, one can define the Gaussian curvature on a coordinate patch of M by using $\vec{N} = \vec{X}_u \times \vec{X}_v / \|\vec{X}_u \times \vec{X}_v\|$ and Equation (6.35) without reference to dn_p, which is not well defined since no Gauss map $n : M \to \mathbb{S}^2$ exists. Consider the Möbius strip

M depicted in Figure 5.15. Show that, using the parametrization of the Möbius strip in Example 5.4.12, the Gaussian curvature is

$$K = -\frac{1}{\left(\frac{1}{4}v^2 + (2 - v\sin\frac{u}{2})^2\right)^2}.$$

10. *Tubes.* Let $\vec{\gamma} : I \to \mathbb{R}^3$ be a regular space curve and let r be a positive real number. Consider the tube of radius r around $\vec{\gamma}(t)$ parametrized by

$$\vec{X}(t, u) = \vec{\gamma}(t) + (r\cos v)\vec{P}(t) + (r\sin v)\vec{B}(t).$$

   (a) Calculate the second fundamental form (L_{ij}) and the matrix for dn_p.

   (b) Calculate the Gaussian curvature function $K(t, v)$ on the tube.

   (c) Prove that all the points with $K = 0$ are either points on curves $\vec{\gamma}(t) \pm r\vec{B}(t)$ for all $t \in I$ or points on circles $\vec{\gamma}(t_0) + (r\cos v)\vec{P}(t_0) + (r\sin v)\vec{B}(t_0)$, where t_0 satisfies $\kappa(t_0) = 0$.

11. Consider the pseudosphere and the parametrization provided in Example 6.6.6.

   (a) Prove the statements about the pseudosphere in Example 6.6.6.

   (b) Find the mean curvature function on the pseudosphere.

   (c) Modify the given parametric equations to find a parametrization of a surface with constant Gaussian curvature $K = -\frac{1}{R^2}$.

   (d) Determine the lines of curvature on the pseudosphere.

12. Consider the monkey saddle $\vec{X}(u, v) = (u, v, u^3 - 3uv^2)$. Calculate the Gaussian curvature function of $\vec{X}$ and the points where $K > 0$, $K < 0$, or $K = 0$.

13. *Surfaces of revolution.* Consider the surface of revolution defined by revolving the parametrized curve $(f(t), h(t))$ in the xy-plane about the y-axis. Its parametrization is

$$\vec{X}(u, v) = (f(v)\cos u, f(v)\sin u, h(v))$$

for $u \in [0, 2\pi)$ and $v \in I$, where I is some interval. Assume that f and h are such that the surface of revolution is a regular surface.

   (a) Calculate the second fundamental form (L_{ij}).

   (b) Calculate the coefficients of the matrix for dn_p.

   (c) Calculate the Gaussian curvature function.

   (d) Determine which points are elliptic, hyperbolic, parabolic, or planar.

   (e) Prove that the lines of curvature are the coordinate lines.

14. *Normal Variations.* Let $\vec{X}(u, v)$ be the parametrization for a coordinate patch V of a regular surface S. Consider the normal variation of S over V that is parametrized by

$$\vec{V}(u, v) = \vec{X}(u, v) + r\vec{N}(u, v)$$

for some constant $r \in \mathbb{R}$ and where $\vec{N}$ is the unit normal vector associated to $\vec{X}$.

(a) Prove that the unit normal $\vec{N}_Y$ associated to $\vec{Y}$ is everywhere equal to $\vec{N}$ (except perhaps up to a sign).

(b) Call K_Y the Gaussian curvature for $\vec{Y}_r$ and K the Gaussian curvature of $\vec{X}$. Prove that

$$K(\vec{X}_u \times \vec{X}_v) = K_Y \frac{\partial \vec{Y}_r}{\partial u} \times \frac{\partial \vec{Y}_r}{\partial v}.$$

(c) Call V_Y the corresponding coordinate patches on the normal variation. Conclude that

$$\iint_V K \, dS = \iint_{V_Y} K_Y \, dS.$$

15. Consider plane curves $\vec{\alpha}(t) = (\alpha_1(s), \alpha_2(s))$ and $\vec{\beta}(t) = (\beta_1(t), \beta_2(t))$ both parametrized by arc length. Assume we are in $\mathbb{R}^3$ with standard basis $\{\vec{i}, \vec{j}, \vec{k}\}$. Consider the parametrized surface S given by

$$\vec{X}(s,t) = \vec{\alpha}(s) + \beta_1(t)\vec{U}(s) + \beta_2(t)\vec{k},$$

where $\vec{U}(s)$ is the usual unit normal vector for plane curves.

(a) Prove that if S is regular, then $1 - \beta_1(t)\kappa_\alpha(s) \neq 0$ for all (s,t), where $\kappa_\alpha(s)$ is the curvature of the plane curve $\vec{\alpha}$.

(b) Calculate the Gaussian curvature of S.

(c) Prove that $\kappa_\alpha(s) = 0$ or $\kappa_\beta(t) = 0$ imply that $K(s,t) = 0$, but explain why the converse is not true.

16. Let S be a regular oriented surface and p a point of S. Let $\vec{u}$ be any fixed unit vector in T_pS. Show that the mean curvature H at p is given by

$$H = \frac{1}{\pi} \int_0^\pi \kappa_n(\theta) \, d\theta,$$

where $\kappa_n(\theta)$ is the normal curvature of S along a direction making an angle θ with $\vec{u}$.

17. (*) *Theorem of Beltrami-Enneper*. Prove that the absolute value of the torsion at any point on an asymptotic curve with nonzero curvature is given by

$$|\tau| = \sqrt{-K},$$

where K is the Gaussian curvature of the surface at that point.

18. Let S be a regular surface in $\mathbb{R}^3$ parametrized by $\vec{X}$, and consider the linear transformation $T : \mathbb{R}^3 \to \mathbb{R}^3$ given by $T(\vec{x}) = A\vec{x}$ with respect to the standard basis, where A is an invertible matrix. It is usually an intractable problem to determine the Gaussian or mean curvature of the image surface $S' = T(S)$ from those of S. Nonetheless, it is possible to answer the following question: Prove that T preserves the sign of the Gaussian curvature of any surface in $\mathbb{R}^3$, more precisely, if $p \in S$ and $q = T(p)$ is the corresponding point on S', then $K(p) = 0 \Leftrightarrow K(q) = 0$ and $\operatorname{sign} K(p) = \operatorname{sign} K(q)$. [Hint: Use Equation (6.18).]

19. (*) Consider a regular surface $S \in \mathbb{R}^3$ defined by $F(x, y, z) = 0$. Use implicit differentiation to show that the Gaussian curvature to a point on S with coordinates (x, y, z) is

$$K = -\frac{1}{(F_x^2 + F_y^2 + F_z^2)^2} \begin{vmatrix} F_{xx} & F_{xy} & F_{xz} & F_x \\ F_{yx} & F_{yy} & F_{yz} & F_y \\ F_{zx} & F_{zy} & F_{zz} & F_z \\ F_x & F_y & F_z & 0 \end{vmatrix}.$$

6.7 Developable Surfaces; Minimal Surfaces

When studying plane curves, we showed in Proposition 1.3.7 that if the curvature $\kappa_g(t)$ of a regular plane curve is always 0, then the curve is a line segment. In this section, we wish to study properties of surfaces with either Gaussian curvature everywhere 0 or mean curvature everywhere 0. Since there exist formulas for the mean curvature and Gaussian curvature in terms of a particular parametrization, the equations $K = 0$ and $H = 0$ are partial differential equations. However, since they are nonlinear differential equations that a priori involve three unknown functions in two variables, they are intractable in general. Nonetheless, we shall study various classes of surfaces that satisfy $K = 0$ or $H = 0$.

Planes satisfy $K = 0$, but other surfaces do as well, for example, cylinders and cones. In a geometric sense, cylinders and cones in fact resemble a plane because, as any elementary school student knows, one can create a cylinder or a cone out of a flat paper without folding, stretching, or crumpling. In the first half of this section, we introduce ruled surfaces and determine the conditions for these ruled surfaces to have Gaussian curvature everywhere 0.

As of yet, we have neither seen a particularly intuitive interpretation of the mean curvature nor have presented surfaces that satisfy $H = 0$. However, as we shall see, surfaces that satisfy $H = 0$ have minimal surface area in the following sense. Given a simple closed curve C in $\mathbb{R}^3$, among surfaces S that have $C = \partial S$ as a boundary, a surface with minimal surface area will have mean curvature H everywhere 0. Consequently, a surface that satisfies $H = 0$ is called a *minimal surface*. Many articles and books are devoted to the study of minimal surfaces (see [22], [29], or [28] to name a few; an Internet search will reveal many more), so in the interest of space, the second half of this section gives only a brief introduction to minimal surfaces.

6.7.1 Developable Surfaces

Geometrically, we define a *ruled surface* as the union of a (differentiable) one-parameter family of straight lines in $\mathbb{R}^3$. One can specify each line in the family by a point $\vec{\alpha}(t)$ and by a direction given by another vector $\vec{w}(t)$. The adjective "differentiable" means that both $\vec{\alpha}$ and $\vec{w}$ are differentiable vector functions over some interval $I \subset \mathbb{R}$. We parametrize a ruled surface by

$$\vec{X}(t, u) = \vec{\alpha}(t) + u\,\vec{w}(t), \qquad \text{with } (t, u) \in I \times \mathbb{R}. \qquad (6.36)$$

We call the lines L_t passing through $\vec{\alpha}(t)$ with direction $\vec{w}(t)$ the *rulings*, and the curve $\vec{\alpha}(t)$ is called the *directrix* of the surface. We should note that this definition does not insist that ruled surfaces be regular; in particular, we allow singular points, that is, points where $\vec{X}_t \times \vec{X}_u = \vec{0}$.

Definition 6.7.1 A *developable surface* is a surface S such that at each point P on S, there is a line (called generator) through P that lies on S and such that S has the same tangent plane at all points on this generator.

A developable surface is a surface that can be formed by bending a portion of the plane into space. Intuitively speaking, each bend line corresponds to a generator. The set of generator lines on the surface define a one-parameter family of lines in $\mathbb{R}^3$ that sweep out the surface. Hence, every developable surface is a ruled surface. Now suppose we can choose a directrix $\vec{\alpha}(t)$ of the developable with the generators as the rulings. Then the normal vector is constant along each ruling and hence the partial derivative in that direction is 0. Then by Proposition 6.6.2, a developable surface has a Gaussian curvature that is identically 0.

Consequently, an equivalent definition of a developable surface is that it is a ruled surface that has Gaussian curvature that is constantly 0.

Example 6.7.2 (Cylinders) The simplest example of a ruled surface is a cylinder. The general definition of a cylinder is a ruled surface that can be given as a one-parameter family of lines $\{\vec{\alpha}(t), \vec{w}(t)\}$, where $\vec{\alpha}(t)$ is planar and $\vec{w}(t)$ is a constant vector in $\mathbb{R}^3$. It is easy to show that this has Gaussian curvature that is identically 0, so the cylinder is a developable surface.

Example 6.7.3 (Cones) The general definition of a cone is a surface that can be given as a one-parameter family of lines $\{\vec{\alpha}(t), \vec{w}(t)\}$, where $\vec{\alpha}(t)$ lies in a plane P and the rulings L_t all pass through some common point $p \notin P$. Therefore, the cone over $\vec{\alpha}(t)$ through p can be parametrized by

$$\vec{X}(t, u) = \vec{\alpha}(t) + u(p - \vec{\alpha}(t)).$$

It is easy to show that the Gaussian curvature is identically 0.

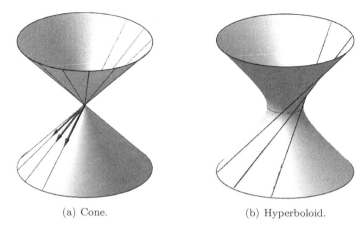

(a) Cone.   (b) Hyperboloid.

Figure 6.12: Ruled surfaces.

Example 6.7.4 As a perhaps initially surprising example, the hyperboloid of one sheet is also a ruled surface. The standard parametrization for the hyperboloid of one sheet is

$$\vec{X}(u, v) = (\cosh v \cos u, \cosh v \sin u, \sinh v).$$

Consider now the ruled surface with directrix $\vec{\alpha}(t) = (\cos t, \sin t, 0)$ and with ruling directions $\vec{w}(t) = \vec{\alpha}'(t) + \vec{k} = (-\sin t, \cos t, 1)$ (see Figure 6.12(b)). We obtain the parametrized surface

$$\vec{Y}(t, u) = (\cos t - u \sin t, \sin t + u \cos t, u).$$

Though it is not particularly easy to express $\vec{Y}$ as a reparametrization of $\vec{X}$, it is not hard to see that both of these parametrizations satisfy the following usual equation that gives the hyperboloid as a conic surface:

$$x^2 + y^2 - z^2 = 1.$$

This ruled surface is not developable since the Gaussian curvature is not identically 0.

In order to identify developable surfaces, we determine the Gaussian curvature for ruled surfaces generally. However, in order to simplify subsequent calculations, we make two additional assumptions that do not lose generality. First, note that one can impose the condition that $\|\vec{w}(t)\| = 1$ without changing the definition or the image of the parametrization In Equation (6.36). In defining particular ruled surfaces, it is usually easier not to make this assumption, but it does simplify calculations since $\|\vec{w}(t)\| = 1$ for all $t \in I$ implies that

$\vec{w}(t) \cdot \vec{w}(t)' = 0$. Second, note that different curves $\vec{\alpha}(t)$ can serve as the directrix for the same ruled surface, so we wish to employ a curve $\vec{\beta}(t)$ as a directrix, which will simplify the algebra in our calculations. We choose a curve $\vec{\beta}(t)$ satisfying $\vec{\beta}'(t) \cdot \vec{w}'(t) = 0$.

Since $\vec{\beta}(t)$ lies on $\vec{X}$, we can write

$$\vec{\beta}(t) = \vec{\alpha}(t) + u(t)\vec{w}(t). \tag{6.37}$$

Then

$$\vec{\beta}'(t) = \vec{\alpha}'(t) + u'(t)\vec{w}(t) + u(t)\vec{w}'(t),$$

and since $\vec{\beta}' \cdot \vec{w}' = 0$, we have

$$\vec{\alpha}'(t) \cdot \vec{w}'(t) + u(t)\vec{w}'(t) \cdot \vec{w}'(t) = 0.$$

Thus, we determine that

$$u(t) = -\frac{\vec{\alpha}'(t) \cdot \vec{w}'(t)}{\|\vec{w}'(t)\|^2}. \tag{6.38}$$

Furthermore, it is easy to prove (see Problem 6.7.4) that, with our present assumptions, $\vec{\beta}(t)$ is unique. We call the curve $\vec{\beta}(t)$ the *line of stricture* of the ruled surface.

We now use the parametrization for the ruled surface $\vec{X} = \vec{\beta}(t) + u\vec{w}(t)$ and proceed to calculate the first and second fundamental forms. Understanding that $\vec{\beta}$ and $\vec{w}$ are functions of t, we find that

$$\vec{X}_t = \vec{\beta}' + u\vec{w}', \qquad\qquad \vec{X}_{tt} = \vec{\beta}'' + u\vec{w}'',$$
$$\vec{X}_u = \vec{w}, \qquad\text{and}\qquad \vec{X}_{tu} = \vec{w}',$$
$$\vec{X}_t \times \vec{X}_u = \vec{\beta}' \times \vec{w} + u\vec{w}' \times \vec{w}, \qquad \vec{X}_{uu} = \vec{0}.$$

One can already notice that if $\vec{w}(t)$ is constant, then $L_{12} = L_{21} = L_{22} = 0$, which leads to $K = 0$ and proves that all cylinders have Gaussian curvature $K = 0$. We will assume now that $\vec{w}'(t)$ is not identically 0.

Because of the conditions $\vec{w} \cdot \vec{w}' = 0$ and $\vec{\beta}' \cdot \vec{w}' = 0$, we can conclude that $\vec{\beta}' \times \vec{w}$ is parallel to $\vec{w}'$. Thus, $\vec{\beta}' \times \vec{w}$ is perpendicular to $\vec{w} \times \vec{w}'$, so

$$\|\vec{X}_t \times \vec{X}_u\|^2 = \|\vec{\beta}' \times \vec{w}\|^2 + u^2\|\vec{w}'\|^2.$$

Furthermore, since $\vec{\beta}' \times \vec{w}$ is parallel to $\vec{w}'$ it is its own projection onto $\vec{w}'$. Hence, $\|\vec{\beta}' \times \vec{w}\| = |(\vec{\beta}' \times \vec{w}) \cdot \vec{w}'|/\|\vec{w}'\|$ and hence,

$$\|\vec{X}_t \times \vec{X}_u\|^2 = (\vec{\beta}'\vec{w}\vec{w}')^2/\|\vec{w}'\|^2 + u^2\|\vec{w}'\|^2 \tag{6.39}$$
$$= \frac{1}{\|\vec{w}'\|^2}\left((\vec{\beta}'\vec{w}\vec{w}')^2 + u^2\|\vec{w}'\|^4\right),$$

where we use the notation $(\vec{\beta}'\vec{w}\vec{w}')$ for $(\vec{\beta}' \times \vec{w}) \cdot \vec{w}'$, the triple-vector product in $\mathbb{R}^3$. Consequently, we can write the unit normal vector $\vec{N}$ as

$$\vec{N}(t, u) = \frac{\|\vec{w}'\|}{\sqrt{(\vec{\beta}'\vec{w}\vec{w}')^2 + u^2\|\vec{w}'\|^4}}(\vec{\beta}' \times \vec{w} + u\,\vec{w}' \times \vec{w}),$$

and therefore, again using the conditions that $\vec{w}\cdot\vec{w}' = 0$ and $\vec{\beta}'\cdot\vec{w}' = 0$, we get

$$(g_{ij}) = \begin{pmatrix} \|\vec{\beta}'\|^2 + u^2\|\vec{w}'\|^2 & \vec{\beta}' \cdot \vec{w} \\ \vec{\beta}' \cdot \vec{w} & 1 \end{pmatrix}$$

and (L_{ij}) is equal to

$$\frac{\|\vec{w}'\|}{\sqrt{(\vec{\beta}'\vec{w}\vec{w}')^2 + u^2\|\vec{w}'\|^4}}\begin{pmatrix} (\vec{\beta}'' + u\vec{w}'') \cdot (\vec{\beta}' \times \vec{w} + u\vec{w}' \times \vec{w}) & (\vec{\beta}'\vec{w}\vec{w}') \\ (\vec{\beta}'\vec{w}\vec{w}') & 0 \end{pmatrix}.$$

We cannot say much for the entry L_{11}, but thanks to Equation (6.35) for the Gaussian curvature and the fact that $\|\vec{X}_u \times \vec{X}_v\|^2 = \det(g_{ij})$ (see Equation (6.6)), we can calculate the Gaussian curvature for a ruled surface as

$$K = \frac{\det(L_{ij})}{\det(g_{ij})} = -\frac{\|\vec{w}'\|^4(\vec{\beta}'\vec{w}\vec{w}')^2}{\left((\vec{\beta}'\vec{w}\vec{w}')^2 + u^2\|\vec{w}'\|^4\right)^2}. \qquad (6.40)$$

This formula for the Gaussian curvature of a ruled surface makes a few facts readily apparent. First, $K \leq 0$ for all points of a ruled surface. Second, by Equation (6.39), all singular points of a ruled surface, i.e., points where $\vec{X}_t \times \vec{X}_u = \vec{0}$, must have $u = 0$ and therefore occur on the line of stricture. Finally, by Equation (6.40), a ruled surface satisfies $K(t, u) = 0$ if and only if $(\vec{\beta}'\vec{w}\vec{w}') = 0$ for all $t \in I$.

Let us return now to the general definition of a ruled surface from Equation (6.36) with regular curves $\vec{\alpha}(t)$ and $\vec{w}(t)$ with no conditions. Since both $\vec{w}$ and $\vec{w}'$ are perpendicular to $\vec{w} \times \vec{w}'$,

$$(\vec{\alpha}'\vec{w}\vec{w}') = \vec{\alpha}' \cdot \vec{w} \times \vec{w}' = \vec{\beta}' \cdot \vec{w} \times \vec{w}' = (\vec{\beta}'\vec{w}\vec{w}').$$

Furthermore, if we call $\hat{w} = \vec{w}/\|\vec{w}\|$, we remark that

$$(\vec{\alpha}'\vec{w}\vec{w}') = \|\vec{w}\|^2(\vec{\alpha}'\hat{w}\hat{w}') = \|\vec{w}\|^2(\vec{\beta}'\hat{w}\hat{w}').$$

Consequently, $(\vec{\beta}'\hat{w}\hat{w}') = 0$ if and only if $(\vec{\alpha}'\vec{w}\vec{w}') = 0$. We have proven the following proposition.

Proposition 6.7.5 *A ruled surface in $\mathbb{R}^3$ with directrix $\vec{\alpha}(t)$ and such that each ruling has direction $\vec{w}(t)$ is a developable surface if and only if $(\vec{\alpha}'\vec{w}\vec{w}') = 0$.*

Essentially by definition, every developable surface has Gaussian curvature that is identically 0. Surprisingly, the converse is true.

Theorem 6.7.6 *A regular surface S of class C^2 in $\mathbb{R}^3$ has $K = 0$ identically if and only if the surface is a developable surface.*

Proof: Suppose that S is a regular surface of class C^2 such that $K = 0$. Let $\vec{X} : U \to \mathbb{R}^3$ be a parametrization $\vec{X}(u, v)$ of a coordinate patch of S. The following reasoning will occur on this coordinate patch, but the result with extend to the whole surface. By Proposition 6.6.2, $K(u, v) = 0$ if and only if the set $\{\vec{N}_u, \vec{N}_v\}$ are everywhere linearly dependent. Also notice that $K = 0$ if and only if $L_{11}L_{22} - L_{12}^2 = 0$. Obviously, in order for this equality to hold L_{11} and L_{22} must have equal signs everywhere. Without loss of generality in what follows, we assume that they are nonnegative.

Consider the asymptotic curves on the surface whose parametrization $(u(t), v(t))$ satisfy the differential equation

$$L_{11}(u')^2 + 2L_{12}u'v' + L_{22}(v')^2 = 0$$
$$\Longleftrightarrow L_{11}(u')^2 + 2\sqrt{L_{11}L_{22}}u'v' + L_{22}(v')^2 = 0$$
$$\Longleftrightarrow (\sqrt{L_{11}}u' + \sqrt{L_{22}}v')^2 = 0.$$

Not both L_{11} and L_{22} can be zero for $L_{11}L_{22} - L_{12}^2 = 0$ to hold. If we suppose that $L_{22} \neq 0$, then by substitution, this gives the differential equation

$$\frac{dv}{du} = -\sqrt{\frac{L_{11}}{L_{22}}}.$$

By the Theorem of Existence and Uniqueness for differential equations, for each point $(u_0, v_0) \in U$, there is an asymptotic curve through (u_0, v_0). Consequently, at each point $p \in S$, there is a neighborhood of p for which it is possible to change parametrizations to use coordinates $(\bar{u}, \bar{v})$ such that the asymptotic curves corresponding to $\bar{v}$ are constant.

Using the $(\bar{u}, \bar{v})$ coordinates, the equation for asymptotic curves is simply $\bar{v}' = 0$ or more precisely $\bar{L}_{22}(\bar{v}')^2 = 0$. From this, we deduce that the components of the second fundamental form have $\bar{L}_{11} = \bar{L}_{12} = 0$. Thus

$$\vec{X}_{\bar{u}} \cdot \vec{N}_{\bar{u}} = \vec{X}_{\bar{v}} \cdot \vec{N}_{\bar{u}} = 0.$$

Thus $\vec{N}_{\bar{u}} = 0$ at all points on the surface and hence the normal vector depends on only one parameter, $\bar{v}$.

Now suppose that the tangent planes of S depend on only one parameter v. Then there exist a differentiable vector function $\vec{a}(v)$ and a differential real-valued function $f(v)$ such that the tangent spaces

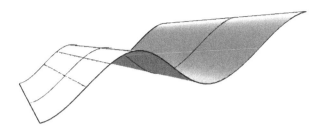

Figure 6.13: A developable surface.

of S (possibly on some open coordinate neighborhood of S) satisfy the equation

$$\vec{x} \cdot \vec{a}(v) = f(v), \qquad \text{where } \vec{x} = (x, y, z).$$

Given a fixed v_0 we have $\vec{x} \cdot \vec{a}(v_0) = f(v_0)$ as well as $\vec{x} \cdot \vec{a}(v_0 + h) = f(v_0 + h)$ as $h \to 0$. In particular, this leads to the fact the normal vectors to the surface S that are also in the tangent plane to S at a point that has $v = v_0$ must also satisfy

$$\vec{x} \cdot \vec{a}'(v_0) = f'(v_0),$$

which is another plane. Hence, such normal vectors are perpendicular to two planes. Consequently, the surface is a ruled surface, such that along the rulings, the normal vectors are constant. Thus the surface S is developable. □

It is interesting to remark that a ruled surface is a cone if $\vec{a}'(t) = 0$ and is a cylinder if and only if $\vec{w}'(t) = 0$, showing again that both cones and cylinders are developable and have $K = 0$. The exercises present examples of developable surfaces that are neither cones nor cylinders. Figure 6.13 also illustrates a developable surface that has $\vec{\alpha} = (0, t, \cos t)$.

Developable surfaces are particularly interesting for design and manufacturing. Because developable surfaces are ruled and have Gaussian curvature 0, they can be created by bending a region of the plane, without stretching it. Consequently, developable surfaces can be made out of sheet metal or any inelastic material that starts out flat. Some postmodern architecture exemplifies the use of developable surfaces (such as Frank Gehry's Guggenheim Museum in Bilbao or City of Wine Hotel Marquès de Riscal).

6.7.2 Minimal Surfaces

Definition 6.7.7 A *minimal surface* is a parametrized surface of class C^2 that satisfies the regularity condition and for which the mean curvature is identically 0.

We first wish to justify the name "minimal."

Let $\vec{X} : U \to \mathbb{R}^3$ be a coordinate neighborhood of a regular parametrized surface of class C^2. Let D' be a connected compact set in U, and let $D = \vec{X}(D')$. Let $h : U \to \mathbb{R}$ be a differentiable function. A normal variation of $\vec{X}$ over D' determined by h is the family of surfaces with $t \in (-\varepsilon, \varepsilon)$ defined by

$$\vec{X}^t : D' \longrightarrow \mathbb{R}^3$$

$$(u, v) \longmapsto \vec{X}(u, v) + th(u, v)\vec{N}(u, v).$$

Proposition 6.7.8 *Let $\vec{X}^t$ be a normal variation of $\vec{X}$ over a compact region D' and determined by some function h. For ε small enough, $\vec{X}^t$ satisfies the regularity condition for all $t \in (-\varepsilon, \varepsilon)$. In this case, the area of $\vec{X}^t$ is*

$$A(t) = \iint\limits_{D'} \sqrt{1 - 4thH + t^2 R} \sqrt{\det(g_{ij})}\, du\, dv \qquad (6.41)$$

for some function $R(u, v, t)$ that is polynomial in t.

Proof: Denote g_{ij}^t as the coefficients of the metric tensor. The surfaces of the normal variation have

$$\vec{X}_u^t = \vec{X}_u + th_u\vec{N} + th\vec{N}_u,$$

$$\vec{X}_v^t = \vec{X}_v + th_v\vec{N} + th\vec{N}_v.$$

Thus, we calculate,

$$g_{11}^t = g_{11} + 2th\vec{X}_u \cdot \vec{N}_u + t^2 h^2 \vec{N}_u \cdot \vec{N}_u + t^2 h_u^2,$$

$$g_{12}^t = g_{12} + th(\vec{X}_u \cdot \vec{N}_v + \vec{X}_v \cdot \vec{N}_u) + t^2 h^2 \vec{N}_u \cdot \vec{N}_v + t^2 h_u h_v,$$

$$g_{22}^t = g_{22} + 2th\vec{X}_v \cdot \vec{N}_v + t^2 h^2 \vec{N}_v \cdot \vec{N}_v + t^2 h_v^2.$$

However, by Equation (6.17), one can summarize the above equations as

$$g_{ij}^t = g_{ij} - 2th L_{ij} + t^2 h^2 \vec{N}_i \cdot \vec{N}_j + t^2 h_i h_j,$$

where we use the notation h_1 (resp. h_2) to indicate the partial derivative h_u (resp. h_v). Therefore, we calculate that

$$\det(g_{ij}^t) = \det(g_{ij}) - 2th(g_{11}L_{22} - 2g_{12}L_{12} + g_{22}L_{11}) + t^2 \bar{R},$$

where $\bar{R}(u, v, t)$ is a function of the form $A_0(u, v) + t A_1(u, v) + t^2 A_2(u, v)$ for continuous functions A_i defined over D'. However, the mean curvature is

$$H = \left(\frac{1}{2}\right) \frac{g_{11}L_{22} - 2g_{12}L_{12} + g_{22}L_{11}}{g_{11}g_{22} - g_{12}^2},$$

which leads to

$$\det(g_{ij}^t) = \det(g_{ij})(1 - 4thH) + t^2\bar{R} = \det(g_{ij})(1 - 4thH + t^2R),$$
$$(6.42)$$

where $R = \bar{R}/\det(g_{ij})$. Since D' is compact, the functions h, H, and R are bounded over D', which shows that

$$\lim_{t\to 0}\det(g_{ij}^t) = \det(g_{ij}) \quad \text{for all } (u,v) \in D'.$$

Hence, if $\varepsilon > 0$ is small enough, then $\det(g_{ij}^t) \neq 0$ for all $t \in (-\varepsilon, \varepsilon)$, and thus, all normal variations satisfy the regularity condition. The rest of the proposition follows from Equation (6.42) since

$$A(t) = \iint_{D'} \sqrt{\det(g_{ij}^t)}\,du\,dv.$$

$$\square$$

Proposition 6.7.9 Let $\vec{X} : U \to \mathbb{R}^3$ be a parametrized surface of class C^2, and let $D' \subseteq U$ be a compact set. Let $A(t)$ be the area function defined in Equation (6.41). Then $\vec{X}$ parametrizes a minimal surface if and only if $A'(0) = 0$ for all D' and all normal variations of $\vec{X}$ over D'.

Proof: We calculate

$$A'(t) = \iint_{D'} \frac{-4hH + 2tR + t^2R_t}{2\sqrt{1 - 4thH + t^2R}} \sqrt{\det(g_{ij})}\,du\,dv,$$

which implies that

$$A'(0) = -2\iint_{D'} hH\sqrt{\det(g_{ij})}\,du\,dv.$$

Obviously, if $\vec{X}$ parametrizes a minimal surface, which means that $H = 0$ for all $(u,v) \in U$, then $A'(0) = 0$ regardless of the function $h(u,v)$ or the compact set D'.

To prove the converse, suppose that $A'(0) = 0$ for all continuous functions $h(u,v)$ and all compact $D' \subset U$. Choosing $h(u,v) = H(u,v)$, since $\det(g_{ij}) > 0$, we have

$$A'(0) = -2\iint_{D'} H^2\sqrt{\det(g_{ij})}\,du\,dv,$$

so $A'(0) \leq 0$ for all choices of D'. Since $H(u,v)$ is continuous, if $H(u_0, v_0) \neq 0$ for any point (u_0, v_0), there is a compact set D' containing (u_0, v_0) such that $H(u,v)^2 > 0$ over D'. Thus, if H is anywhere nonzero, there exists a compact subset D' such that $A'(0) < 0$.

This is a contradiction, so we conclude that $A'(0) = 0$ for all h and all D' implies that $H(u, v) = 0$ for all $(u, v) \in U$.   □

Proposition 6.7.9 shows that a parametrized surface $\vec{X}$ that satisfies the regularity condition $\|\vec{X}_u \times \vec{X}_v\| \neq 0$ is minimal (has $H = 0$ everywhere) if it is a surface such that over every patch of surface there is no way to deform it along normal vectors to obtain a surface of lesser surface area.

Minimal surfaces have also enjoyed considerable popular attention, especially in museums of science, as soap films on a wire frame. When a wire frame is dipped into a soapy liquid and then pulled out, the surface tension on the soap film pulls the film into the state of least potential energy, which turns out to be the surface such that no normal variation can decrease surface area. When experimenting with soap film surfaces, we can use wire frames that are nonregular curves or not even curves at all, e.g., the skeleton of a cube.

The problem of determining a minimal surface with a given closed regular space curve C as a boundary was raised by Lagrange in 1760. However, the mathematical problem became known as Plateau's problem, after Joseph Plateau who specifically studied soap film surfaces. The 19th century saw a few specialized solutions to the problem, but it was not solved until 1930. (See [28] for a more complete history of Plateau's problem.)

Though we can easily construct a minimal surface with a soap film on a wire frame, either checking that a surface is minimal or finding a parametrization for a minimal surface is quite difficult. The study of minimal surfaces continues to provide new areas of research and connections with other branches of analysis. Perhaps among the most interesting results are a connection between complex analytic functions and minimal surfaces [11, p. 206] or the use of elliptic integrals to parametrize special minimal surfaces [28, Chapter 1].

PROBLEMS

1. Prove that the surface given by $z = kxy$, where k is a constant, is a ruled surface and give it as a one-parameter family of lines.

2. Figure 6.13 shows a developable surface that has the line of stricture $\vec{\alpha}(t) = (0, t, \cos t)$. Suppose the rulings are $\vec{w}(t) = (1, w_2(t), w_3(t))$. Find the equations or differential equation required by $w_2(t)$ and $w_3(t)$ to produce a developable surface. [In Figure 6.13, the rulings have $\vec{w}(t) = (1, -0.3t, -0.3 \cos t)$.]

3. The tangential surface to a space curve $\vec{\alpha}(t)$ was presented in Problem 6.6.6. Show that the curve $\vec{\alpha}$ is the line of stricture for the tangential surface. Show that the tangential surface to a regular space curve is a developable surface.

4. Suppose that $\{\vec{\alpha}_1(t), \vec{w}(t)\}$ and $\{\vec{\alpha}_2(t), \vec{w}(t)\}$, where $\|\vec{w}(t)\| = 1$, are two one-parameter families of lines that trace out the same ruled surface. Prove that Equation (6.37) with (6.38), using either $\vec{\alpha}_1$ or $\vec{\alpha}_2$, produce the same line of stricture.

5. Suppose that a ruled surface has $\vec{\alpha}(t) = (\cos t, \sin t, 0)$ as the line of stricture. Suppose also that $w(t) = (w_1(t), w_2(t), 1)$. Find algebraic or differential equations that $w_1(t)$ and $w_2(t)$ must satisfy so that the ruled surface is a developable surface.

6. Let S be an orientable surface, and let $\vec{\alpha}(s)$ be a curve on S parametrized by arc length. Assume that $\vec{\alpha}$ is nowhere tangent to the asymptotic direction on S, and define $\vec{N}(s)$ as the unit normal vector to S along $\vec{\alpha}(s)$. Consider the ruled surface

$$\vec{X}(s, u) = \vec{\alpha}(s) + u\,\frac{\vec{N}(s) \times \vec{N}'(s)}{\|\vec{N}'(s)\|}.$$

The assumption that $\vec{\alpha}'(s)$ is not an asymptotic direction ensures that $\vec{N}'(s) \neq \vec{0}$. Prove that $\vec{X}(s, v)$ is a developable surface. (This kind of surface is called the *envelope* of a family of tangent planes along a curve of a surface.)

7. Let $\vec{X}(t, v) = \vec{\alpha}(t) + v\vec{w}(t)$ be a developable surface. Prove that at a regular point

$$\vec{N}_v \cdot \vec{X}_t = 0 \qquad \text{and} \qquad \vec{N}_v \cdot \vec{X}_v = 0.$$

Use this result to prove that the tangent plane of a developable surface is constant along a line of ruling.

8. Show that a regular surface in $\mathbb{R}^3$ defined by an equation $F(x, y, z) = 0$ is developable if and only if

$$\begin{vmatrix} F_{xx} & F_{xy} & F_{xz} & F_x \\ F_{yx} & F_{yy} & F_{yz} & F_y \\ F_{zx} & F_{zy} & F_{zz} & F_z \\ F_x & F_y & F_z & 0 \end{vmatrix} = 0.$$

9. Consider the helicoid parametrized by $\vec{X}(u, v) = (v \cos u, v \sin u, cu)$, where c is a fixed number.
   (a) Determine the asymptotic curves.
   (b) Determine the lines of curvature.
   (c) Show that the mean curvature of the helicoid is 0.

10. Prove that the catenoid parametrized by

$$\vec{X}(u, v) = (a \cosh v \cos u, a \cosh v \sin u, av)$$

is a minimal surface.

11. Prove that Enneper's minimal surface $\vec{X}(u, v) = (u - \frac{u^3}{3} + uv^2, v - \frac{v^3}{3} + vu^2, u^2 - v^2)$ is indeed a minimal surface.

12. Show that for all constants D_1 and D_2, the surface parametrized by the function graph

$$\vec{X}(u,v) = (u, v, -\ln(\cos(u + D_1)) + \ln(\cos(-v + D_2)))$$

where it is defined is a minimal surface.

13. (**ODE**) Prove that the *only* nonplanar minimal surfaces of the form $\vec{X}(u,v) = (u, v, h(u) + k(v))$ have

$$h(u) = -\frac{1}{C}\ln(\cos(Cu + D_1)) \quad \text{and} \quad k(v) = \frac{1}{C}\ln(\cos(-Cv + D_2))$$

for some nonzero constant C and any constants D_1 and D_2.

14. Let $U \subset \mathbb{R}^2$, and suppose that $\vec{X} : U \to \mathbb{R}^3$ and $\vec{Y} : U \to \mathbb{R}^3$ parametrize two minimal surfaces defined over the same domain. Prove that for all $t \in [0, 1]$, the surfaces parametrized by $\vec{Z}^t(u,v) = (1-t)\vec{X}(u,v) + t\vec{Y}(u,v)$ are also minimal surfaces.

15. (**ODE**) Prove that the *only* surface of revolution that is a minimal surface is a catenoid (see Problem 6.7.10).

16. Suppose that S is a minimal regular surface with no planar points. Prove that at all points $p \in S$, the Gauss map $n : S \to \mathbb{S}^2$ satisfies

$$dn_p(\vec{w}_1) \cdot dn_p(\vec{w}_2) = -K(p)\vec{w}_1 \cdot \vec{w}_2$$

for all $\vec{w}_1, \vec{w}_2 \in T_pS$, where $K(p)$ is the Gaussian curvature of S at p. Use this result to show that on a minimal surface the angle between two intersecting curves on S is the angle between their images on $\mathbb{S}^2$ under the Gauss map n.

17. Let $\vec{X}(u,v)$ be a parametrization of a regular, orientable surface, and let $\vec{N}(u, v) = \vec{X}_u \times \vec{X}_v / \|\vec{X}_u \times \vec{X}_v\|$ be the orientation. A *parallel surface* to $\vec{X}$ is a surface parametrized by

$$\vec{Y}(u, v) = \vec{X}(u, v) + a\vec{N}(u, v).$$

(a) If K and H are the Gaussian and mean curvatures of $\vec{X}$, prove that

$$\vec{Y}_u \times \vec{Y}_v = (1 - 2Ha + Ka^2)\vec{X}_u \times \vec{X}_v.$$

(b) Prove that at regular points, the Gaussian curvature of $\vec{Y}$ is

$$\frac{K}{1 - 2Ha + Ka^2},$$

and the mean curvature is

$$\frac{H - Ka}{1 - 2Ha + Ka^2}.$$

(c) Use the above to prove that if $\vec{X}$ is a surface with constant mean curvature $H = c \neq 0$, then there is a parallel surface to $\vec{X}$ that has constant Gaussian curvature.

CHAPTER 7

Fundamental Equations of Surfaces

In Chapter 6, we began to study the local geometry of a surface and saw how many computations depend on the coefficients of the first and second fundamental forms. In fact, in Section 6.1 we saw that many metric calculations (angles between curves, arc length, area, etc.) depend only on the first fundamental form. We first approached such concepts from the perspective of objects in $\mathbb{R}^3$, but with the first fundamental form, one can perform all the calculations by using only the coordinates that parametrize the surface, without referring to the ambient space. In fact, it is not at all difficult to imagine a regular surface as a subset of $\mathbb{R}^n$, with $n > 3$, and the formulas that depend on the first fundamental form would remain unchanged. See Chapter 9. Concepts that rely only on the first fundamental form are called *intrinsic properties* of a surface.

On the other hand, concepts such as the second fundamental form, the Gauss map, Gaussian curvature, and principal curvatures were defined in reference to the unit normal vector, and these are not necessarily intrinsic properties. To illustrate this point, consider a parametrized surface S that is a subset of $\mathbb{R}^4$, and let p be a point of S. One can still define the tangent plane as the span of two linearly independent tangent vectors, but there no longer exists a unique (up to sign) unit normal vector to S at p. In this situation, one cannot define the Gauss map. Properties that we presented as depending on the Gauss map either cannot be defined in this situation or need to be defined in an alternate way.

This chapter studies what kind of information about a surface one can know from just the first fundamental form versus that which can be determined from knowledge of both the first and the second fundamental forms.

Section 7.1 introduces the Christoffel symbols and studies relations between the coefficients of the first and second fundamental forms. In Section 7.2, we present the famous Theorema Egregium,

DOI: 10.1201/9781003295341-7

which proves that the Gaussian curvature of a surface is in fact an intrinsic property. More precisely, the theorem expresses the Gaussian curvature in terms of the g_{ij} functions. In Section 7.3, we present another landmark result for surfaces in $\mathbb{R}^3$, the Fundamental Theorem of Surface Theory, which proves that under appropriate conditions, the coefficients of the first and second fundamental forms determine the surface up to position and orientation in space.

Some of the quantities we encounter in this chapter involve multiple indices. Some of these quantities represent a new mathematical object called a tensor. Tensors simultaneously generalize vectors, matrices, inner products, and many other objects that arise in linear algebra. Furthermore, just as it is possible to define a vector field that to each point in a region of the plane (or space) one associates a vector, so it is possible to define a tensor field.

An introduction to tensor notation and the classical description of a tensor is given in the appendix in Section A.1. The only notational convention we underscore here is the Einstein summation convention: In a tensor notation expression, whenever a superscript index also appears as a subscript index, it is understood that we sum over that index. For example, suppose that A_{kl}^{ij} and $B_{\beta\gamma}^{\alpha}$ are different collections of quantities with each index showing running from 1 to n, where n is the dimension of the ambient space $\mathbb{R}^n$. Then by the expression $A_{kl}^{ij}B_{im}^{k}$ we mean the set of quantities

$$C_{lm}^{j} = A_{kl}^{ij}B_{im}^{k} \overset{\text{def}}{=} \sum_{i=1}^{n}\sum_{k=1}^{n} A_{kl}^{ij}B_{im}^{k}.$$

The collection A_{kl}^{ij} of quantities consists of n^4 quantities, the collection of $B_{\beta\gamma}^{\alpha}$ consists of n^3 quantities, and the collection C_{lm}^{j} consists again of n^3 quantities.

7.1   Gauss's Equations; Christoffel Symbols

We now return to the study of surfaces. As we shall see, the results of this section assume that one can take the third derivative of a coordinate parametrization. Therefore, in this section and in the remainder of the chapter, unless otherwise stated we consider only regular surfaces of class C^3.

If one compares the theory of surfaces developed so far to the theory of curves, one will point out one major gap in the presentation of the former. In the theory of space curves, we discussed natural equations, namely the curvature $\kappa(s)$ and torsion $\tau(s)$ with respect to arc length, and we proved that these two functions locally define a unique curve up to its position in space. Implicit in the proof of this result for natural equations of curves was the fact that in general there do not exist algebraic relations between the functions $\kappa(s)$ and $\tau(s)$.

In the theory of surfaces, the problem of finding natural equations cannot be quite so simple. For example, even when restricting our attention to the first fundamental form, we know that given any three functions $E(u,v)$, $F(u,v)$, and $G(u,v)$, there does not necessarily exist a surface with

$$\begin{pmatrix} g_{11} & g_{12} \\ g_{21} & g_{22} \end{pmatrix} = \begin{pmatrix} E & F \\ F & G \end{pmatrix}$$

because we need the nontrivial requirement that $EG - F^2 > 0$. Furthermore, one might suspect that, because of the smoothness conditions that imply that $\vec{X}_{uuv} = \vec{X}_{uvu} = \vec{X}_{vuu}$, the (g_{ij}) and (L_{ij}) coefficients may satisfy some inherent relations.

Also, when discussing natural equations for space curves, one often uses the Frenet frame as a basis for $\mathbb{R}^3$, with origin at a point p of the curve. In particular, one must use the Frenet frame to obtain a parametrization of a curve in the neighborhood of p from the knowledge of the curvature $\kappa(s)$ and torsion $\tau(s)$. When performing calculations in the neighborhood of a point p on a surface S parametrized by $\vec{X} : U \to \mathbb{R}^3$, the basis $\{\vec{X}_1, \vec{X}_2, \vec{N}\}$ is the most natural, but, unlike the Frenet frame, this basis is not orthonormal. In addition, the basis $\{\vec{X}_1, \vec{X}_2, \vec{N}\}$ depends significantly on the parametrization $\vec{X}$, while a reparametrization of a curve can at most change the sign of $\vec{T}$ and $\vec{B}$.

One might propose $\{\vec{e}_1, \vec{e}_2, \vec{N}\}$, where $\vec{e}_1$ and $\vec{e}_2$ are the principal directions, as a basis more related to the geometry of a surface. The orthonormality has advantages, but calculations using this basis quickly become intractable. Furthermore, the eigenvectors $\vec{e}_1$ and $\vec{e}_2$ are not well-defined at umbilic points since the eigenspaces of dn_p at umbilic points consist of the whole tangent space T_pS. For these reasons, it remains more natural to use the basis $\{\vec{X}_1, \vec{X}_2, \vec{N}\}$ for $\mathbb{R}^3$ in a neighborhood of p and study the relations that arise between the (g_{ij}) and (L_{ij}) coefficients.

Recapping earlier definitions, we have

$$\begin{pmatrix} \vec{X}_1 \cdot \vec{X}_1 & \vec{X}_1 \cdot \vec{X}_2 & \vec{X}_1 \cdot \vec{N} \\ \vec{X}_2 \cdot \vec{X}_1 & \vec{X}_2 \cdot \vec{X}_2 & \vec{X}_2 \cdot \vec{N} \\ \vec{N} \cdot \vec{X}_1 & \vec{N} \cdot \vec{X}_2 & \vec{N} \cdot \vec{N} \end{pmatrix} = \begin{pmatrix} g_{11} & g_{12} & 0 \\ g_{21} & g_{22} & 0 \\ 0 & 0 & 1 \end{pmatrix}.$$

Every vector in $\mathbb{R}^3$ can be expressed as a linear combination in this basis. In particular, one would like to express the second derivatives $\vec{X}_{11}$, $\vec{X}_{12}$, $\vec{X}_{21}$, and $\vec{X}_{22}$ as linear combinations of this basis. From Equation (6.18), we know that $L_{ij} = \vec{X}_{ij} \cdot \vec{N}$, and since $\vec{N} \cdot \vec{X}_i = 0$, we deduce that L_{ij} is the coordinate of $\vec{X}_{ij}$ along basis vector $\vec{N}$.

However, we do not know the coordinates of $\vec{X}_{ij}$ along the other two basis vectors $\vec{X}_1$ and $\vec{X}_2$.

Definition 7.1.1 The collection of eight functions $\Gamma^i_{jk} : U \to \mathbb{R}$ with indices $1 \leq i, j, k \leq 2$ are defined as the unique functions that satisfy

$$\vec{X}_{jk} = \Gamma^1_{jk}\vec{X}_1 + \Gamma^2_{jk}\vec{X}_2 + L_{jk}\vec{N}. \tag{7.1}$$

Definition 7.1.1 names the functions Γ^i_{jk} implicitly, but we now proceed to find formulas for them in terms of the metric coefficients.

Note that though we have eight combinations for three indices, each ranging between 1 and 2, we do not in fact have eight distinct functions because $\vec{X}_{jk} = \vec{X}_{kj}$. Thus, we already know that

$$\Gamma^i_{jk} = \Gamma^i_{kj}.$$

Before determining formulas for the Γ^i_{jk} functions, we will first establish expressions for $\vec{X}_{ij} \cdot \vec{X}_k$ that are easier to find. Since these quantities occur frequently, there is a common shorthand symbol for them, namely,

$$[ij, k] = \vec{X}_{ij} \cdot \vec{X}_k.$$

Begin by fixing $j = k = 1$. Using the formulas

$$\frac{\partial g_{11}}{\partial x^1} = \frac{\partial}{\partial x^1}(\vec{X}_1 \cdot \vec{X}_1) = 2\vec{X}_{11} \cdot \vec{X}_1,$$

$$\frac{\partial g_{12}}{\partial x^1} = \frac{\partial}{\partial x^1}(\vec{X}_1 \cdot \vec{X}_2) = \vec{X}_{11} \cdot \vec{X}_2 + \vec{X}_{21} \cdot \vec{X}_1,$$

$$\frac{\partial g_{11}}{\partial x^2} = \frac{\partial}{\partial x^2}(\vec{X}_1 \cdot \vec{X}_1) = 2\vec{X}_{12} \cdot \vec{X}_1,$$

we deduce that

$$[11, 1] = \frac{1}{2}\frac{\partial g_{11}}{\partial x^1} \quad \text{and} \quad [11, 2] = \frac{\partial g_{12}}{\partial x^1} - \frac{1}{2}\frac{\partial g_{11}}{\partial x^2}.$$

Pursuing the calculations for the remaining cases, using the fact that (g_{ij}) is symmetric, and rewriting the results in a convenient manner, one can prove the following lemma.

Lemma 7.1.2 *For all indices $1 \leq i, j, k \leq 2$, we have*

$$[ij, k] = \vec{X}_{ij} \cdot \vec{X}_k = \frac{1}{2}\left(\frac{\partial g_{jk}}{\partial x^i} + \frac{\partial g_{ki}}{\partial x^j} - \frac{\partial g_{ij}}{\partial x^k}\right). \tag{7.2}$$

Using this lemma, one can easily establish a formula for the functions Γ^i_{jk}. We remind the reader that we shall often use the Einstein summation convention when an index is repeated in one superscript and one subscript position. (See Section A.1 for a more accurate explanation of the convention.) Before we give the proof, we also mention the useful artifice in tensorial notation called the *Kronecker delta* δ_i^j with $1 \le i \le n$ and $1 \le j \le n$ defined as

$$\delta_i^j = \begin{cases} 1 & \text{if } i = j \\ 0 & \text{if } i \ne j. \end{cases}$$

The Kronecker delta is a way to represent the identity matrix.

Proposition 7.1.3 *Let $\vec{X} : U \to \mathbb{R}^3$ be the parametrization of a regular surface of class C^2 in the neighborhood of some point $p = \vec{X}(q)$. Then the coefficients Γ_{jk}^i satisfy*

$$\Gamma_{jk}^i = \sum_{l=1}^{2} g^{il} [jk, l] = \sum_{l=1}^{2} g^{il} \frac{1}{2} \left(\frac{\partial g_{kl}}{\partial x^j} + \frac{\partial g_{lj}}{\partial x^k} - \frac{\partial g_{jk}}{\partial x^l}, \right) \qquad (7.3)$$

where (g^{ij}), with the indices in superscript, is the inverse matrix $(g_{ij})^{-1}$.

Proof: We have already determined formulas for $[ij, k] = \vec{X}_{ij} \cdot \vec{X}_k$. However, from Equation (7.1), taking a dot product with respect to the derivative vectors $\vec{X}_i$, we obtain

$$[ij, k] = \Gamma_{ij}^l g_{lk}.$$

Multiplying the matrix of the first fundamental form (g_{lk}) by its inverse, we can write
$$g_{lk} g^{k\alpha} = \delta_l^\alpha$$
where δ_l^α is the Kronecker delta (defined in Equation (A.3)). Then we get

$$\sum_{k=1}^{2} g^{k\alpha} [ij, k] = \Gamma_{ij}^l g_{lk} g^{k\alpha} = \Gamma_{ij}^l \delta_l^\alpha = \Gamma_{ij}^\alpha.$$

The proposition follows from the symmetry of the (g^{ij}) matrix. $\square$

Definition 7.1.4 The symbols $[ij, k]$ are called the Christoffel symbols of the first kind, while the functions Γ_{jk}^i are called the Christoffel symbols of the second kind or, more simply, just the Christoffel symbols.

The formula in Proposition 7.1.3 along with Definition 7.1.1 is called *Gauss's formula* for surfaces.

One of the first uses of Gauss's formula is that knowing the functions L_{jk} and Γ^i_{jk} allows us to write not just the second partial derivatives of the parametrization $\vec{X}$ with coordinates with respect to the basis ordered $(\vec{X}_1, \vec{X}_2, \vec{N})$, but also all higher derivatives of $\vec{X}(x^1, x^2)$. We point out that knowing g_{ij} in the neighborhood of a point allows us to determine the lengths of and angle between $\vec{X}_1$ and $\vec{X}_2$. Furthermore, as we shall see later, knowing the three distinct functions L_{jk} and the six distinct functions Γ^i_{jk} for a particular surface S in the neighborhood of p allows one to use a Taylor series expansion in the two variables x^1 and x^2 to write an infinite sum that provides a parametrization of S in a neighborhood of p once p, $\vec{X}_1$, and $\vec{N}$ are given. However, we still have not identified any relations that must exist between L_{ij} and g_{ij}, so we cannot yet state an equivalent to the natural equations theorem for space curves.

We point out that though we used a superscript and subscripts for the Christoffel symbol Γ^i_{jk}, these functions do *not* form the components of a tensor but rather transform according to the following proposition.

Proposition 7.1.5 *Let (x^1, x^2) and $(\bar{x}^1, \bar{x}^2)$ be two coordinate systems for a neighborhood of a point p on a surface S. If we denote by Γ^m_{ij} and $\bar{\Gamma}^\mu_{\alpha\beta}$ the Christoffel symbols in the respective coordinate systems, then they are related by*

$$\bar{\Gamma}^\mu_{\alpha\beta} = \frac{\partial x^i}{\partial \bar{x}^\alpha} \frac{\partial x^j}{\partial \bar{x}^\beta} \frac{\partial \bar{x}^\mu}{\partial x^m} \Gamma^m_{ij} + \frac{\partial^2 x^m}{\partial \bar{x}^\alpha \partial \bar{x}^\beta} \frac{\partial \bar{x}^\mu}{\partial x^m}.$$

Proof: We leave some of the details of this proof for the reader but present an outline here.

Consider two systems of coordinates (x^1, x^2) and $(\bar{x}^1, \bar{x}^2)$ for a neighborhood of a point p on a surface S. We know that the metric coefficients change according to

$$\bar{g}_{\alpha\beta} = \frac{\partial x^i}{\partial \bar{x}^\alpha} \frac{\partial x^j}{\partial \bar{x}^\beta} g_{ij} \quad \text{and} \quad \bar{g}^{\alpha\beta} = \frac{\partial \bar{x}^\alpha}{\partial x^i} \frac{\partial \bar{x}^\beta}{\partial x^j} g^{ij}. \qquad (7.4)$$

In the $(\bar{x}^1, \bar{x}^2)$ coordinate system, we denote the Christoffel symbols of the first kind by

$$\overline{[\alpha\beta, \nu]} = \frac{1}{2}\left(\frac{\partial \bar{g}_{\beta\nu}}{\partial \bar{x}^\alpha} + \frac{\partial \bar{g}_{\nu\alpha}}{\partial \bar{x}^\beta} - \frac{\partial \bar{g}_{\alpha\beta}}{\partial \bar{x}^\nu}\right), \qquad (7.5)$$

and we must first relate this to the Christoffel symbols $[ij, k]$ in the (x^1, x^2) coordinate system. Note that in this proof, we make indices (i, j, k, m) in the (x^1, x^2) coordinate system correspond to indices $(\alpha, \beta, \nu, \mu)$. Using Equation (7.4), the first term in Equation

(7.5) transforms according to

$$\frac{\partial \bar{g}_{\beta\nu}}{\partial \bar{x}^\alpha} = \frac{\partial^2 x^j}{\partial \bar{x}^\alpha \partial \bar{x}^\beta} \frac{\partial x^k}{\partial \bar{x}^\nu} g_{jk} + \frac{\partial x^j}{\partial \bar{x}^\beta} \frac{\partial^2 x^k}{\partial \bar{x}^\alpha \partial \bar{x}^\nu} g_{jk} + \frac{\partial x^j}{\partial \bar{x}^\beta} \frac{\partial x^k}{\partial \bar{x}^\nu} \frac{\partial x^i}{\partial \bar{x}^\alpha} \frac{\partial g_{jk}}{\partial x^i}.$$

(7.6)

Equation (7.6) and the corresponding results of the other two terms in the sum in Equation (7.5) produce an expression for the transformation property of the Christoffel symbol of the first kind. The expression is not pleasing, especially since it involves the coefficients g_{ij} explicitly. One must now calculate

$$\bar{\Gamma}^\mu_{\alpha\beta} = \sum_{\nu=1}^{2} \bar{g}^{\mu\nu} \overline{[\alpha\beta, \nu]} = \sum_{\nu=1}^{2} \frac{\partial \bar{x}^\mu}{\partial x^m} \frac{\partial \bar{x}^\nu}{\partial x^l} g^{ml} \overline{[\alpha\beta, \nu]},$$

where we replace $\overline{[\alpha\beta, \nu]}$ with the terms found in Equation (7.6) and similar equalities. After appropriate simplifications, we find that

$$\begin{aligned}
\bar{\Gamma}^\mu_{\alpha\beta} = {} & \frac{\partial \bar{x}^\mu}{\partial x^m} \frac{\partial x^i}{\partial \bar{x}^\alpha} \frac{\partial x^j}{\partial \bar{x}^\beta} \Gamma^m_{ij} + \frac{1}{2} \left(\frac{\partial \bar{x}^\mu}{\partial x^j} \frac{\partial^2 x^j}{\partial \bar{x}^\alpha \partial \bar{x}^\beta} + \frac{\partial \bar{x}^\mu}{\partial x^i} \frac{\partial^2 x^i}{\partial \bar{x}^\alpha \partial \bar{x}^\beta} \right) \\
& + \frac{1}{2} \frac{\partial \bar{x}^\nu}{\partial x^l} \frac{\partial \bar{x}^\mu}{\partial x^m} \left(\frac{\partial x^j}{\partial \bar{x}^\alpha} \frac{\partial^2 x^k}{\partial \bar{x}^\alpha \partial \bar{x}^\nu} g^{ml} g_{jk} - \frac{\partial x^j}{\partial \bar{x}^\alpha} \frac{\partial^2 x^i}{\partial \bar{x}^\alpha \partial \bar{x}^\nu} g^{ml} g_{ij} \right) \\
& + \frac{1}{2} \frac{\partial \bar{x}^\nu}{\partial x^l} \frac{\partial \bar{x}^\mu}{\partial x^m} \left(\frac{\partial x^i}{\partial \bar{x}^\alpha} \frac{\partial^2 x^k}{\partial \bar{x}^\beta \partial \bar{x}^\nu} g^{ml} g_{ki} - \frac{\partial x^i}{\partial \bar{x}^\alpha} \frac{\partial^2 x^j}{\partial \bar{x}^\nu \partial \bar{x}^\beta} g^{ml} g_{ji} \right).
\end{aligned}$$

(7.7)

However, it is important to remember that when one sums over an index, the actual name of the index does not change the result of the summation. Applying this observation to Equation (7.7) and remembering that the components g_{ij} are symmetric in their indices finishes the proof of the proposition. $\square$

One should not view the fact that the Christoffel symbols Γ^i_{jk} do not form the components of a tensor as just a small annoyance; it is of fundamental importance in the theory of manifolds. From an intuitive perspective, one could understand tensors as objects that are related to the tangent space to a surface at a point. However, since the Γ^i_{jk} functions explicitly involve the second derivatives of a parametrization of a neighborhood of a point on a surface, one might have been able to predict that the Γ^i_{jk} functions would not necessarily form the components of a tensor.

Example 7.1.6 (Sphere) As a simple example, consider the usual parametrization of the sphere,

$$\vec{X}(x^1, x^2) = (R \cos x^1 \sin x^2, R \sin x^1 \sin x^2, R \cos x^2).$$

In spherical coordinates, the variable names we used correspond to $x^1 = \theta$ as the meridian and $x^2 = \varphi$ the latitude, measured as the angle down from the positive z-axis. A simple calculation leads to

$$g_{ij} = \begin{pmatrix} R^2 \sin^2(x^2) & 0 \\ 0 & R^2 \end{pmatrix} \quad \text{and} \quad g^{ij} = \frac{1}{R^2} \begin{pmatrix} \frac{1}{\sin^2(x^2)} & 0 \\ 0 & 1 \end{pmatrix}.$$

We first note that $[ij, k] = [ji, k]$. Then by (7.2), we find that

$$[11, 1] = \frac{1}{2}\frac{\partial g_{11}}{\partial x^1} = 0,$$

$$[12, 1] = [21, 1] = \frac{1}{2}\left(\frac{\partial g_{21}}{\partial x^1} + \frac{\partial g_{11}}{\partial x^2} - \frac{\partial g_{12}}{\partial x^1}\right) = R^2 \sin(x^2)\cos(x^2),$$

$$[22, 1] = \frac{1}{2}\left(\frac{\partial g_{21}}{\partial x^2} + \frac{\partial g_{12}}{\partial x^2} - \frac{\partial g_{22}}{\partial x^1}\right) = 0,$$

$$[11, 2] = \frac{1}{2}\left(\frac{\partial g_{12}}{\partial x^1} + \frac{\partial g_{21}}{\partial x^2} - \frac{\partial g_{11}}{\partial x^2}\right) = -R^2 \sin(x^2)\cos(x^2),$$

$$[12, 2] = [21, 2] = \frac{1}{2}\left(\frac{\partial g_{22}}{\partial x^1} + \frac{\partial g_{21}}{\partial x^2} - \frac{\partial g_{12}}{\partial x^2}\right) = 0,$$

$$[22, 2] = \frac{1}{2}\frac{\partial g_{22}}{\partial x^2} = 0.$$

Then, using (7.3), we get

$$\Gamma_{11}^1 = g^{11}[11, 1] + g^{12}[11, 2] = 0,$$
$$\Gamma_{12}^1 = \Gamma_{21}^1 = g^{11}[12, 1] + g^{12}[12, 2] = \cot(x^2),$$
$$\Gamma_{22}^1 = g^{11}[22, 1] + g^{12}[22, 2] = 0,$$
$$\Gamma_{11}^2 = g^{21}[11, 1] + g^{22}[11, 2] = -\sin(x^2)\cos(x^2),$$
$$\Gamma_{12}^2 = \Gamma_{21}^2 = g^{21}[12, 1] + g^{22}[12, 2] = 0,$$
$$\Gamma_{22}^2 = g^{21}[22, 1] + g^{22}[22, 2] = 0.$$

As an example of an application of Gauss's equation, recall from Equation (6.34) that two vectors $\vec{u}_1$ and $\vec{u}_2$ in the tangent space T_pS to a surface S at p are called conjugate if

$$I_p(dn_p(\vec{u}_1), \vec{u}_2) = I_p(\vec{u}_1, dn_p(\vec{u}_2)) = 0. \qquad (7.8)$$

Intuitively speaking, Equation (7.8) states that conjugate directions are such that the direction of change of the unit normal vector $\vec{N}$ along $\vec{u}_1$ is perpendicular to $\vec{u}_2$ and vice versa. Given a parametrization $\vec{X}$ of a regular surface in the neighborhood of a point p, we say that $\vec{X}$ produces a conjugate set of coordinate lines if

$$dn_p(\vec{X}_1) \cdot \vec{X}_2 = 0 = dn_p(\vec{X}_2) \cdot \vec{X}_1$$

for all (x^1, x^2) in the domain of $\vec{X}$. One can state this alternatively as

$$\vec{N}_1 \cdot \vec{X}_2 = 0 = \vec{N}_2 \cdot \vec{X}_1,$$

so $\vec{X}$ produces a set of conjugate coordinate lines if and only if $L_{12} = 0$. In this situation, Gauss's equation for $\vec{X}_{12}$ reduces to

$$\vec{X}_{12} = \Gamma^1_{12}\vec{X}_1 + \Gamma^2_{12}\vec{X}_2, \tag{7.9}$$

so the parametrization $\vec{X}$ has a conjugate set of coordinate lines if and only if Equation (7.9) holds.

In addition, Equation (7.9) encompasses three independent linear differential equations in the rectangular coordinate functions of the parametrization. Therefore, given functions $P(x^1, x^2)$ and $Q(x^1, x^2)$, any three linearly independent solutions to the equation

$$\frac{\partial^2 f}{\partial x^1 \partial x^2} - P(x^1, x^2)\frac{\partial f}{\partial x^1} - Q(x^1, x^2)\frac{\partial f}{\partial x^2} = 0, \tag{7.10}$$

when taken as the coordinate functions, give a parametrization of a surface with conjugate coordinate lines in which $\Gamma^1_{12} = P(x^1, x^2)$ and $\Gamma^2_{12} = Q(x^1, x^2)$. Equation (7.10) is, in general, not simple to solve but since $\vec{N}$ does not appear in Equation (7.10), the coordinate function solutions are independent of each other.

Example 7.1.7 (Intrinsic Geometry) Since the Christoffel symbols are defined using the metric coefficients and their partial derivatives, these functions are intrinsic. We can use the metric coefficients $g_{11}(u, v) = 1/v^2$, $g_{12}(u, v) = 0 = g_{21}(u, v)$, and $g_{22}(u, v) = 1/v^2$ from Example 6.1.9 to compute the Christoffel symbols for the points in the Poincaré upper half-plane. Note that

$$0 = \frac{\partial g_{11}}{\partial u} = \frac{\partial g_{12}}{\partial u} = \frac{\partial g_{22}}{\partial u}$$

since the coefficients only depend on v. Also

$$\frac{\partial g_{11}}{\partial v} = \frac{\partial g_{22}}{\partial v} = -\frac{2}{v^3} \text{ and } \frac{\partial g_{12}}{\partial v} = 0.$$

From this information we can calculate the Christoffel symbols for this metric.

Using Equation (7.2) we find

$$[11, 1] = 0, \qquad [11, 2] = \frac{1}{v^3},$$

$$[12, 1] = -\frac{1}{v^3}, \qquad [12, 2] = 0,$$

$$[21, 1] = -\frac{1}{v^3}, \qquad [21, 2] = 0,$$

$$[22, 1] = 0, \qquad [22, 2] = -\frac{1}{v^3}.$$

Then using Equation (7.2), we get

$$\Gamma^1_{11} = 0, \qquad \Gamma^2_{11} = 1/v,$$
$$\Gamma^1_{12} = -1/v, \qquad \Gamma^2_{12} = 0,$$
$$\Gamma^1_{21} = -1/v, \qquad \Gamma^2_{21} = 0,$$
$$\Gamma^1_{22} = 0, \qquad \Gamma^2_{22} = -1/v.$$

PROBLEMS

1. Fill in the details of the proof for Proposition 7.1.5.

2. Calculate the Christoffel symbols for the torus parametrized by
$$\vec{X}(u,v) = \big((a + b\cos v)\cos u, (a + b\cos v)\sin u, b\sin v\big),$$
   where we assume that $b < a$.

3. Calculate the Christoffel symbols for functions graphs, i.e., surfaces parametrized by $\vec{X}(u,v) = (u, v, f(u,v))$, where f is a function from $U \subset \mathbb{R}^2$ to $\mathbb{R}$.

4. Calculate the Christoffel symbols for surfaces of revolution (see Problem 5.2.14).

5. Calculate the Christoffel symbols for the pseudosphere (see Example 6.6.6).

6. *Tubes.* Let $\vec{\alpha}(t)$ be a regular space curve and let r be small enough that the tube
$$\vec{X}(t,u) = \vec{\alpha}(t) + (r\cos u)\vec{P}(t) + (r\sin u)\vec{B}(t)$$
   is a regular surface. Calculate the Christoffel symbols for this tube.

7. Prove that a parametrization $\vec{X}$ of a neighborhood on a surface S is such that all the coordinate lines $x^2 = c$ are asymptotic lines if and only if $L_{11} = 0$.

8. Defining the function $D(x^1, x^2)$ as $D^2 = \det(g_{ij})$, show that
$$\frac{\partial}{\partial x^1}(\ln D) = \Gamma^1_{11} + \Gamma^2_{12} \qquad \text{and} \qquad \frac{\partial}{\partial x^2}(\ln D) = \Gamma^1_{12} + \Gamma^2_{22}.$$

9. Suppose that S is a surface with a parametrization that satisfies $\Gamma^1_{12} = \Gamma^2_{12} = 0$. Prove that S is a translation surface.

10. Using D as defined in Problem 7.1.8, calling $\theta(x^1, x^2)$ the angle between the coordinate lines, prove that
$$\frac{\partial\theta}{\partial x^1} = -\frac{D}{g_{11}}\Gamma^2_{11} - \frac{D}{g_{22}}\Gamma^2_{11} \qquad \text{and} \qquad \frac{\partial\theta}{\partial x^2} = -\frac{D}{g_{11}}\Gamma^2_{12} - \frac{D}{g_{22}}\Gamma^1_{22}.$$
$$\text{(7.11)}$$
    Show that if the parametrization is orthogonal, i.e., $g_{12} = 0$, Equation (7.11) becomes
$$g_{22}\Gamma^2_{11} + g_{11}\Gamma^1_{12} = 0 \qquad \text{and} \qquad g_{22}\Gamma^2_{12} + g_{11}\Gamma^1_{22} = 0.$$

11. As in Example 7.1.7, calculate the Christoffel symbols of the first and second kind for a general metric of the form $g_{11}(u,v) = f(v)$, $g_{22}(u,v) = f(v)$ and $g_{12}(u,v) = 0$.

7.2 Codazzi Equations; Theorema Egregium

Gauss' formula in Equation (7.1) defines expressions for any second derivative $\vec{X}_{ij}$ in terms of the coefficients of the first and second fundamental forms and the basis $\{\vec{X}_1, \vec{X}_2, \vec{N}\}$. We remind the reader that the symmetry of the dot product imposes $g_{ij} = g_{ji}$. Furthermore, from $\vec{X}_{ij} = \vec{X}_{ji}$, we also deduced that $L_{ij} = L_{ji}$.

Definition 5.3.1 of a regular surface imposes the condition that all the higher derivatives of any parametrization $\vec{X}$ of a coordinate patch be continuous. Therefore, for a regular parametrization, the order in which one takes derivatives with respect to given variables is irrelevant. In particular, for the mixed third derivatives, we have $\vec{X}_{112} = \vec{X}_{121}$ and $\vec{X}_{221} = \vec{X}_{122}$, which can be listed in the following slightly more suggestive manner:

$$\frac{\partial \vec{X}_{11}}{\partial x^2} = \frac{\partial \vec{X}_{12}}{\partial x^1} \quad \text{and} \quad \frac{\partial \vec{X}_{22}}{\partial x^1} = \frac{\partial \vec{X}_{12}}{\partial x^2}.$$

Consequently, because of the nontrivial expressions in Gauss's equations, we obtain

$$\frac{\partial}{\partial x^2}\left(\Gamma_{11}^1 \vec{X}_1 + \Gamma_{11}^2 \vec{X}_2 + L_{11}\vec{N}\right) = \frac{\partial}{\partial x^1}\left(\Gamma_{12}^1 \vec{X}_1 + \Gamma_{12}^2 \vec{X}_2 + L_{12}\vec{N}\right),$$

$$\frac{\partial}{\partial x^2}\left(\Gamma_{12}^1 \vec{X}_1 + \Gamma_{12}^2 \vec{X}_2 + L_{12}\vec{N}\right) = \frac{\partial}{\partial x^1}\left(\Gamma_{22}^1 \vec{X}_1 + \Gamma_{22}^2 \vec{X}_2 + L_{22}\vec{N}\right).$$

$$(7.12)$$

It is natural to equate the basis components of each equation. However, when we take the partial derivatives, we will obtain expressions involving second derivatives of $\vec{X}$ and derivatives of $\vec{N}$ that we need to put back into the usual $\{\vec{X}_1, \vec{X}_2, \vec{N}\}$ basis using the Weingarten equations given in Equation (6.27) or Gauss's formula in Equation (7.1). Doing this transforms Equation (7.12) into six distinct equations from which we deduce two significant theorems in the theory of surfaces.

Theorem 7.2.1 (Codazzi Equations) *For any parametrized surface of class C^3, the following hold:*

$$\frac{\partial L_{11}}{\partial x^2} - \frac{\partial L_{12}}{\partial x^1} = L_{11}\Gamma_{12}^1 + L_{12}\Gamma_{12}^2 - L_{12}\Gamma_{11}^1 - L_{22}\Gamma_{11}^2,$$

$$\frac{\partial L_{12}}{\partial x^2} - \frac{\partial L_{22}}{\partial x^1} = L_{11}\Gamma_{22}^1 + L_{12}\Gamma_{22}^2 - L_{12}\Gamma_{12}^1 - L_{22}\Gamma_{12}^2.$$

$$(7.13)$$

We can summarize these two equations in one by

$$\frac{\partial L_{ij}}{\partial x^l} - \Gamma_{il}^k L_{kj} = \frac{\partial L_{il}}{\partial x^j} - \Gamma_{ij}^k L_{kl}$$

for all $1 \le i, j, l \le 2$, where Einstein summation is implied.

Proof: Equate the $\vec{N}$ coefficient functions in both equations from Equation (7.12).   □

The classical formulation of the above relationships in Equation (7.13) is collectively called the Codazzi equations or sometimes the Mainardi-Codazzi equations. These equations present a relationship that must hold between the coefficients of the first and second fundamental forms that must hold for all regular surfaces. They are a crucial part of the Fundamental Theorem of Surface Theory that we will present in Section 7.3.

The following theorem is another result that stems from equating components along $\vec{X}_1$ and $\vec{X}_2$ in Equation (7.12).

Theorem 7.2.2 (Theorema Egregium) *The Gaussian curvature of a surface is an intrinsic property of the surface, that is, it depends only on the coefficients of the metric tensor and higher derivatives thereof.*

Proof: Equating the $\vec{X}_1$ and $\vec{X}_2$ coefficient functions in the first equation of Equation (7.12), we obtain

$$\frac{\partial \Gamma^1_{11}}{\partial x^2} + \Gamma^1_{11}\Gamma^1_{12} + \Gamma^2_{11}\Gamma^1_{22} + L_{11}a^1_2 = \frac{\partial \Gamma^1_{12}}{\partial x^1} + \Gamma^1_{12}\Gamma^1_{11} + \Gamma^2_{12}\Gamma^1_{12} + L_{12}a^1_1,$$

$$\frac{\partial \Gamma^2_{11}}{\partial x^2} + \Gamma^1_{11}\Gamma^2_{12} + \Gamma^2_{11}\Gamma^2_{22} + L_{11}a^2_2 = \frac{\partial \Gamma^2_{12}}{\partial x^1} + \Gamma^1_{12}\Gamma^2_{11} + \Gamma^2_{12}\Gamma^2_{12} + L_{12}a^2_1.$$

Since $a^i_j = -\sum_{l=1}^2 L_{jl}g^{li}$, where $(g^{ij}) = (g_{ij})^{-1}$, after some simplification, we can write these two equations as

$$g_{21}\frac{L_{11}L_{22} - (L_{12})^2}{\det g} = \frac{\partial \Gamma^1_{12}}{\partial x^1} - \frac{\partial \Gamma^1_{11}}{\partial x^2} + \Gamma^2_{12}\Gamma^1_{12} - \Gamma^2_{11}\Gamma^1_{22}, \quad (7.14)$$

$$-g_{11}\frac{L_{11}L_{22} - (L_{12})^2}{\det g} = \frac{\partial \Gamma^2_{12}}{\partial x^1} - \frac{\partial \Gamma^2_{11}}{\partial x^2} + \Gamma^1_{12}\Gamma^2_{11} + \Gamma^2_{12}\Gamma^2_{12}$$
$$- \Gamma^1_{11}\Gamma^2_{12} - \Gamma^2_{11}\Gamma^2_{22}. \quad (7.15)$$

Since $K = \det(L_{ij})/\det(g_{ij})$, we can use either one of these equations to obtain a formula for K in terms of the function g_{ij} and Γ^i_{jk}. However, since the Christoffel symbols Γ^i_{jk} are themselves determined by the coefficients of the metric tensor, then these equations provide formulas for the Gaussian curvature exclusively in terms of the first fundamental form.   □

Gauss coined the name "Theorema Egregium," which means "an excellent theorem," and indeed this result should seem rather surprising. A priori, the Gaussian curvature depends on the components of the first fundamental form and the second fundamental form.

After all, the Gaussian curvature at a point p on the surface is $K = \det(dn_p)$, where dn_p is the differential of the Gauss map. Since the coefficients of the matrix dn_p with respect to the basis $\{\vec{X}_1, \vec{X}_2\}$ involve the L_{ij} coefficients, one would naturally expect that we would need the second fundamental form to calculate K. However, the Theorema Egregium shows that only the coefficient functions of the first fundamental form are necessary.

In the study of surfaces, we have described a geometric quantity as a "local" or "global property" of a curve or surface if it does not change under the orientation and position of the curve or surface in space. On the other hand, topology studies properties of point sets that are invariant under any homeomorphism – a bijective function that is continuous in both directions. The concept of an intrinsic property lies somewhere between these two extremes. Any bijective function between two regular surfaces preserving the metric tensor defines an *isometry* between the two surfaces. Then intrinsic geometry will treat these two surfaces as the same.

In the above proof, Equations (7.14) and (7.15) motivate the following definition of the *Riemann symbols*:

$$R^l_{ijk} = \frac{\partial \Gamma^l_{ik}}{\partial x^j} - \frac{\partial \Gamma^l_{jk}}{\partial x^i} + \Gamma^m_{ik}\Gamma^l_{mj} - \Gamma^m_{jk}\Gamma^l_{mi}. \qquad (7.16)$$

(Note that by Einstein summation convention, the term $\Gamma^m_{ik}\Gamma^l_{mj}$ means that we sum over $m = 1, 2$, and likewise for the term $\Gamma^m_{jk}\Gamma^l_{mi}$.) As it turns out (see Problem 7.2.3), the Riemann symbols form the components of a $(1, 3)$-tensor. Then, a closely associated tensor denoted by

$$R_{ijkl} = R^m_{ijk} g_{ml} \qquad (7.17)$$

has the interesting property that $R_{1212} = \det(L_{ij})$, and hence, the Gaussian curvature of a surface at a point is given by the component function

$$K = \frac{R_{1212}}{\det(g_{ij})}. \qquad (7.18)$$

The tensor associated to the components R^l_{ijk} is called the *Riemann curvature tensor* and plays an important role in the analysis on manifolds.

Example 7.2.3 (Cone) As a simple example, consider the right circular cone defined by the parametrization

$$\vec{X}(u, v) = (v \cos u, v \sin u, v),$$

where $u \in [0, 2\pi]$ and $v > 0$, which we know, as a developable surface, has Gaussian curvature $K = 0$ everywhere. Simple calculations give, for the metric tensor,

$$g_{11} = v^2, \qquad g_{12} = 0, \qquad g_{22} = 2.$$

For the Christoffel symbols, one obtains

$$\Gamma^2_{11} = -\frac{1}{2}v, \qquad \Gamma^1_{12} = \Gamma^1_{21} = \frac{1}{v}$$

with all the other symbols of the second kind $\Gamma^i_{jk} = 0$. With these data, an application of Equation (7.16) shows that all but two of the Riemann symbols vanish simply because all the terms vanish. However, for the two nontrivial terms, we calculate

$$
\begin{aligned}
R^1_{122} &= \frac{\partial \Gamma^1_{12}}{\partial v} - \frac{\partial \Gamma^1_{22}}{\partial u} + \Gamma^1_{12}\Gamma^1_{12} + \Gamma^2_{12}\Gamma^1_{22} - \Gamma^1_{22}\Gamma^1_{11} - \Gamma^2_{22}\Gamma^1_{21} \\
&= -\frac{1}{v^2} + \frac{1}{v^2} = 0, \\
R^2_{121} &= \frac{\partial \Gamma^2_{11}}{\partial v} - \frac{\partial \Gamma^2_{12}}{\partial u} + \Gamma^1_{11}\Gamma^2_{12} + \Gamma^2_{11}\Gamma^2_{22} - \Gamma^1_{21}\Gamma^2_{11} - \Gamma^2_{21}\Gamma^2_{21} \\
&= -\frac{1}{2} - \left(\frac{1}{v}\right)\left(-\frac{1}{2}v\right) = 0.
\end{aligned}
$$

Consequently, though the Christoffel symbols are not identically 0, the Riemann curvature tensor is identically 0. We find then that $R_{1212} = 0$ as a function of u and v, and hence, using Equation (7.18), we recover the fact that the cone has Gaussian curvature identically 0 everywhere.

The Theorema Egregium (Theorem 7.2.2) excited the mathematics community in the middle of the 19th century and sparked a search for a variety of alternative formulations for the Gaussian curvature of a surface as a function of the g_{ij} metric coefficients. We leave a few such formulas for K for the exercises, but we present a few here.

Recall that $L_{ij} = \vec{X}_{ij} \cdot \vec{N}$, $\det(g_{ij}) = \|\vec{X}_u \times \vec{X}_v\|^2$, and $\vec{N} = \vec{X}_u \times \vec{X}_v / \|\vec{X}_u \times \vec{X}_v\|$. From these facts, one can easily see that

$$K = \frac{L_{11}L_{22} - L^2_{12}}{\det(g_{ij})} = \frac{(\vec{X}_{uu}\vec{X}_u\vec{X}_v)(\vec{X}_{vv}\vec{X}_u\vec{X}_v) - (\vec{X}_{uv}\vec{X}_u\vec{X}_v)^2}{\det(g_{ij})^2},$$

$$(7.19)$$

where $(\vec{A}\vec{B}\vec{C})$ is the triple-vector product in $\mathbb{R}^3$. However, the triple-vector product is a determinant and we know that for all matrices M_1 and M_2, one has $\det(M_1M_2) = \det(M_1)\det(M_2)$, and so Equation

(7.19) becomes

$$K = \frac{1}{\det(g_{ij})^2} \left(\begin{vmatrix} \vec{X}_{uu} \cdot \vec{X}_{vv} & \vec{X}_{uu} \cdot \vec{X}_u & \vec{X}_{uu} \cdot \vec{X}_v \\ \vec{X}_u \cdot \vec{X}_{vv} & \vec{X}_u \cdot \vec{X}_u & \vec{X}_u \cdot \vec{X}_v \\ \vec{X}_v \cdot \vec{X}_{vv} & \vec{X}_v \cdot \vec{X}_u & \vec{X}_v \cdot \vec{X}_v \end{vmatrix} \right.$$
$$\left. - \begin{vmatrix} \vec{X}_{uv} \cdot \vec{X}_{uv} & \vec{X}_{uv} \cdot \vec{X}_u & \vec{X}_{uv} \cdot \vec{X}_v \\ \vec{X}_u \cdot \vec{X}_{uv} & \vec{X}_u \cdot \vec{X}_u & \vec{X}_u \cdot \vec{X}_v \\ \vec{X}_v \cdot \vec{X}_{uv} & \vec{X}_v \cdot \vec{X}_u & \vec{X}_v \cdot \vec{X}_v \end{vmatrix} \right). \qquad (7.20)$$

This equation involves only the metric tensor coefficients and Γ^i_{jk} symbols, except in the inner products $\vec{X}_{uu} \cdot \vec{X}_{vv}$ and $\vec{X}_{uv} \cdot \vec{X}_{uv}$. At first glance, this problem seems insurmountable, but if one performs the Laplace expansion on the determinants along the first row, factors terms involving $\det(g_{ij})$, and collects into determinants again, one can show that Equation (7.20) is equivalent to

$$K = \frac{1}{\det(g_{ij})^2} \left(\begin{vmatrix} \vec{X}_{uu} \cdot \vec{X}_{vv} - \vec{X}_{uv} \cdot \vec{X}_{uv} & \vec{X}_{uu} \cdot \vec{X}_u & \vec{X}_{uu} \cdot \vec{X}_v \\ \vec{X}_u \cdot \vec{X}_{vv} & \vec{X}_u \cdot \vec{X}_u & \vec{X}_u \cdot \vec{X}_v \\ \vec{X}_v \cdot \vec{X}_{vv} & \vec{X}_v \cdot \vec{X}_u & \vec{X}_v \cdot \vec{X}_v \end{vmatrix} \right.$$
$$\left. - \begin{vmatrix} 0 & \vec{X}_{uv} \cdot \vec{X}_u & \vec{X}_{uv} \cdot \vec{X}_v \\ \vec{X}_u \cdot \vec{X}_{uv} & \vec{X}_u \cdot \vec{X}_u & \vec{X}_u \cdot \vec{X}_v \\ \vec{X}_v \cdot \vec{X}_{uv} & \vec{X}_v \cdot \vec{X}_u & \vec{X}_v \cdot \vec{X}_v \end{vmatrix} \right).$$

The value in this is that though one cannot express $\vec{X}_{uu} \cdot \vec{X}_{vv}$ or $\vec{X}_{uv} \cdot \vec{X}_{uv}$ using only the metric coefficients, it can be shown (Problem 7.2.1) that

$$\vec{X}_{uu} \cdot \vec{X}_{vv} - \vec{X}_{uv} \cdot \vec{X}_{uv} = -\frac{1}{2} g_{11,22} + g_{12,12} - \frac{1}{2} g_{22,11}, \qquad (7.21)$$

where by $g_{ij,kl}$ we mean the function

$$\frac{\partial^2 g_{ij}}{\partial x^k \partial x^l}.$$

This produces the following formula for K that is appealing on the grounds that it illustrates symmetry of the metric tensor coefficients in determining K:

$$K = \frac{1}{\det(g_{ij})^2} \left(\begin{vmatrix} -\frac{1}{2} g_{11,22} + g_{12,12} - \frac{1}{2} g_{11,22} & \frac{1}{2} g_{11,1} & g_{12,1} - \frac{1}{2} g_{11,2} \\ g_{12,2} - \frac{1}{2} g_{22,1} & g_{11} & g_{12} \\ \frac{1}{2} g_{22,2} & g_{21} & g_{22} \end{vmatrix} \right.$$
$$\left. - \begin{vmatrix} 0 & \frac{1}{2} g_{11,2} & \frac{1}{2} g_{22,1} \\ \frac{1}{2} g_{11,2} & g_{11} & g_{12} \\ \frac{1}{2} g_{22,1} & g_{21} & g_{22} \end{vmatrix} \right). \qquad (7.22)$$

In the situation where one considers an orthogonal parametrization of a neighborhood of a surface, namely, when $g_{12} = 0$, Equation
(7.22) takes on the particularly interesting form of

$$K = -\frac{1}{\sqrt{g_{11}g_{22}}}\left(\frac{\partial}{\partial x^1}\left(\frac{1}{\sqrt{g_{11}}}\frac{\partial\sqrt{g_{22}}}{\partial x^1}\right) + \frac{\partial}{\partial x^2}\left(\frac{1}{\sqrt{g_{22}}}\frac{\partial\sqrt{g_{11}}}{\partial x^2}\right)\right). \quad (7.23)$$

By historical habit, Equations (7.22) and (7.23) are referred to as
Gauss's equations.

We now prove a theorem that we could have introduced much
earlier, but it arises in part as an application of the Codazzi equations.

Definition 7.2.4 A set $S \subset \mathbb{R}^n$ is called *path-connected* if for any
pair of points $p, q \in S$, there exists a continuous path (curve) $\alpha :$
$[0, 1] \to \mathbb{R}^n$, with $\alpha(0) = p$, $\alpha(1) = q$, and $\alpha([0, 1]) \subset S$.

Proposition 7.2.5 *If S is a path-connected regular surface in which
all its points are umbilical, then S is either contained in a plane or
in a sphere.*

Proof: Let $U \subset \mathbb{R}^2$ be an open set, and let $\vec{X} : U \to \mathbb{R}^3$ be the
parametrization of a neighborhood $V = \vec{X}(U)$ of S. Since all points
of S are umbilical, then the eigenvalues of dn_p are equal to a value λ
that *a priori* is a function of the coordinates (u, v) of the patch V.
We first show that $\lambda(u, v)$ is a constant over U.

By Proposition 6.5.4, there always exist two linearly independent
eigenvectors to dn_p, so dn_p is always diagonalizable. Thus, at any
point p of S, the whole tangent plane T_pS is the eigenspace of the
eigenvalue λ. Thus, in particular, we have

$$\vec{N}_u = dn_p(\vec{X}_u) = \lambda\vec{X}_u,$$
$$\vec{N}_v = dn_p(\vec{X}_v) = \lambda\vec{X}_v.$$

Since $\vec{N}_{uv} = \vec{N}_{vu}$, a fact that is equivalent to the Codazzi equations
(see Problem 7.2.2), we have

$$\lambda_v\vec{X}_u + \lambda\vec{X}_{uv} = \lambda_u\vec{X}_v + \lambda\vec{X}_{vu}. \qquad '$$

Since $\vec{X}_u$ and $\vec{X}_v$ are linearly independent, we deduce that $\lambda_u = \lambda_v =$
0, and therefore λ is a constant function.

If $\lambda = 0$, then $\vec{N}_u = \vec{N}_v = \vec{0}$, so the unit normal vector $\vec{N}$ is a
constant vector $\vec{N}_0$ over the domain U. Then

$$\frac{\partial}{\partial u}(\vec{X} \cdot \vec{N}_0) = \vec{X}_u \cdot \vec{N}_0 + \vec{0} = \vec{0}$$

and similarly for $\frac{\partial}{\partial v}(\vec{X} \cdot \vec{N}_0) = \vec{0}$, which shows that $\vec{X} \cdot \vec{N}_0 = \vec{0}$, which further proves that $V = \vec{X}(U)$ lies in the plane through p perpendicular to $\vec{N}_0$.

On the other hand, if $\lambda \neq 0$, then consider the vector function

$$\vec{Y}(u, v) = \vec{X}(u, v) - \frac{1}{\lambda}\vec{N}(u, v).$$

By a similar calculation as above, we see that the function $\vec{Y}(u, v)$ is a constant vector. Then we deduce that

$$\|\vec{X}(u, v) - \vec{Y}\| = \left\|\frac{1}{\lambda}\vec{N}\right\| = \frac{1}{|\lambda|},$$

which shows that $V = \vec{X}(U)$ lies on a sphere of center $\vec{Y}$ and radius $\frac{1}{|\lambda|}$.

So far, the above proof has established that any parametrization of a neighborhood V of S either lies on a sphere or lies in a plane but not that all of S necessarily lies on a sphere or in a plane. The assumption that S is path-connected extends the result to the whole surface as follows.

By Example A.1.10, if V_1 and V_2 are two coordinate patches of S with respective coordinate systems (x^1, x^2) and $(\bar{x}^1, \bar{x}^2)$ such that $V_1 \cap V_2 \neq \emptyset$, then on $V_1 \cap V_2$ the coefficients of the Gauss map with respect to the coordinate systems are related by

$$\bar{a}^i_j = \frac{\partial \bar{x}^i}{\partial x^k} \frac{\partial x^l}{\partial \bar{x}^j} a^k_l.$$

This is tantamount to saying that the matrices for the Gauss map relative to different coordinate systems are similar matrices. Hence, if V_1 and V_2 are two overlapping coordinate patches, then the eigenvalue λ is constant over $V_1 \cup V_2$, and then all of $V_1 \cup V_2$ either lies in a plane or on a sphere.

Since S is path-connected, given any two points p and q on S, there exists a path $\alpha : [0, 1] \rightarrow S$, with $\alpha(0) = p$ and $\alpha(1) = q$. From basic facts in topology, we know that $\alpha([0, 1])$ is a compact set and that any open cover of a compact set has a finite subcover. Thus, given a collection of coordinate patches that cover $\alpha([0, 1])$, then only a finite subcollection of these coordinate patches is necessary to cover $\alpha([0, 1])$. Call this collection $\{V_1, V_2, \ldots, V_r\}$. By the reasoning in the above paragraph, we see that $V_1 \cup V_2 \cup \cdots \cup V_r$ lies either in a plane or on a sphere. Thus all points of S either lie in the same plane or lie on the same sphere.   □

PROBLEMS

1. Prove Equation (7.21).

2. Show that one can obtain the Codazzi equations from the equation $\vec{N}_{12} = \vec{N}_{21}$ and by using the Gauss equations in Equation (7.1).

3. Prove that the Riemann symbols defined in Equation (7.16) form the components of a $(1,3)$-tensor. [Requires Appendix A.]

4. Prove Equation (7.18) and the claim that supports it.

5. Suppose that $g_{12} = L_{12} = 0$ (i.e., the coordinate lines are lines of curvature), so that the principal curvatures satisfy

$$\kappa_1 = \frac{L_{11}}{g_{11}} \quad \text{and} \quad \kappa_2 = \frac{L_{22}}{g_{22}}.$$

Prove that the Codazzi equations are equivalent to

$$\frac{\partial \kappa_1}{\partial x^2} = \frac{g_{11,2}}{2g_{11}}(\kappa_2 - \kappa_1) \quad \text{and} \quad \frac{\partial \kappa_2}{\partial x^1} = \frac{g_{22,1}}{2g_{22}}(\kappa_1 - \kappa_2),$$

where by $g_{11,2}$ we mean $\dfrac{\partial g_{11}}{\partial x^2}$ and similarly for $g_{22,1}$.

6. Prove that the Riemann symbols R^l_{ijk} defined in Equation (7.16) are antisymmetric in the indices $\{i, j\}$. Conclude that for surfaces, R^l_{ijk} represent at most four distinct functions.

7. Prove Equation (7.23).

8. Prove the following formula by Blaschke for the Gaussian curvature:

$$K = -\frac{1}{4\det(g_{ij})^2} \begin{vmatrix} g_{11} & g_{12} & g_{22} \\ g_{11,1} & g_{12,1} & g_{22,1} \\ g_{11,2} & g_{12,2} & g_{22,2} \end{vmatrix}$$

$$- \frac{1}{2\sqrt{\det(g_{ij})}} \left(\frac{\partial}{\partial x^1} \left(\frac{g_{22,1} - g_{12,2}}{\sqrt{\det(g_{ij})}} \right) - \frac{\partial}{\partial x^2} \left(\frac{g_{12,1} - g_{11,2}}{\sqrt{\det(g_{ij})}} \right) \right).$$

9. Consider the two surfaces parametrized by

$$\vec{X}(u, v) = (v \cos u, v \sin u, \ln v),$$
$$\vec{Y}(u, v) = (v \cos u, v \sin u, u).$$

Prove that these two surfaces have equal Gaussian curvature functions over the same domain but that they do not possess the same metric tensor. (This gives an example of two surfaces with given coordinate systems over which different metric tensor components lead to the same Gaussian curvature function.)

10. Since the Gaussian curvature is intrinsic, depending only on the metric coefficients and their derivatives, compute that Gaussian curvature for the metric with $g_{11}(u, v) = g_{22}(u, v) = 1/v^2$ and $g_{12}(u, v) = 0$ on the Poincaré upper half-plane, as in Examples 6.1.9 and 7.1.7.

11. Carry out the same computation as in the previous problem for the general metric $g_{11}(u,v) = g_{22}(u,v) = f(v)$ and $g_{12}(u,v) = 0$ on the upper half-plane. [See Problems 6.1.27 and 7.1.11.]

12. Consider a surface with metric coefficients $g_{11}(u,v) = 1$, $g_{12}(u,v) = 0$, and $g_{22}(u,v) = f(u,v)$, for some positive function f. Find the Christoffel symbols of both kinds for this metric and compute the Gaussian curvature K.

13. *Liouville surface.* Consider a surface with metric coefficients defined by $g_{11}(u,v) = g_{22}(u,v) = U(u) + V(v)$ and $g_{12}(u,v) = 0$, where $U(u)$ is a function of u and $V(v)$ is a function of v. Find the Christoffel symbols of both kinds for this metric and compute the Gaussian curvature K. Such a surface is called a *Liouville surface.*

14. The coordinate curves of a parametrization $\vec{X}(u,v)$ form a *Tchebysheff net* if the lengths of the opposite sides of any quadrilateral formed by them are equal.

    (a) Prove that a necessary and sufficient condition for a parametrization to be a Tchebysheff net is
    $$\frac{\partial g_{11}}{\partial v} = \frac{\partial g_{22}}{\partial u} = 0.$$

    (b) **(ODE)** Prove that when a parametrization constitutes a Tchebysheff net, there exists a reparametrization of the coordinate neighborhood so that the new components of the metric tensor are
    $$g_{11} = 1, \qquad g_{12} = \cos\theta, \qquad g_{22} = 1,$$
    where θ is the angle between the coordinate lines at the given point on the surface.

    (c) Show that in this case,
    $$K = -\frac{\theta_{uv}}{\sin\theta}.$$

7.3 Fundamental Theorem of Surface Theory

The original proof of this theorem was provided by Bonnet in 1855 in [5]. In more recent texts, we can find the proof in the Appendix to Chapter 4 in [11] or in Chapter VI of [32]. We will not provide a complete proof of the Fundamental Theorem of Surface Theory here since it involves solving a system of partial differential equations, but we will sketch the main points behind it.

Suppose we consider a regular oriented surface S and coordinate patch V parametrized by $\vec{X} : U \to \mathbb{R}^3$. We have seen that the coefficients (g_{ij}) and (L_{ij}) of the first and second fundamental forms satisfy $\det(g_{ij}) > 0$ and the Gauss-Codazzi equations. Given these conditions, there exists an essentially unique surface with specified first and second fundamental forms is a profound result, called the Fundamental Theorem of Surface Theory.

Theorem 7.3.1 *If E, F, G and e, f, g are sufficiently differentiable functions of (u, v) that satisfy the Gauss-Codazzi Equations (7.22) and (7.13) and $EG - F^2 > 0$, then there exists a parametrization $\vec{X}$ of a regular orientable surface that admits*

$$g_{11} = E, \qquad g_{12} = F, \qquad g_{22} = G,$$
$$L_{11} = e, \qquad L_{12} = f, \qquad L_{22} = g.$$

Furthermore, this surface is uniquely determined up to its position in space.

The setup for the proof is to consider nine functions $\xi_i(u, v)$, $\varphi_i(u, v)$, and $\psi_i(u, v)$ with $1 \leq i \leq 3$, and think of these functions as the components of the vector functions $\vec{X}_u$, $\vec{X}_v$, and $\vec{N}$ so that $\vec{X}_u = (\xi_1, \xi_2, \xi_3)$, $\vec{X}_v = (\varphi_1, \varphi_2, \varphi_3)$, and $\vec{N} = (\psi_1, \psi_2, \psi_3)$. With this setup, the equations that define Gauss's and Weingarten equations, namely, Equations (7.1) and (6.27), become the following system of 18 partial differential equations: for $i = 1, 2, 3$,

$$\frac{\partial \xi_i}{\partial u} = \Gamma_{11}^1 \xi_i + \Gamma_{11}^2 \varphi_i + L_{11} \psi_i, \qquad \frac{\partial \xi_i}{\partial v} = \Gamma_{12}^1 \xi_i + \Gamma_{12}^2 \varphi_i + L_{12} \psi_i,$$

$$\frac{\partial \varphi_i}{\partial u} = \Gamma_{21}^1 \xi_i + \Gamma_{21}^2 \varphi_i + L_{21} \psi_i, \qquad \frac{\partial \varphi_i}{\partial v} = \Gamma_{22}^1 \xi_i + \Gamma_{22}^2 \varphi_i + L_{22} \psi_i,$$

$$\tag{7.24}$$

$$\frac{\partial \psi_i}{\partial u} = a_1^1 \xi_i + a_1^2 \varphi_i, \qquad\qquad \frac{\partial \psi_i}{\partial v} = a_2^1 \xi_i + a_2^2 \varphi_i.$$

In general, when a system of partial differential equations involving n functions $u_i(x_1, \ldots, x_m)$ has $n < m$, the solutions may involve not only constants of integration but also unknown functions that can be any continuous function from $\mathbb{R}$ to $\mathbb{R}$ (or some appropriate interval). However, when $n > m$, i.e., when there are more functions in the system than there are independent variables, the system may be "overdetermined" and may either have less freedom in its solution set or have no solutions at all. In fact, we cannot expect the above system to have solutions if the mixed partial derivatives of $\vec{\xi} = (\xi_1, \xi_2, \xi_3)$, $\vec{\varphi} = (\varphi_1, \varphi_2, \varphi_3)$, and $\vec{\psi} = (\psi_1, \psi_2, \psi_3)$ are not equal. This is usually called the *compatibility condition* for systems of partial differential equations, and, as we see in the above system, this condition imposes relations between the functions $\Gamma_{jk}^i(u, v)$ and $L_{jk}(u, v)$.

The key ingredient behind the Fundamental Theorem of Surface Theory is Theorem V in Appendix B of [32] that, applied to our context, states that if all second derivatives of the Γ_{jk}^i and L_{jk} functions are continuous and if the compatibility condition holds in Equation (7.24), solutions to the system exist and are unique once values for $\vec{\xi}(u_0, v_0)$, $\vec{\varphi}(u_0, v_0)$, and $\vec{\psi}(u_0, v_0)$ are given, where (u_0, v_0) is a point in the common domain of $\Gamma^i_{jk}(u, v)$ and $L_{jk}(u, v)$. The compatibility condition required in this theorem is satisfied if and only if the functions $g_{11} = E$, $g_{12} = F$, $g_{22} = G$, $L_{11} = e$, $L_{12} = f$, and $L_{22} = g$ satisfy the Gauss-Codazzi equations.

Solutions to Equation (7.24) can be chosen in such a way that

$$\vec{\xi}(u_0, v_0) \cdot \vec{\xi}(u_0, v_0) = E(u_0, v_0), \quad \vec{\varphi}(u_0, v_0) \cdot \vec{\varphi}(u_0, v_0) = G(u_0, v_0),$$
$$\vec{\xi}(u_0, v_0) \cdot \vec{\varphi}(u_0, v_0) = F(u_0, v_0), \quad \vec{\psi}(u_0, v_0) \cdot \vec{\psi}(u_0, v_0) = 1,$$
$$\vec{\psi}(u_0, v_0) \cdot \vec{\xi}(u_0, v_0) = 0, \qquad\qquad \vec{\psi}(u_0, v_0) \cdot \vec{\varphi}(u_0, v_0) = 0,$$
$$\frac{\vec{\xi}(u_0, v_0) \times \vec{\varphi}(u_0, v_0)}{\|\vec{\xi}(u_0, v_0) \times \vec{\varphi}(u_0, v_0)\|} = \vec{\psi}(u_0, v_0). \tag{7.25}$$

The next step of the proof is to show that, given the above initial conditions, the following equations hold for all (u, v) where the solutions are defined:

$$\vec{\xi}(u, v) \cdot \vec{\xi}(u, v) = E(u, v), \qquad \vec{\varphi}(u, v) \cdot \vec{\varphi}(u, v) = G(u, v),$$
$$\vec{\xi}(u, v) \cdot \vec{\varphi}(u, v) = F(u, v), \qquad \vec{\psi}(u, v) \cdot \vec{\psi}(u, v) = 1,$$
$$\vec{\psi}(u, v) \cdot \vec{\xi}(u, v) = 0, \qquad\qquad \vec{\psi}(u, v) \cdot \vec{\varphi}(u, v) = 0,$$
$$\frac{\vec{\xi}(u, v) \times \vec{\varphi}(u, v)}{\|\vec{\xi}(u, v) \times \vec{\varphi}(u, v)\|} = \vec{\psi}(u, v).$$

From the solutions for $\vec{\xi}$, $\vec{\varphi}$, and $\vec{\psi}$, we form the new system of differential equations

$$\begin{cases} \vec{X}_u = \vec{\xi}, \\ \vec{X}_v = \vec{\varphi}. \end{cases}$$

One obtains a solution for the function $\vec{X}$ over appropriate (u, v) by

$$\vec{X}(u, v) = \int_{u_0}^{u} \vec{\xi}(u, v)\, du + \int_{v_0}^{v} \vec{\varphi}(u_0, v)\, dv. \tag{7.26}$$

The resulting vector function $\vec{X}$ is defined over an open set $U \subset \mathbb{R}^2$ containing (u_0, v_0), and $\vec{X}$ parametrizes a regular surface S. By construction, the coefficients of the first fundamental form for this surface are

$$g_{11} = E, \qquad g_{12} = F, \qquad g_{22} = G.$$

One then proves that it is also true that the coefficients of the second fundamental form satisfy

$$L_{11} = e, \qquad L_{12} = f, \qquad L_{22} = g.$$

It remains to show that this surface is unique up to a rigid motion in $\mathbb{R}^3$. It is not hard to see that the equalities in Equation (7.25) imposed on the initial conditions still allow the freedom to choose the unit vector $\hat{\xi}(u_0, v_0) = \vec{\xi}(u_0, v_0)/\|\vec{\xi}(u_0, v_0)\|$ and the vector $\vec{\psi}(u_0, v_0)$, which must be perpendicular to $\hat{\xi}(u_0, v_0)$. The vectors

$$\hat{\xi}(u_0, v_0), \quad \vec{\psi}(u_0, v_0), \quad \hat{\xi}(u_0, v_0) \times \vec{\psi}(u_0, v_0)$$

form a positive orthonormal frame, so any two choices allowed by (7.25) differ from each other by a rotation in $\mathbb{R}^3$. Finally, the integration in Equation (7.26) introduces a constant vector of integration. Thus, two solutions to Gauss's and Weingarten's equations differ from each other by a rotation and a translation, namely, any rigid motion in $\mathbb{R}^3$.

PROBLEMS

1. **(ODE)** Consider solutions to Gauss's and Weingarten's equations for which the coefficients of the first and second fundamental forms are constant.

   (a) Let E, F, and G be constants such that $EG - F^2 > 0$ and view them as constant functions. Prove that the Gauss-Codazzi equations impose $L_{jk} = 0$.

   (b) Prove that all solutions to the Gauss-Weingarten equations in this situation are planes.

2. **(ODE)** Find all regular parametrized surfaces that have

$$g_{11} = 1, \qquad g_{12} = 0, \qquad g_{22} = \cos^2 u$$
$$L_{11} = 1, \qquad L_{12} = 0, \qquad L_{22} = \cos^2 u.$$

3. Does there exist a surface $\vec{X}(u, v)$ with

$$g_{11} = 1, \qquad g_{12} = 0, \qquad g_{22} = \cos^2 u,$$
$$L_{11} = \cos^2 u, \qquad L_{12} = 0, \qquad L_{22} = 1?$$

CHAPTER 8

Gauss-Bonnet Theorem; Geodesics

Historians of mathematics often point to Euclid as the inventor, or at least the father in some metaphorical sense, of the synthetic methods of mathematical proofs. Euclid's *Elements*, which comprises 13 books, treats a wide variety of topics but focuses heavily on geometry. In his *Elements*, Euclid presents 23 geometric definitions along with five postulates (or axioms), and in Books I–VI and XI–XIII, he proves around 250 propositions about lines, circles, angles, and ratios of quantities in the plane and in space.

Most popular texts about the nature of mathematics agree that Euclid's geometric methods held a foundational importance in the development of mathematics (see, e.g., [12, pp. 76–77] or [18, p. 7]). Many such texts also retell the story of the discovery in the 19th century of geometries that remain consistent yet do not satisfy the fifth and most debatable of Euclid's postulates ([6], [9, pp. 214–227], [10], [18, pp. 217–223]). One can readily list elliptical geometry and hyperbolic geometry as examples of such geometries.

At this point in this book, with the methods from the theory of curves and surfaces now at our disposal, we stand in a position to study geometry on any regular surface. As we shall see, on a general regular surface the notion of "straightness" is not intuitive, and one might debate whether such a notion should exist at all. Hence, instead of only trying to consider lines and circles, we first study regular curves on a regular surface in general and only later define notions of shortest distance, straightness, and parallelism. Arguably, the most important theorem of this book is the Gauss-Bonnet Theorem, which relates the Gaussian curvature on a region $\mathcal{R}$ of the surface to a specific function along the boundary $\partial\mathcal{R}$. One of the central themes of this chapter is to present the Gauss-Bonnet Theorem and introduce geodesics (curves that generalize the concept of shortest distance and straightness). In the last two sections, we discuss applications of the Gauss-Bonnet Theorem to non-Euclidean geometry.

DOI: 10.1201/9781003295341-8

8.1 Curvatures and Torsion

8.1.1 Natural Frames

Throughout this chapter, we let S be a regular surface of class C^2 and let V be an open set of S parametrized by a vector function $\vec{X} : U \to \mathbb{R}^3$, where U is an open set in $\mathbb{R}^2$. We consider a curve C of class C^2 of the form $\vec{\gamma} = \vec{X} \circ \vec{\alpha}$, where $\vec{\alpha} : I \to U$ is a curve in the domain of $\vec{X}$, with $\vec{\alpha}(t) = (u(t), v(t))$.

Let $p = \vec{\gamma}(t_0)$ be a point on the curve and on the surface. Using the theory of curves in $\mathbb{R}^3$, one typically performs calculations on quantities related to $\vec{\gamma}$ in the Frenet frame $(\vec{T}, \vec{P}, \vec{B})$. (Since order matters when discussing coordinates, we use the triple notation as opposed to set notation to describe an ordered basis.) This frame is orthonormal and hence has many nice properties. On the other hand, using the local theory of surfaces, one would typically perform calculations in the $(\vec{X}_u, \vec{X}_v, \vec{N})$ reference frame. This frame is not orthonormal but is often the most practical. If a point p on a surface is not an umbilical point, the frame $(\vec{e}_1, \vec{e}_2, \vec{N})$, where the vectors $\vec{e}_i$ are principal directions (well defined up to a change in sign), is a natural orthonormal frame associated to S at p. We could construct yet another orthonormal frame by applying the Gram-Schmidt orthonormalization to $(\vec{X}_u, \vec{X}_v, \vec{N})$.

Though more geometric in nature than $(\vec{X}_u, \vec{X}_v, \vec{N})$, the frame $(\vec{e}_1, \vec{e}_2, \vec{N})$ does not lend itself well to calculations with specific parametrizations. Studying curves on surfaces, one could choose between these three reference frames, but it turns out that a combination will be most helpful.

Borrowing first from surface theory, we use the unit normal vector $\vec{N}$, which is invariant up to a sign under parametrization of S. We will often consider $\vec{N}(t)$ to be a single-variable vector function over the interval I, by which we mean explicitly $\vec{N}(t) = \vec{N} \circ \vec{\alpha}(t)$. Borrowing from the theory of space curves, we use $\vec{T}$, the unit tangent vector to the curve C at p, which is also invariant up to a sign under reparametrization of the curve. By construction, $\vec{N}$ and $\vec{T}$ are perpendicular to each other, and we just need to choose a third vector to complete the frame. We cannot use $\vec{P}$ as a third vector in the frame because there is no guarantee that the principal normal vector $\vec{P}(t)$ is perpendicular to $\vec{N}$. In fact, Section 6.5 discusses how the second fundamental form measures the relationship between $\vec{N}$ and $\vec{P}$. Consequently, to complete a natural orthonormal frame related to a curve on a surface, we define the new vector function

$$\vec{U}(t) = \vec{N}(t) \times \vec{T}(t).$$

We remind the reader that in the theory of plane curves, the unit normal $\vec{U}_{\text{plane}}$ to a curve at a point is defined as the counterclockwise rotation of the unit tangent $\vec{T}$. We can rephrase this definition as

$$\vec{U}_{\text{plane}} = \vec{k} \times \vec{T}$$

Therefore, our definition of a normal vector in the theory of plane curves matches the above definition if we considers the xy-plane a surface in $\mathbb{R}^3$, with $\vec{k}$ as the unit surface normal vector. In this context, what we called $\vec{U}_{\text{plane}}$ we now denote as $\vec{U}$. This observation motivates the use of the oriented basis $(\vec{T}, \vec{U}, \vec{N})$ where studying local properties of curves on surfaces.

Definition 8.1.1 Let S be a regular surface and let $\mathcal{C}$ be a regular curve on S. At each point on $\mathcal{C}$, the oriented basis $(\vec{T}, \vec{U}, \vec{N})$, is called the *Darboux frame*.

8.1.2 Normal Curvature

As in the theory of space curves, we begin the study of curves on surfaces with the derivative $\vec{T}' = s'\kappa\vec{P}$. Since the principal normal vector is perpendicular to $\vec{T}$, this vector $\vec{T}'$, often called the *curvature vector*, is perpendicular to $\vec{T}$, and thus, in the $(\vec{T}, \vec{U}, \vec{N})$ frame, decomposes into a $\vec{U}$ component and an $\vec{N}$ component. (Alternately, the "curvature vector" sometimes refers to just $\kappa\vec{P}$.) In Definition 6.5.1, we already introduced the notion of the normal curvature $\kappa_n(t)$ as the component of $\vec{T}'$ along $\vec{N}$. We now define an additional function $\kappa_g : I \to \mathbb{R}$ such that

$$\vec{T}' = s'(t)\kappa(t)\vec{P} = s'(t)\kappa_g(t)\vec{U} + s'(t)\kappa_n(t)\vec{N} \qquad (8.1)$$

for all $t \in I$. We call κ_g the *geodesic curvature* of C at p on S.

Since $(\vec{N}, \vec{T}, \vec{U})$ is an orthonormal frame, we can already provide one way to calculate these curvatures, namely, using

$$\kappa_g = \kappa\vec{P} \cdot \vec{U} \qquad \text{and} \qquad \kappa_n = \kappa\vec{P} \cdot \vec{N}. \qquad (8.2)$$

If one is given parametric equations for the curve $\vec{\gamma}$ and the surface $\vec{X}$, then Equation (8.2) is often the most direct way of calculating κ_n. If we are given the second fundamental form to the surface S, there is an alternative way to calculate κ_n. We remind the reader of Proposition 6.5.2, which relates the second fundamental form of S to the normal curvature by

$$\kappa_n(t_0) = II_p(\vec{T}(t_0)).$$

Therefore, the second fundamental form of a unit vector $\vec{T}$ is the component of a curve's curvature vector $\vec{T}'$ in the normal direction.

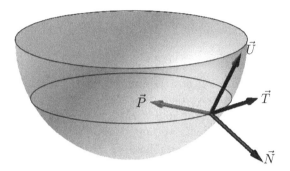

Figure 8.1: $\vec{N}$, $\vec{T}$, $\vec{U}$, and $\vec{P}$ of a curve on a surface.

If $\vec{T}(t_0)$ makes an angle θ with $\vec{e}_1$, the principal direction associated to the maximum principal curvature, then by Euler's curvature formula in Equation (6.31),

$$\kappa_n(t_0) = \kappa_1 \cos^2 \theta + \kappa_2 \sin^2 \theta.$$

We remind the reader that $\vec{T}$ at a point p on a regular curve C may change sign under a reparametrization and thus the principal normal $\vec{P}$ may as well. For surfaces, the unit normal $\vec{N}$, which is calculated by

$$\vec{N} = \frac{\vec{X}_u \times \vec{X}_v}{\|\vec{X}_u \times \vec{X}_v\|},$$

may change sign under a reparametrization. Thus, in Equation (8.2) one notices that κ_n and κ_g are unique only up to a possible sign change under a reparametrization of the curve or of the surface. On the other hand, none of these vectors changes direction under positively oriented reparametrizations.

Figure 8.1 illustrates the principal normal $\vec{P}$ in the $(\vec{T}, \vec{U}, \vec{N})$ reference frame of a circle on the surface of a sphere. In this figure, however, since the surface is a sphere, all the points are umbilical, and there does not exist a unique basis $\{\vec{e}_1, \vec{e}_2\}$ of principal directions (even up to sign).

As an application of Euler's formula in the context of curves on surfaces, let us consider any two curves $\vec{\gamma}_1$ and $\vec{\gamma}_2$ on the surface S, each parametrized by arc length in such a way that they intersect orthogonally at the point p at $s = s_0$. If we write

$$\vec{\gamma}_1'(s_0) = (\cos \theta)\vec{e}_1 + (\sin \theta)\vec{e}_2,$$

then

$$\vec{\gamma}_2'(s_0) = \pm((\sin \theta)\vec{e}_1 - (\cos \theta)\vec{e}_2)$$

for either $-$ or $+$ signs as a choice on $\pm$. If we denote $\kappa_n^{(1)}$ and $\kappa_n^{(2)}$ as the normal curvatures of $\vec{\gamma}_1$ and $\vec{\gamma}_2$, respectively, at p, then one obtains the interesting fact that

$$\kappa_n^{(1)} + \kappa_n^{(2)} = \cos^2\theta\kappa_1 + \sin^2\theta\kappa_2 + \sin^2\theta\kappa_1 + \cos^2\theta\kappa_2$$
$$= \kappa_1 + \kappa_2 = 2H.$$

In other words, for any two orthogonal unit tangent directions $\vec{v}_1$ and $\vec{v}_2$ at a point p, the average of the associated normal curvatures is equal to the mean curvature and does not depend on the particular directions of $\vec{v}_1$ and $\vec{v}_2$.

Note that since $\vec{T} \cdot \vec{N} = 0$, it follows that $\vec{N}' \cdot \vec{T} = -\vec{T}' \cdot \vec{N}$, and hence, we deduce that

$$\vec{N}' \cdot \vec{T} = -s'(t)\kappa_n(t).$$

We remind the reader that an asymptotic curve on a surface is a curve that satisfies $\kappa_n(t) = 0$ (see Definition 6.5.11) at all points on the curve. Geometrically, this means that an asymptotic curve is a curve on which $\vec{N}'$ has no $\vec{T}$ component or, vice versa, a curve on which $\vec{T}'$ has no $\vec{N}$ component. Also, from an intuitive perspective, though we introduced κ_n as a measure of how much $\vec{T}$ changes in the normal direction $\vec{N}$, $-\kappa_n$ gives a measure of how much $\vec{N}$ changes in the tangent direction $\vec{T}$.

8.1.3 Geodesic Curvature

The geodesic curvature κ_g of a curve on a surface corresponds to the component of the curvature vector occurring in the tangent plane. Equation (8.2) states this relationship as

$$\kappa_g = \kappa\vec{P} \cdot \vec{U},$$

which is equivalent to

$$s'(t)\kappa_g = \vec{T}' \cdot \vec{U} = \vec{T}' \cdot (\vec{N} \times \vec{T}) = (\vec{T}'\vec{N}\vec{T}) = (\vec{T}\vec{T}'\vec{N}), \qquad (8.3)$$

where $(\vec{u}\vec{v}\vec{w})$ is the triple-vector product in $\mathbb{R}^3$. This already provides a formula to calculate $\kappa_g(t)$ in specific situations. It turns out, however, that the geodesic curvature is an intrinsic quantity, a fact that we show now.

Let us write $\vec{\alpha}(t) = (u(t), v(t))$ and $\vec{\gamma} = \vec{X} \circ \vec{\alpha}$ for the curve on the surface. (To simplify the expression of certain formulas as we did in Chapter 7, we will refer to the coordinates u and v as x^1 and x^2, and we will use the expressions $\dot{x}^1, \dot{x}^2, \ddot{x}^1, \ddot{x}^2$ to refer to the derivatives of u', v', u'', v''.) For the tangent vector, we have

$$s'(t)\vec{T} = \vec{X}_u u'(t) + \vec{X}_v v'(t) = \sum_{i=1}^{2} \dot{x}^i \vec{X}_i. \qquad (8.4)$$

Taking another derivative of this expression, we get

$$s''\vec{T} + s'\vec{T}'$$

$$= \vec{X}_{uu}(u')^2 + 2\vec{X}_{uv}u'v' + \vec{X}_{vv}(v')^2 + \vec{X}_u u'' + \vec{X}_v v'' \qquad (8.5)$$

$$= \sum_{i=1}^{2}\sum_{j=1}^{2} \dot{x}^i \dot{x}^j \vec{X}_{ij} + \sum_{k=1}^{2} \ddot{x}^k \vec{X}_k.$$

The triple-vector product $(\vec{T}\vec{T}'\vec{N})$ can also be written as $(\vec{T} \times \vec{T}') \cdot \vec{N}$. Therefore, since $\vec{T} \times \vec{T} = \vec{0}$, taking the cross product of the two expressions in Equations (8.4) and (8.5) and then taking the dot product with $\vec{N}$ leads to

$$(s')^2(\vec{T}\vec{T}'\vec{N}) = \Big[(\vec{X}_u \times \vec{X}_{uu})(u')^3 + (2\vec{X}_u \times \vec{X}_{uv} + \vec{X}_v \times \vec{X}_{uu})(u')^2 v'$$

$$+ (\vec{X}_u \times \vec{X}_{vv} + 2\vec{X}_v \times \vec{X}_{uv})u'(v')^2 + (\vec{X}_v \times \vec{X}_{vv})(v')^3\Big] \cdot \vec{N}$$

$$+ (\vec{X}_u \times \vec{X}_v) \cdot \vec{N}(u'v'' - u''v'). \qquad (8.6)$$

It is possible to express all the coefficients of the terms $(u')^3$, $(u')^2 v'$, and so forth using the metric tensor coefficients g_{ij} or their derivatives. For example, using the result of Problem 3.1.6,

$$(\vec{X}_u \times \vec{X}_{uu}) \cdot \vec{N} = (\vec{X}_u \times \vec{X}_{uu}) \cdot \frac{\vec{X}_u \times \vec{X}_v}{\sqrt{\det(g)}}$$

$$= \frac{(\vec{X}_u \cdot \vec{X}_u)(\vec{X}_{uu} \cdot \vec{X}_v) - (\vec{X}_u \cdot \vec{X}_v)(\vec{X}_{uu} \cdot \vec{X}_u)}{\sqrt{\det(g)}}$$

$$= \frac{g_{11}[11,2] - g_{12}[11,1]}{\sqrt{\det(g)}}$$

$$= \frac{\det(g)g^{22}[11,2] + \det(g)g^{21}[11,1]}{\sqrt{\det(g)}}$$

$$= \Gamma_{11}^2 \sqrt{\det(g)}.$$

Repeating the calculations for all the relevant terms, we get

$$(\vec{X}_u \times \vec{X}_{uu}) \cdot \vec{N} = \Gamma_{11}^2 \sqrt{\det(g)}, \quad (\vec{X}_v \times \vec{X}_{uu}) \cdot \vec{N} = -\Gamma_{11}^1 \sqrt{\det(g)},$$

$$(\vec{X}_u \times \vec{X}_{uv}) \cdot \vec{N} = \Gamma_{12}^2 \sqrt{\det(g)}, \quad (\vec{X}_v \times \vec{X}_{uv}) \cdot \vec{N} = -\Gamma_{12}^1 \sqrt{\det(g)},$$

$$(\vec{X}_u \times \vec{X}_{vv}) \cdot \vec{N} = \Gamma_{22}^2 \sqrt{\det(g)}, \quad (\vec{X}_v \times \vec{X}_{vv}) \cdot \vec{N} = -\Gamma_{22}^1 \sqrt{\det(g)}.$$

Putting these expressions, along with $(\vec{X}_u \times \vec{X}_v) \cdot \vec{N} = \sqrt{\det(g)}$, into Equation (8.6) and using Equation (8.3) gives

$$(s')^3 \kappa_g = \Big(\Gamma_{11}^2(u')^3 + (2\Gamma_{12}^2 - \Gamma_{11}^1)(u')^2 v' + (\Gamma_{22}^2 - 2\Gamma_{12}^1)u'(v')^2$$

$$- \Gamma_{22}^1(v')^3 + u'v'' - u''v'\Big)\sqrt{g_{11}g_{22} - (g_{12})^2}. \qquad (8.7)$$

This shows that, as opposed to the normal curvature, the geodesic curvature is an intrinsic quantity, depending only on the metric tensor and the parametric equations $\vec{\alpha}(t) = (u(t), v(t))$.

Though complete, Equation (8.7) is not written as concisely as it could be using tensor notation. We define the permutation symbols

$$
\varepsilon_{h_1 h_2 \cdots h_n} = \begin{cases} +1, & \text{if } h_1 h_2 \cdots h_n \text{ is an even permutation of } 1, 2, \ldots, n, \\ -1, & \text{if } h_1 h_2 \cdots h_n \text{ is an odd permutation of } 1, 2, \ldots, n, \\ 0, & \text{if } h_1 h_2 \cdots h_n \text{ is not a permutation of } 1, 2, \ldots, n. \end{cases}
$$

(See also Section A.1 for more details on this tensor notation.) We shall use this symbol in the simple case with $n = 2$, so with only two indices. Then our indices satisfy $1 \leq i, j \leq 2$ and $\varepsilon_{ii} = 0$, $\varepsilon_{12} = 1$, and $\varepsilon_{21} = -1$. We can now summarize (8.7) as

$$
\kappa_g = \frac{\sqrt{\det(g)}}{(s')^3} \left(\sum_{i,j,k,l=1}^{2} \varepsilon_{il} \Gamma_{jk}^l \dot{x}^i \dot{x}^j \dot{x}^k + \sum_{i,j=1}^{2} \varepsilon_{ij} \dot{x}^i \ddot{x}^j \right)
$$

or, in other words,

$$
\kappa_g = \frac{\sqrt{\det(g)}}{(s')^3} \left(\varepsilon_{il} \Gamma_{jk}^l \dot{x}^i \dot{x}^j \dot{x}^k + \varepsilon_{ij} \dot{x}^i \ddot{x}^j \right),
$$

where we use the Einstein summation notation convention.

Example 8.1.2 (Plane Curves) We modeled the $(\vec{T}, \vec{U}, \vec{N})$ frame off the frame $(\vec{T}, \vec{U}, \vec{k})$, viewing the plane as a surface in $\mathbb{R}^3$ with normal vector $\vec{k}$. Consider then curves in the plane as curves on a surface. Using an orthonormal basis in the plane, i.e., the trivial parametrization $\vec{X}(u, v) = (u, v)$, we get $g_{11} = g_{22} = 1$ and $g_{12} = 0$. In this case, all the Christoffel symbols are 0 and of course $\sqrt{\det(g)} = 1$. Then Equation (8.7) gives

$$
\kappa_g = \frac{u'v'' - u''v'}{(s')^3},
$$

which is precisely the Equation (1.12) we obtained for the curvature of a plane curve in Section 1.3.

It is an interesting fact, the proof of which we leave as an exercise to the reader (Problem 8.1.8), that the geodesic curvature at a point p on a curve C on a surface S is equal to the geodesic curvature of the plane curve obtained by projecting C orthogonally onto T_pS. From an intuitive perspective, this property indicates why the geodesic curvature should be an intrinsic quantity. Intrinsic properties depend on the metric tensor, which is an inner product on the tangent space T_pS, so any quantity that measures something within the tangent space is usually an intrinsic property.

8.1.4   Geodesic Torsion

Similar to how we viewed the Frenet frame in Chapter 3, we consider the triple $(\vec{T}, \vec{U}, \vec{N})$ as a moving frame based at a point $\vec{\gamma}(t)$. Since $\vec{T}$, $\vec{U}$, and $\vec{N}$, are unit vectors, using the same reasoning as we did to establish Equation (3.5), we can determine that

$$\frac{d}{dt} \begin{pmatrix} \vec{T} & \vec{U} & \vec{N} \end{pmatrix} = \begin{pmatrix} \vec{T} & \vec{U} & \vec{N} \end{pmatrix} A(t),$$

where $A(t)$ is an antisymmetric matrix. We already know most of the coefficient functions in $A(t)$. In Equation (8.1), our definitions of the normal and geodesic curvature gave

$$\vec{T}' = s'(t)\kappa_g(t)\vec{U} + s'(t)\kappa_n(t)\vec{N} \qquad (8.8)$$

and using the usual dot product relations among the vectors in an orthonormal basis, namely

$$\vec{T} \cdot \vec{T} = 1, \qquad \vec{U} \cdot \vec{U} = 1, \qquad \vec{N} \cdot \vec{N} = 1,$$
$$\vec{N} \cdot \vec{T} = 0, \qquad \vec{N} \cdot \vec{U} = 0, \qquad \vec{T} \cdot \vec{U} = 0,$$

we deduce the following equalities:

$$\vec{T}' \cdot \vec{T} = 0, \qquad \vec{U}' \cdot \vec{U} = 0, \qquad \vec{N}' \cdot \vec{N} = 0,$$
$$\vec{N}' \cdot \vec{T} = -\vec{T}' \cdot \vec{N}, \qquad \vec{N}' \cdot \vec{U} = -\vec{U}' \cdot \vec{N}, \qquad \vec{T}' \cdot \vec{U} = -\vec{U}' \cdot \vec{T}.$$

Consequently, from Equation (8.8),

$$\vec{N}' \cdot \vec{T} = -s'\kappa_n \qquad \text{and} \qquad \vec{U}' \cdot \vec{T} = -s'\kappa_g.$$

    Therefore, in order to describe the coefficients of $A(t)$, and thereby express the derivatives $\vec{T}'$, $\vec{U}'$, $\vec{N}'$ in the $(\vec{T}, \vec{U}, \vec{N})$ frame, we need to label one more coefficient function. Define the *geodesic torsion* $\tau_g : I \to \mathbb{R}$ to be the unique function such that

$$\vec{U}' \cdot \vec{N} = s'(t)\tau_g(t) \qquad \text{for all } t \in I.$$

With this function defined, we can imitate Equation (3.5) for the change of the Frenet frame, and write

$$\frac{d}{dt} \begin{pmatrix} \vec{T} & \vec{U} & \vec{N} \end{pmatrix} = \begin{pmatrix} \vec{T} & \vec{U} & \vec{N} \end{pmatrix} s'(t) \begin{pmatrix} 0 & -\kappa_g(t) & -\kappa_n(t) \\ \kappa_g(t) & 0 & -\tau_g(t) \\ \kappa_n(t) & \tau_g(t) & 0 \end{pmatrix}.$$

The normal and geodesic curvatures both possess fairly intuitive geometric explanations as to what they measure. However, it is harder to say precisely what the geodesic torsion measures. Using the formula

$$\vec{N}' = s'(-\kappa_n \vec{T} - \tau_g \vec{U}),$$

we could say that the geodesic torsion measures the rate of change of $\vec{N}$ in the $\vec{U}$ direction, which would mean the rate of change of $\vec{N}$ twisting around the direction of motion along the curve (the direction in the tangent plane perpendicular to $\vec{T}$). This intuition is not particularly instructive. In Section 8.4, we study geodesic curves of a surface, which in some sense generalize the notion of straight lines on the surface. Problem 8.4.1 shows that the geodesic torsion is the torsion function of the geodesic curves.

PROBLEMS

1. Consider the cylinder parametrized by $\vec{X}(u, v) = (\cos u, \sin u, v)$ for $(u, v) \in [0, 2\pi] \times \mathbb{R}$ and consider the curve $\mathcal{C}$ on the cylinder with coordinate functions $(u(t), v(t)) = (t, \sin mt)$. Calculate the normal curvature, the geodesic curvature, and the geodesic torsion functions of $\mathcal{C}$ on the cylinder.

2. Let $\vec{X} = (R\cos u \sin v, R\sin u \sin v, R\cos v)$, with $(u, v) \in [0, 2\pi] \times [0, \pi]$, be a parametrization for a sphere. Consider the circle on the sphere given by $\vec{\gamma}(t) = \vec{X}(t, \varphi_0)$, where φ_0 is a fixed constant. Calculate the normal curvature, the geodesic curvature, and the geodesic torsion of $\vec{\gamma}$ on $\vec{X}$ (see Figure 8.1 for an illustration with $\varphi_0 = 2\pi/3$).

3. On the circular paraboloid parametrized by $\vec{X}(u, v) = (u, v, u^2 + v^2)$, calculate the normal curvature, the geodesic curvature, and the geodesic torsion of the curve on the surface with $u(t) = 1 + 2t$ and $v(t) = 3t$. [Hint: Use (8.7) for the geodesic curvature.]

4. Parametrize the circular paraboloid with polar coordinates with $u = \theta$ and $v = r$ so that $\vec{X}(u, v) = (v \cos u, v \sin u, v^2)$.

   (a) Show that a circle of center (a, b) and radius R in the domain has the equation $v^2 = a^2 + b^2 + R^2 + 2aR\cos u + 2bR\sin u$.

   (b) Use (8.7) to calculate the geodesic curvature of the curve on the circular paraboloid given as the circle in part (a) mapped onto the circular paraboloid.

5. Consider the cylinder $x^2 + y^2 = 1$, and consider the curve $\mathcal{C}$ on the cylinder obtained by intersecting the cylinder with the plane through the x-axis that makes an angle of θ with the xy-plane.

   (a) Show that $\mathcal{C}$ is an ellipse.

   (b) Find the normal curvature, geodesic curvature, and geodesic torsion functions of $\mathcal{C}$ on the cylinder.

   (c) Is the geodesic torsion an intrinsic quantity?

6. Let $\vec{\gamma}(s)$ be a curve parametrized by arc length on a surface parametrized by $\vec{X}(u, v)$. Prove that

$$\kappa_g \sqrt{\det(g)} = \frac{\partial}{\partial u}(\vec{T} \cdot \vec{X}_v) - \frac{\partial}{\partial v}(\vec{T} \cdot \vec{X}_u).$$

7. Consider the torus parametrized by $\vec{X}(u,v) = ((a+b\sin v)\cos u, (a + b\sin v)\sin u, b\cos v)$, where $a > b$. The curve $\vec{\gamma}(mt, nt)$, where m and n are relatively prime, is called the (m,n)-torus knot. Calculate the geodesic curvature of the (m,n)-torus knot on the torus.

8. Let $\vec{X}$ be the parametrization for a coordinate patch of a regular surface S, and let $\vec{\gamma} = \vec{X} \circ \vec{\alpha}$ be the parametrization for a curve $\mathcal{C}$ on S. Consider a point p on S, and let $\mathcal{C}'$ be the orthogonal projection of $\mathcal{C}$ onto the tangent plane T_pS. Prove that the geodesic curvature κ_g at p of $\mathcal{C}$ on S is equal to the curvature κ_g of $\mathcal{C}'$ at p as a plane curve in T_pS.

9. Let Γ be a regular space curve that is a component of the intersection of two regular surfaces S_1 and S_2. Suppose that near a point P on the curve Γ, the curve has the Frenet frame $(\vec{T}, \vec{P}, \vec{B})$, and the Darboux frames $(\vec{T}, \vec{U}_1, \vec{N}_1)$ and $(\vec{T}, \vec{U}_2, \vec{N}_2)$ corresponding to S_1 and S_2 respectively. Let $\theta(t)$ be the function along Γ of the angle from $\vec{N}_1$ to $\vec{N}_2$, so that $\sin\theta\vec{T} = \vec{N}_1 \times \vec{N}_2$. Let κ_{n1} and κ_{n2} be the normal curvatures of Γ on S_1 and S_2 respectively; let τ_{g1} and τ_{g2} be the geodesic torsions of Γ on S_1 and S_2 respectively. Prove the following two identities:

(a) $\dfrac{d\theta}{ds} = \tau_{g2} - \tau_{g1}$

(b) $\kappa^2 \sin^2\theta = \kappa_{n1}^2 + \kappa_{n2}^2 - 2\kappa_{n1}\kappa_{n2}\cos\theta$

10. *Bonnet's Formula.* Suppose that a curve on a surface is given by $\varphi(u, v) = c$, where c is a constant. Prove that the geodesic curvature is given by

$$\kappa_g = \frac{1}{\sqrt{\det(g)}}\left[\frac{\partial}{\partial u}\left(\frac{g_{12}\varphi_v - g_{22}\varphi_u}{g_{11}\varphi_v^2 - 2g_{12}\varphi_u\varphi_v + g_{22}\varphi_u^2}\right) + \frac{\partial}{\partial v}\left(\frac{g_{12}\varphi_u - g_{11}\varphi_v}{g_{11}\varphi_v^2 - 2g_{12}\varphi_u\varphi_v + g_{22}\varphi_u^2}\right)\right].$$

11. *Liouville's Formula.* (Comment: This result is of vital importance in the proof of the Gauss-Bonnet Theorem in the next section.) Let $\vec{X}(u, v)$ be an orthogonal parametrization of a patch on a regular surface S. Let $\mathcal{C}$ be a curve on this patch parametrized by arc length by $\vec{\gamma}(s) = \vec{X}(u(s), v(s))$. Let $\theta(s)$ be a function defined along $\mathcal{C}$ that gives the angle between $\vec{T}$ and $\vec{X}_u$. Prove that the geodesic curvature of $\mathcal{C}$ is given by

$$\kappa_g = \frac{d\theta}{ds} + \kappa_{(u)}\cos\theta + \kappa_{(v)}\sin\theta,$$

where $\kappa_{(u)}$ is the geodesic curvature along the u-parameter curve (i.e., $v = v_0$) and similarly for $\kappa_{(v)}$.

12. (*) Consider the surface S that is a function graph $z = f(x, y)$ and let $\mathcal{C}$ be a level curve given by $f(x, y) = c$. Using implicit differentiation, find a function $G(x, y)$ such that $G(x_0, y_0)$ is the geodesic curvature of $\mathcal{C}$ on S at a point (x_0, y_0, c).

13. Let S be a regular surface parametrized by $\vec{X}(u,v)$, and let $\mathcal{C}$ be a regular curve on S parametrized by $\vec{X}(t) = \vec{X}(u(t),v(t))$ for $t \in [a,b]$. Consider the normal tube around $\mathcal{C}$ parametrized by

$$\vec{Y}_r(t,v) = \vec{X}(t) + r\cos v\,\vec{U}(t) + r\sin v\,\vec{N}(t)$$

for some $r > 0$.

   (a) Find the metric tensor associated to $\vec{Y}(t,v)$, and conclude that

   $$\det g = (s')^2 r^2 (1 - r\kappa_g \cos v - r\kappa_n \sin v)^2.$$

   (b) Show that if r is small enough but still positive, then $\vec{Y}_r$ is a regular parametrization.

   (c) Assuming that $\vec{Y}_r$ is regular, calculate the coefficients L_{ij} of the second fundamental form of the normal tube around $\mathcal{C}$. Prove that the Gaussian curvature K satisfies

   $$K(t,v) = -\frac{s'(t)}{\sqrt{\det(g)}}(\kappa_n(t)\sin v + \kappa_g(t)\cos v).$$

8.2   Gauss-Bonnet Theorem, Local Form

No course on classical differential geometry is complete without the Gauss-Bonnet Theorem, arguably the most profound theorem in the differential geometry study of surfaces. The Gauss-Bonnet Theorem simultaneously encompasses a total curvature theorem for surfaces, the total geodesic curvature formula for plane curves, and other famous results, such as the sum of angles formula for a triangle in plane, spherical, or hyperbolic geometry. Other applications of the theorem extend much further and lead to deep connections between topological invariants and differential geometric quantities, such as the Gaussian curvature.

In his landmark paper [15], Gauss proved an initial version of what is now called the Gauss-Bonnet Theorem. The form in which we present it was first published by Bonnet in 1848 in [4]. Various alternative proofs exist (e.g., [34]) and a search of the literature turns up a large variety of generalizations (e.g., to polyhedral surfaces and to higher-dimensional manifolds). Despite the far-reaching consequences of the theorem, the difficulty of the proof resides in one essentially topological property of curves on surfaces. When we present this theorem, we will simply provide a reference. The theorem then follows easily from Green's Theorem for simple closed curves in the plane.

First, however, we motivate the Gauss-Bonnet Theorem with the following intuitive example.

Example 8.2.1 (The Moldy Potato Chip) We particularly encourage the reader to consult the demo applet for this example.

Consider a region $\mathcal{R}$ on a regular surface S such that the boundary curve $\partial\mathcal{R}$ is a regular curve that is simple and simply connected (can be shrunk continuously to a point) on the surface S. We now create the "moldy potato chip" as the surface that consists of taking the region $\mathcal{R}$ and spreading it out over every possible normal direction by a distance of r, where r is a fixed real number. As the demo shows, the surface of the moldy potato chip (MPC) consists of three pieces: two pieces that are the parallel surfaces of S "above" and "below" $\mathcal{R}$, and one region that consists of a half-tube around $\partial\mathcal{R}$. (In the applet, these portions are colored by magenta and green, respectively.)

The total Gaussian curvature $\iint_{\text{MPC}} K \, dS$ of the MPC is the same as the area of the portion of the unit sphere covered by the Gauss map of the MPC. Consider this area as the boundary $\partial\mathcal{R}$ is shrunk continuously to a point; it must vary continuously with the shrinking of $\partial\mathcal{R}$. However, this area must always be an integral multiple of 4π since the Gauss map of a surface without boundary covers the unit sphere a fixed number of times. The only function that is continuous and discrete is a constant function. In the limit, the moldy surface around a point is simply a sphere whose Gauss map image is the unit sphere with area 4π. Hence, the total curvature of the moldy potato chip is always 4π.

However, Problem 6.6.14 showed that if $r > 0$ is small enough, the total Gaussian curvature for each of the two parallel surfaces above and below $\mathcal{R}$ is $\iint_{\mathcal{R}} K \, dS$. Furthermore (and we leave the full details of this until Example 8.2.6), the total Gaussian curvature of the half-tube around $\partial\mathcal{R}$ is $2\int_{\partial\mathcal{R}} \kappa_g \, ds$. Hence, we conclude that

$$\iint_{\text{MPC}} K \, dS = 4\pi \iff 2\iint_{\mathcal{R}} K \, dS + 2\int_{\partial\mathcal{R}} \kappa_g \, ds = 4\pi$$

$$\iff \iint_{\mathcal{R}} K \, dS + \int_{\partial\mathcal{R}} \kappa_g \, ds = 2\pi. \qquad (8.9)$$

The applet for this example colors the magenta and green regions in a lighter shade when the Gaussian curvature of the MPC is positive and in a darker shade when the Gaussian curvature is negative. In this way, the user can see how, even if certain regions of the Gauss map are covered more than once, the signed area cancels portions out so that the total signed area covered is 4π.)

In what follows, we work to establish Equation (8.9) rigorously and expand the hypotheses under which it holds, the end result being the celebrated global Gauss-Bonnet Theorem. Furthermore, we provide an intrinsic proof of the Gauss-Bonnet Theorem, which establishes it as long as we have a metric tensor and without the assumption that the surface is in an ambient Euclidean three-space.

Certain types of regions on surfaces play an important role in what follows. We call a region $\mathcal{R}$ on a regular surface S *simple* if its boundary can be parametrized by a simple, closed, piecewise regular curve.

Using theorems of existence and uniqueness to certain differential equations, there are a variety of ways to see that near every point p on S there exists a neighborhood of p that is parametrized by an orthogonal parametrization. In Section 8.5 we will see such an orthogonal parametrization using what are called geodesic coordinate systems on S.

On a regular surface of class C^3 in $\mathbb{R}^3$, we can utilize extrinsic properties to realize an orthogonal parametrization. As discussed in Section 6.5, if $p = \vec{X}(u_0, v_0)$ is not an umbilical point of S, then the lines of curvature provide orthogonal coordinate lines for a parametrization of a neighborhood of p. More precisely, by the theorem on existence and uniqueness of solutions to differential equations, it is possible to find a solution to the first-order differential equation that results from Equation (6.30) with the initial condition of (u_0, v_0) and for each point, say $(u_0 + h, v_0)$ with $-\varepsilon < h < \varepsilon$, along the resulting line of curvature we can calculate the perpendicular line of curvature as the solution to the other branch resulting from Equation (6.30). This is not a tractable problem for specific surfaces but what matters for what follows is that at every point p, we can parametrize an open neighborhood of p on S with an orthogonal parametrization.

Let S be a regular surface of class C^3, and let $\vec{X} : U \to \mathbb{R}^3$ be an orthogonal parametrization of a neighborhood $V = \vec{X}(U)$ in S. Consider a regular curve $\vec{\gamma}(s)$ parametrized by arc length whose image lies in V. Liouville's Formula (see Problem 8.1.11) states that the geodesic curvature of $\vec{\gamma}(s)$ satisfies

$$\kappa_g(s) = \frac{d\varphi}{ds} + \kappa_{(u)} \cos\varphi + \kappa_{(v)} \sin\varphi,$$

where $\kappa_{(u)}$ is the geodesic curvature along the u-parameter curve (i.e., $v = v_0$) and similarly for $\kappa_{(v)}$ and $\varphi(s)$ is the angle $\vec{\gamma}'(s)$ makes with $\vec{X}_u$. Using Equation (8.7) to calculate $\kappa_{(u)}$ and $\kappa_{(v)}$ and writing $\cos\varphi$ and $\sin\varphi$ in terms of the metric tensor, one can rewrite Liouville's Formula as

$$\kappa_g(s) = \frac{d\varphi}{ds} + \frac{1}{2\sqrt{g_{11}g_{22}}} \frac{\partial g_{22}}{\partial u} \frac{dv}{ds} - \frac{1}{2\sqrt{g_{11}g_{22}}} \frac{\partial g_{11}}{\partial v} \frac{du}{ds}. \tag{8.10}$$

Now consider a simple region $\mathcal{R}$ in V with a boundary $\partial\mathcal{R}$ that is parametrized by arc length by a regular curve $\vec{\alpha}(s)$ defined over the interval $[0, \ell]$. Additionally, suppose that $\vec{\alpha}(s) = \vec{X}(u(s), v(s))$ for coordinate functions $u(s)$ and $v(s)$. Finally, suppose that U' is the subset of U such that $\vec{X}(U') = \mathcal{R}$.

Integrating the geodesic curvature around the curve $\partial \mathcal{R}$, we get

$$\int_{\partial \mathcal{R}} \kappa_g \, ds$$
$$= \int_0^\ell \left(\frac{1}{2\sqrt{g_{11}g_{22}}} \frac{\partial g_{22}}{\partial u} \frac{dv}{ds} - \frac{1}{2\sqrt{g_{11}g_{22}}} \frac{\partial g_{11}}{\partial v} \frac{du}{ds} \right) ds + \int_0^\ell \frac{d\varphi}{ds} \, ds,$$

where ℓ is the length of $\partial \mathcal{R}$. Applying Green's Theorem (Theorem 2.1.6) to the first term, this is equal to

$$\iint_{U'} \frac{1}{2} \left(\frac{\partial}{\partial u} \left(\frac{1}{\sqrt{g_{11}g_{22}}} \frac{\partial g_{22}}{\partial u} \right) + \frac{\partial}{\partial v} \left(\frac{1}{\sqrt{g_{11}g_{22}}} \frac{\partial g_{11}}{\partial v} \right) \right) du \, dv + \int_0^\ell \frac{d\varphi}{ds} \, ds.$$

By (7.23), this becomes

$$\int_{\partial \mathcal{R}} \kappa_g \, ds = - \iint_{U'} K \sqrt{g_{11}g_{22}} \, du \, dv + \int_0^\ell \frac{d\varphi}{ds} \, ds,$$

which leads to

$$\int_{\partial \mathcal{R}} \kappa_g \, ds = - \iint_{U'} K \sqrt{g_{11}g_{22}} \, du \, dv + 2\pi. \qquad (8.11)$$

The last term in Equation (8.11) is the content of the Theorem of Turning Tangents, proved by H. Hopf in [19], which we state without proving.

Lemma 8.2.2 (Theorem of Turning Tangents, Regular)
Let C be a simple, closed, regular curve on a regular surface S of class C^3 parametrized by $\vec{\alpha}$. Let $\varphi(s)$ be defined as above. Then

$$\int_C \frac{d\varphi}{ds} \, ds = \pm 2\pi,$$

where the sign of 2π is positive if the orientation of $\vec{\alpha}$ is such that the normal vector $\vec{U}(s)$ to $\vec{\alpha}(s)$ points into the region enclosed by the curve and negative otherwise.

Putting together Equation (8.11) and Lemma 8.2.2 establishes a first local version of the Gauss-Bonnet Theorem.

Theorem 8.2.3 (Local Gauss-Bonnet Theorem, Regular)
Let $\vec{X} : U \to \mathbb{R}^3$ be an orthogonal parametrization of a neighborhood $V = \vec{X}(U)$ of an oriented surface S of class C^3. Let $\mathcal{R} \subset V$ be a

region of S such that the boundary is $\partial \mathcal{R} = \vec{\alpha}([0, \ell])$ for some simple, closed, regular, positively oriented curve $\vec{\alpha} : [0, \ell] \to S$ of class C^3 parametrized by arc length. Then

$$\int_{\partial \mathcal{R}} \kappa_g \, ds + \iint_{\mathcal{R}} K \, dS = 2\pi.$$

We have called the above formulation of the Gauss-Bonnet Theorem a local version since, as stated, it requires that the region $\mathcal{R}$ of S be inside a coordinate neighborhood of S that admits an orthogonal parametrization. Furthermore, the above theorem assumes that $\partial \mathcal{R}$ is a regular curve. By Problem 6.1.22, we know that if S is a regular surface of class C^2, then at every point $p \in S$ there exists a neighborhood of p that can be parametrized by a regular orthogonal parametrization. We will first generalize this theorem to include regions whose boundaries satisfy a looser condition than bring regular. Then, to obtain a global version of the theorem, we will "piece together" local instances of the above theorem.

The above proof of the local Gauss-Bonnet Theorem is *almost* an intrinsic proof but not quite. In order to use Liouville's formula, we needed to refer to an orthogonal parametrization of a patch of the surface. At that point, we chose to refer to a parametrization of a patch that has curvature lines as coordinate lines. However, curvature lines are along the direction of principal curvatures, which are extrinsic properties. Proposition 8.5.5 proves the existence of geodesic coordinate systems on patches of a regular surface. Such coordinate systems do depend entirely on the components of the metric tensor and its derivatives and they induce orthogonal parametrizations. Citing this proposition that we will encounter later establishes the Gauss-Bonnet Theorem as a purely intrinsic result.

We now extend the local Gauss-Bonnet Theorem to the broader class of piecewise regular curves in order to set the scene for the global theorem.

In order to present the Gauss-Bonnet Theorem for piecewise regular curves, we need to first establish a few definitions about the geometry of such curves on surfaces.

We call a set of points $\{t_i\}_{i \in \mathcal{K}}$ in $\mathbb{R}$ *discrete* if

$$\inf\{|t_i - t_j| \, \big| \, i, j \in \mathcal{K} \text{ and } i \neq j\} > 0,$$

i.e., if any two distinct points are separated by at least some fixed, positive real number. The indexing set $\mathcal{K}$ may be finite, say $\mathcal{K} = \{1, \ldots, k\}$, or, if it is infinite, it may be taken as either the set of nonnegative integers $\mathbb{N}$ or the set of integers $\mathbb{Z}$.

Suppose that a curve $\vec{X} : I \to \mathbb{R}^m$ is regular near $t_0 \in I$, i.e., on interval $(t_0 - \varepsilon, t_0 + \varepsilon)$, where $\varepsilon > 0$ and suppose also that it is regular on $(t_0 - \varepsilon, t_0)$ and $(t_0, t_0 + \varepsilon)$. Then we can consider the limits of the unit tangent $\vec{T}(t)$ as t approaches t_0 from the right and from the left. Call

$$\vec{T}(t_0^-) = \lim_{t \to t_0^-} \vec{T}(t) \qquad \text{and} \qquad \vec{T}(t_0^+) = \lim_{t \to t_0^+} \vec{T}(t).$$

To be precise, we call a curve $\vec{\alpha} : I \to \mathbb{R}^n$ *piecewise regular* if

- $\vec{\alpha}$ is continuous over I;

- there exists a discrete set of points $\{t_i\}_{i \in \mathcal{K}}$ such that $\vec{\alpha}$ is regular over each open interval in the set difference $I - \{t_i\}_{i \in \mathcal{K}}$; and

- $\vec{T}(t_i^-)$ and $\vec{T}(t_i^+)$ exist for all $t_i \in \{t_i\}_{i \in \mathcal{K}}$.

A regular curve is also piecewise regular, by virtue of considering the situation in which the set $\{t_i\}_{i \in \mathcal{K}}$ is empty. If $\vec{\alpha}$ is regular over an interval I', then the trace $\vec{\alpha}(I')$ is called a *regular arc* of the curve. If $\vec{T}(t_i^-) = -\vec{T}(t_i^+)$, then a point $\vec{\alpha}(t_i)$ is called a *cusp*. Otherwise, if $\vec{T}(t_i^-)$ and $\vec{T}(t_i^+)$ are not collinear, then $\vec{\alpha}(t_i)$ is called a *corner*.

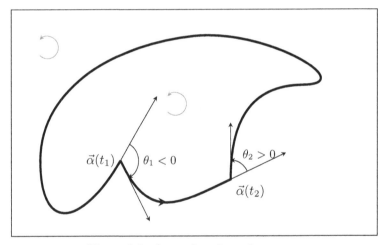

Figure 8.2: A regular piecewise curve.

We point out that our definition for piecewise regular curves applies to any curve in $\mathbb{R}^m$. However we now also assume that $\vec{\alpha}$ has the properties that will interest us for the Gauss-Bonnet Theorem, namely that $\vec{\alpha} : I \to \mathbb{R}^3$ is a simple, closed, piecewise regular curve on a regular oriented surface S with orientation n. In this case, I is a closed and bounded interval, and there can be at most a finite number of vertices. Furthermore, we impose the criterion that the curve $\vec{\alpha}$ traces out the image $\vec{\alpha}(I)$ in the same direction as the orientation of S. See the surface and curve orientations in Figure 8.2.

For all corners $\vec{\alpha}(t_i)$, we define the *external angle* θ_i of the vertex as the angle $-\pi < \theta_i < \pi$ swept out from $\vec{T}(t_i^-)$ to $\vec{T}'(t_i^+)$ in the plane through the point $\vec{\alpha}(t_i)$ and spanned by $\vec{T}'(t_i^-)$ and $\vec{T}'(t_i^+)$ (see Figure 8.2). Note that this external angle may be positive or negative.

We are now in a position to approach the local Gauss-Bonnet Theorem for piecewise regular curves.

Let S be a regular surface and let V be a neighborhood of S parametrized orthogonally by $\vec{X} : U \to \mathbb{R}^3$ with $U \subseteq \mathbb{R}^2$. We assume that $\vec{X} : U \to V$ is a homeomorphism. Consider a simple region $\mathcal{R}$ in the neighborhood V on S such that its boundary $\partial\mathcal{R}$ is parametrized by a simple, closed, piecewise regular curve $\vec{\alpha} : I \to \mathbb{R}^3$. Suppose that $\vec{\alpha}$ is parametrized by arc length so that $I = [0, \ell]$ and has vertices at $s_1 < s_2 < \cdots < s_k$. Set $s_0 = 0$ and $s_{k+1} = \ell$. Also call C_i the image of $\vec{\alpha}([s_{i-1}, s_i])$ for $1 \le i \le k+1$; these are the regular arcs of $\partial\mathcal{R}$. Suppose, additionally, that $\vec{\alpha}(s) = \vec{X}(u(s), v(s))$. Let U' be the subset of U such that $\mathcal{R} = \vec{X}(U')$.

As in the previous case, integrating the geodesic curvature around curve C, we get

$$\int_{\partial\mathcal{R}} \kappa_g \, ds = \sum_{i=0}^{k} \int_{C_i} \kappa_g \, ds = \sum_{i=0}^{k} \int_{s_i}^{s_{i+1}} \kappa_g(s) \, ds$$

$$= \sum_{i=0}^{k} \int_{s_i}^{s_{i+1}} \left(\frac{1}{2\sqrt{g_{11}g_{22}}} \frac{\partial g_{22}}{\partial u} \frac{dv}{ds} - \frac{1}{2\sqrt{g_{11}g_{22}}} \frac{\partial g_{11}}{\partial v} \frac{du}{ds} \right) ds$$

$$+ \sum_{i=0}^{k} \int_{s_i}^{s_{i+1}} \frac{d\varphi}{ds} \, ds.$$

Applying Green's Theorem (generalized to simple closed piecewise regular curves) to the first term, we get

$$\int_{\partial\mathcal{R}} \kappa_g \, ds = \iint_{U'} \frac{1}{2} \left(\frac{\partial}{\partial u} \left(\frac{1}{\sqrt{g_{11}g_{22}}} \frac{\partial g_{22}}{\partial u} \right) + \frac{\partial}{\partial v} \left(\frac{1}{\sqrt{g_{11}g_{22}}} \frac{\partial g_{11}}{\partial v} \right) \right) du \, dv$$

$$+ \sum_{i=0}^{k} \int_{s_i}^{s_{i+1}} \frac{d\varphi}{ds} \, ds.$$

By Equation (7.23), this becomes

$$\int_{\partial\mathcal{R}} \kappa_g \, ds = - \iint_{U'} K\sqrt{g_{11}g_{22}} \, du \, dv + \sum_{i=0}^{k} \int_{s_i}^{s_{i+1}} \frac{d\varphi}{ds} \, ds,$$

which leads to

$$\int_{\partial\mathcal{R}} \kappa_g \, ds = - \iint_{\vec{X}(U')} K \, dS + \sum_{i=0}^{k} \left(\varphi(s_{i+1}) - \varphi(s_i) \right). \qquad (8.12)$$

It remains for us to interpret the last term in Equation (8.12). As before, is precisely the more general version of the Theorem of Turning Tangents [19].

Lemma 8.2.4 (Theorem of Turning Tangents) *Let $\vec{\alpha}$ be a simple, closed, piecewise regular curve on a regular surface S of class C^3. Let $\vec{\alpha}(s_i)$ be the vertices with external angles θ_i, and let $\varphi(s)$ be as defined above. Then*

$$\sum_{i=0}^{k} \big(\varphi(s_{i+1}) - \varphi(s_i)\big) = \pm 2\pi - \sum_{i=1}^{k} \theta_i,$$

where the sign of 2π is positive if the orientation of $\vec{\alpha}$ is such that the normal vector $\vec{U}(s)$ to $\vec{\alpha}(s)$ points into the region enclosed by the curve and negative otherwise.

Putting together Equation (8.12) and Lemma 8.2.4 establishes a local version of the Gauss-Bonnet Theorem.

Theorem 8.2.5 (Local Gauss-Bonnet Theorem) *Let $\vec{X} : U \to \mathbb{R}^3$ be an orthogonal parametrization of a region $V = \vec{X}(U)$ of an oriented surface S of class C^3. Let $\mathcal{R} \subset V$ be a simple region of S, and suppose that the boundary is $\partial\mathcal{R} = \vec{\alpha}([0, \ell])$ for some simple, closed, piecewise regular, positively oriented curve $\vec{\alpha} : [0, \ell] \to S$ of class C^2, parametrized by arc length. Let $\vec{\alpha}(s_i)$, with $1 \leq i \leq k$, be the vertices of $\partial\mathcal{R}$, and let θ_i be their external angles. Call C_i the regular arcs of $\partial\mathcal{R}$. Then*

$$\int_{\partial\mathcal{R}} \kappa_g \, ds + \iint_{\mathcal{R}} K \, dS + \sum_{i=1}^{k} \theta_i = 2\pi.$$

Note that in this statement of the theorem, there are k vertices and $k + 1$ regular arcs because the formulation assumes that $\vec{\alpha}(0)$ is not a vertex. Consequently, we must remember to interpret the integral on the left as an integral over $k + 1$ regular arcs as

$$\int_{\partial\mathcal{R}} \kappa_g \, ds = \sum_{i=0}^{k} \int_{C_i} \kappa_g \, ds.$$

We typically call the more general Theorem 8.2.5 the Local Gauss-Bonnet Theorem and understand that Theorem 8.2.3 is a subcase of this theorem.

We illustrate this theorem in a similar vein as the Moldy Potato Chip example but now with a patch of surface bounded by a piecewise regular curve.

Example 8.2.6 (The Moldy Patch) The motivating example, Example 8.2.1 with the moldy potato chip, falls just shy of giving an extrinsic (assumes we have/know a normal vector to the surface) proof of the local Gauss-Bonnet Theorem. We provide the details here.

Consider a simply connected region $\mathcal{R}$ on a regular surface S as described in Theorem 8.2.5. Suppose that $\mathcal{R}$ is parametrized by $\vec{X}$: $U \to \mathbb{R}^3$ for some $U \subset \mathbb{R}^2$, and call $\vec{N}$ the associated normal vector.

Define the surface $\mathcal{T}_r$ as the tubular neighborhood of $\mathcal{R}$ with radius r. This means that $\mathcal{T}_r$ consists of

1) two pieces for the normal variation to $\mathcal{R}$ parametrized, respectively, by $\vec{X} + r\vec{N}$ and $\vec{X} - r\vec{N}$ over U (which we call respectively $U_{(+r)}$ and $U_{(-r)}$);

2) k half-tubes of radius r around the smooth pieces of the boundary $\partial \mathcal{R}$ pointing "away" from the region $\mathcal{R}$;

3) k lunes of spheres (of radius r) at the k vertices of $\partial \mathcal{R}$.

Figure 8.3 shows the pieces of $\mathcal{T}_r$ for a patch on a torus. We assume from now on that r is small enough so that each of the pieces of $\mathcal{T}_r$ is a regular surface.

Let K be the Gaussian curvature of S over $\mathcal{R}$, and call $K_{\mathcal{T}}$ the Gaussian curvature of the moldy patch $\mathcal{T}_r$. By Proposition 6.6.2, the quantity $K_{\mathcal{T}} \, dS_{\mathcal{T}}$ is a signed area element of the image on the unit sphere of $\mathcal{T}_r$ under its Gauss map. So in calculating the total curvature of the moldy patch, one is adding or subtracting area of the sphere depending on the sign of $K_{\mathcal{T}}$. We now reason why

$$\iint_{\mathcal{T}_r} K_{\mathcal{T}} \, dS_{\mathcal{T}} = 4\pi. \qquad (8.13)$$

Since $\mathcal{T}_r$ has no boundary, then the Gauss map of $\mathcal{T}_r$ covers the unit sphere an integer number of times, where we add area for positive curvature and subtract for negative curvature. Thus, the integral in Equation (8.13) must be equal to $4\pi h$, where h is an integer. Suppose that the region $\mathcal{R}$ is *contractible*, which means it can be shrunk continuously to a point while remaining on the surface. Now if $\mathcal{R}$ is a point, then $\mathcal{T}_r$ is a sphere of radius r. In this case, the Gauss map for $\mathcal{T}_r$ is a bijective map onto the unit sphere, and hence, the integral in Equation (8.13) gives precisely the surface of the unit sphere, so is equal to 4π. However, as one "uncontracts" from a point to $\mathcal{R}$, the integral in Equation (8.13) must vary continuously. However, a continuous function to the set of integer multiples of 4π is a constant function. Though we have not spelled out the whole topological background, this reasoning justifies Equation (8.13).

We now break down Equation (8.13) according to various pieces

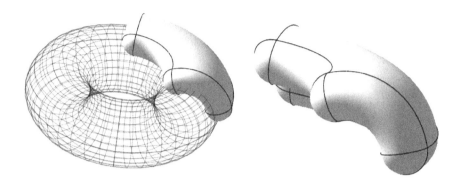

Figure 8.3: A "moldy patch."

of $\mathcal{T}_r$: the normal variation patches, the half-tubes, and the lunes of spheres.

By Problem 6.6.14, if we call $K_{(+r)}$ and $K_{(-r)}$ the respective Gaussian curvatures of $\vec{X} + r\vec{N}$ and $\vec{X} - r\vec{N}$ over U, then

$$2 \iint_U K \, dS = \iint_{U_{(+r)}} K_{(+r)} \, dS + \iint_{U_{(-r)}} K_{(-r)} \, dS.$$

In Problem 8.1.13, one calculates the Gaussian curvature of the normal tube around a curve on the surface. Consider a regular arc C of $\partial\mathcal{R}$, and assume it is parametrized by $t \in [a, b]$. As a consequence of Problem 8.1.13, over the normal half-tube HT pointing away from $\vec{U}$ (i.e., away from the inside of $\mathcal{R}$), the total Gaussian curvature is

$$\iint_{HT} K \, dS = \int_a^b \int_{\pi/2}^{3\pi/2} -s'(t)\big((\sin v)\kappa_n(t) + (\cos v)\kappa_g(t)\big) \, dt$$

$$= 2 \int_a^b s'(t)\kappa_g(t) \, dt = 2 \int_C \kappa_g \, ds.$$

Finally, we consider the lunes of $\mathcal{T}_r$ that are around the vertices of $\partial\mathcal{R}$. Under the image of the Gauss map, the lunes map to the same corresponding lune on the unit sphere. Thus, in Equation (8.13), a lune around a vertex with exterior angle θ_i contributes $(\theta_i/2\pi)4\pi = 2\theta_i$ to the surface of the unit sphere.

Combining each of these results, we find that

$$\iint_{\mathcal{T}_r} K_{\mathcal{T}} \, dS_{\mathcal{T}} = 2 \iint_U K \, dS + 2 \sum_{i=0}^{k} \int_{C_i} \kappa_g \, ds + 2 \sum_{i=1}^{k} \theta_i,$$

and the local Gauss-Bonnet Theorem follows immediately from Equation (8.13).

In our presentation of the local Gauss-Bonnet Theorem, we did not allow corners to be cusps. The main difficulty lies in deciding whether to assign a value of π or $-\pi$ to the angle θ_i of any given cusp in order to retain the validity of the local Gauss-Bonnet Theorem. We can allow for cusps on the boundary $\partial \mathcal{R}$ if we employ the following sign convention for the angles of cusps: if the cusp $\vec{\alpha}(t_i)$ points into the interior of the closed curve (i.e., $\vec{\alpha}(t_i^-)$ points into the interior) then $\theta_i = -\pi$; and if the cusp $\vec{\alpha}(t_i)$ points away from the interior of the closed curve (i.e., $\vec{\alpha}(t_i^-)$ points away from the interior) then $\theta_i = \pi$. (See Figure 8.4. The orientation of the surface matters since it determines the direction of travel around the boundary curve $\partial \mathcal{R}$.)

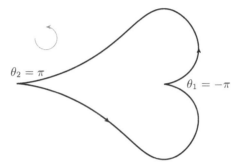

Figure 8.4: External angles at cusps.

8.3 Gauss-Bonnet Theorem, Global Form

In order to extend the Gauss-Bonnet Theorem to a global presentation (i.e., outside of a single coordinate patch on the surface), we need to briefly discuss triangulations of surfaces, the classification of orientable surfaces, and the Euler characteristic of regions of surfaces in $\mathbb{R}^3$. These topics are typically considered in the area of topology, but we summarize the results that we need in order to give a full treatment to the global Gauss-Bonnet Theorem without insisting that the reader have mastery of the supporting topology. (The authors include technical details behind these concepts in Appendices A.5 and A.6 in [24]. Otherwise, the interested reader could consult Chapters 6 and 7 of [1].)

In intuitive terms, a *triangulation* of a surface consists of a network of a finite number of regular curve segments on the surface such that any point on the surface either lies on one of the curves or lies in a region that is bounded by precisely three curve segments. The first picture in Figure 8.5 depicts a triangulation on a torus. As an additional technical requirement, one should be able to continuously

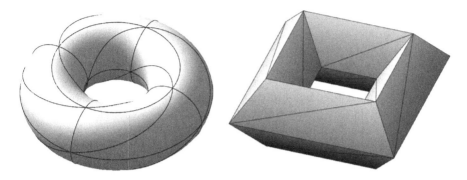

Figure 8.5: Torus triangulation.

deform the surface with its triangulation so that each "triangle" becomes a true triangle without changing the topological nature of the surface. (Compare the two pictures in Figure 8.5.)

In a triangulation, a *vertex* is an endpoint of one of the curve segments on the surface. We call the curve segments *edges*, and the regions enclosed by edges (the "triangles") we call *faces*. An interesting and useful result, first proved by Rado in 1925, is that every regular compact surface admits a triangulation.

A result of basic topology is that given a compact regular surface S, the quantity

$$\#(\text{vertices}) - \#(\text{edges}) + \#(\text{faces})$$

is the same regardless of any triangulation of S. This number is called the *Euler characteristic* of S and is denoted by $\chi(S)$. Furthermore, from the definition of triangulation, one can deduce that the Euler characteristic does not change if the surface is deformed continuously (no cutting or pinching). One often restates this last property by saying that the Euler characteristic is a *topological invariant*.

For example, the torus triangulation in Figure 8.5 has 16 vertices, 48 edges, and 32 faces. Thus, the Euler characteristic of the torus is 0. As another example consider the tetrahedron, which is homeomorphic to the sphere. A tetrahedron has four vertices, six edges, and four faces, so its Euler characteristic, and therefore the Euler characteristic of the sphere, is $\chi = 4 - 6 + 4 = 2$.

A profound theorem in topology, the Classification Theorem of Surfaces states that every orientable surface without boundary is homeomorphic to a sphere or to a sphere with a finite number of "handles" added to it. Figure 8.6(a) shows a torus while Figure 8.6(b) shows a sphere with one handle added to it. These two surfaces are in fact the same under a continuous deformation, i.e., they are homeomorphic. Figure 8.6(c) depicts a two-holed torus that, in the language

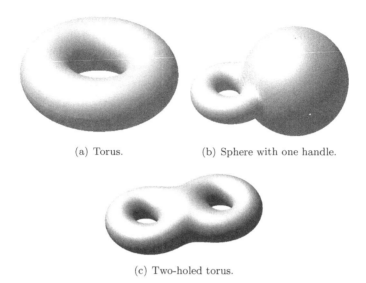

(a) Torus.　　　　(b) Sphere with one handle.

(c) Two-holed torus.

Figure 8.6: Tori.

of the Classification Theorem of Surfaces, is called a sphere with two handles.

It is not hard to show that the Euler characteristic of a sphere with g handles added is

$$\chi(S) = 2 - 2g. \tag{8.14}$$

The notion of the Euler characteristic applies equally well to a surface with boundary as long as the boundary is completely covered by edges and vertices of the triangulation. For example, we encourage the reader to verify that a sphere with a small disk removed has Euler characteristic of 1.

We must now discuss orientations on a triangulation. When considering adjacent triangles, we can think of the orientation of a triangle as a direction of travel around the edges. Two adjacent triangles have a compatible orientation if the orientation of the first leads one to travel along the common edge in the opposite direction of the orientation on the second triangle (see Figure 8.7). It turns out that if a surface is orientable, then it is possible to choose an orientation of each triangle in the triangulation such that adjacent triangles have compatible orientations.

A compact regular surface S is covered by a finite number of coordinate neighborhoods given by regular parametrizations. In general, if $\mathcal{R}$ is a regular region of S, it may not lie entirely in one coordinate patch. However, it is possible to show that not only does a regular region $\mathcal{R}$ admit a triangulation, but every regular region $\mathcal{R}$ admits a triangulation such that each triangle is contained in a coordinate neighborhood. This comment and two lemmas show that it makes sense to talk about the surface integral over the whole region $\mathcal{R} \subset S$.

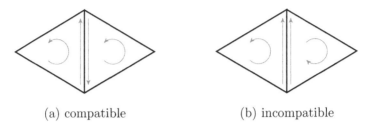

(a) compatible   (b) incompatible

Figure 8.7: Adjacent oriented triangles.

Lemma 8.3.1 *Suppose that $\vec{X}_1 : U_1 \to \mathbb{R}^3$ and $\vec{X}_2 : U_2 \to \mathbb{R}^3$ are two systems of coordinates of a regular surface S. Call (u_i, v_i) the coordinates of $\vec{X}_i$, and call $g^{(i)}$ the corresponding metric tensor. Suppose that $T \subset \vec{X}_1(U_1) \cap \vec{X}_2(U_2)$. Then for any function $f : T \to \mathbb{R}$, we have*

$$\iint_{\vec{X}_1^{-1}(T)} f(u_1, v_1) \sqrt{\det(g^{(1)})}\, du_1\, dv_1$$
$$= \iint_{\vec{X}_2^{-1}(T)} f(u_2, v_2) \sqrt{\det(g^{(2)})}\, du_2\, dv_2. \quad (8.15)$$

Proof: By Equation (6.7), one deduces that

$$\det(g^{(1)}) = \left(\frac{\partial(u_2, v_2)}{\partial(u_1, v_1)} \right)^2 \det(g^{(2)}).$$

However, $\frac{\partial(u_2, v_2)}{\partial(u_1, v_1)}$ is the Jacobian of the coordinate transformation $F : \vec{X}_1^{-1}(T) \to \vec{X}_2^{-1}(T)$ defined by $\vec{X}_2^{-1} \circ \vec{X}_1$ restricted to $\vec{X}_1^{-1}(T)$. The result follows as an application of the change of variables formula in double integrals. $\square$

Lemma 8.3.2 *Let S be a regular oriented surface, and let $\mathcal{R}$ be a regular compact region of S, possibly with a boundary. Given a collection $\{\vec{X}_i\}_{i \in I}$ of coordinate neighborhoods that cover S, $\{T_j\}_{j \in J}$ triangles of a triangulation of $\mathcal{R}$, and $i : J \to I$ such that T_j is in the image of $\vec{X}_{i(j)}$, define the sum*

$$\sum_{j \in J} \iint_{\vec{X}_{i(j)}^{-1}(T_j)} f(u_{i(j)}, v_{i(j)}) \sqrt{\det g^{(i(j))}}\, du_{i(j)}\, dv_{i(j)}.$$

This sum is independent of the choice of triangulation of $\mathcal{R}$, collection of coordinate patches, and function i.

Proof: That the sum is independent of the collection of coordinate neighborhoods follows from Lemma 8.3.1. That the sum does not depend on the choice of triangulation is a little tedious to prove and is left as an exercise for the reader. □

   This leads to a definition of a surface integral over any region on a regular surface.

Definition 8.3.3 Let S be a regular oriented surface, and let $\mathcal{R}$ be a regular compact region of S, possibly with a boundary. We call the common sum described in Lemma 8.3.2 the surface integral of f over $\mathcal{R}$ and denote it by

$$\iint_{\mathcal{R}} f \, dS.$$

We point out that if $\vec{X} : U \to \mathbb{R}^3$ is a regular parametrization of a region of S such that $\vec{X}(U)$ is dense in $\mathcal{R}$, then

$$\iint_{\mathcal{R}} f \, dS = \iint_{U} f(\vec{X}(u,v)) \|\vec{X}_u \times \vec{X}_v\| \, dA,$$

where the right-hand side is the usual double integral.

   We can now state the main theorem of this section.

Theorem 8.3.4 (Global Gauss-Bonnet Theorem)   *Let S be a regular oriented surface of class C^3, and let $\mathcal{R}$ be a compact region of S with boundary $\partial \mathcal{R}$. Suppose that $\partial \mathcal{R}$ is a simple, closed, piecewise regular, positively oriented curve. Suppose that $\partial \mathcal{R}$ has k regular arcs C_i of class C^2, and let θ_i be the external angles of the vertices of $\partial \mathcal{R}$. Then*

$$\sum_{i=1}^{k} \int_{C_i} \kappa_g \, ds + \iint_{\mathcal{R}} K \, dS + \sum_{i=1}^{k} \theta_i = 2\pi \chi(\mathcal{R}),$$

where $\chi(\mathcal{R})$ is the Euler characteristic of $\mathcal{R}$.

Proof: Let $\mathcal{R}$ be covered by a collection of coordinate patches. Let $\{T_j\}_{j \in J}$ be the triangles of a triangulation of $\mathcal{R}$ in which all the triangles T_j on S are subsets of some coordinate patch. Suppose also that every triangle in the set $\{T_j\}$ is equipped with an orientation that is compatible with the orientation of S.

   For each triangle T_j, for $1 \leq l \leq 3$, call E_{jl} the edges of T_j (as curves on S), call V_{jl} the vertices of T_j, and let β_{jl} be the interior angle of T_j at V_{jl}. For this triangulation, call a_0 the number of vertices, a_1 the number of edges, and a_2 the number of triangles. By construction,

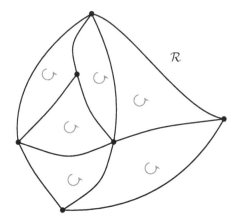

Figure 8.8: A triangulation of a region $\mathcal{R}$.

the local Gauss-Bonnet Theorem applies to each triangle T_j on S, so on each T_j, we have

$$\sum_{l=1}^{3} \int_{E_{jl}} \kappa_g \, ds + \iint_{T_j} K \, dS = 2\pi - \sum_{l=1}^{3} (\pi - \beta_{jl}) = -\pi + \sum_{l=1}^{3} \beta_{jl}. \quad (8.16)$$

Now consider the sum of Equation (8.16) over all the triangles T_j. Since each triangle has an orientation compatible with the orientation of S, then whenever two triangles share an edge, the edge is traversed in opposite orientations on the adjacent triangles (see Figure 8.8). Consequently, in the sum of Equation (8.16), each integral $\displaystyle\int_{E_{jl}} \kappa_g \, ds$ cancels out another similar integral along any edge that is not a part of the boundary of $\mathcal{R}$. Therefore, applying Lemma 8.3.2, the left-hand side of the sum of Equation (8.16) is precisely

$$\sum_{i=1}^{k} \int_{C_i} \kappa_g \, ds + \iint_{\mathcal{R}} K \, dS. \quad (8.17)$$

The right-hand side of the sum of Equation (8.16) is

$$-\pi a_2 + \sum_{j} \sum_{i=1}^{3} \beta_{jl}.$$

In the double sum $\sum \sum \beta_{jl}$, the sum of interior angles associated to a vertex on the interior of $\mathcal{R}$ contributes 2π, and the sum of angles associated to a vertex V on the boundary $\partial\mathcal{R}$ contributes

$\pi - ($exterior angle of V). Thus,

$$\sum_j \left(-\pi + \sum_{l=1}^{3} \beta_{jl} \right) = -\sum_{i=1}^{k} \theta_i - \pi a_2 + 2\pi(\#\text{interior vertices})$$
$$+ \pi(\#\text{exterior vertices}).$$

Since there are as many vertices on the boundary as there are edges, we rewrite this as

$$\sum_j \left(-\pi + \sum_{l=1}^{3} \beta_{jl} \right) = -\sum_{i=1}^{k} \theta_i - \pi a_2 + 2\pi a_0 - \pi(\#\text{exterior edges}).$$
(8.18)

However, since each triangle has three edges,

$$-a_2 = 2a_2 - 3a_2 = 2a_2 - 2(\#\text{interior edges}) - (\#\text{exterior edges}).$$
(8.19)

Consequently, taking the sum of (8.16) and combining (8.17), (8.18), and (8.19) one obtains

$$\sum_i \int_{C_i} \kappa_g \, ds + \iint_{\mathcal{R}} K \, dS = -\sum_{i=1}^{k} \theta_i + 2\pi(a_2 - a_1 + a_0).$$

By definition of the Euler characteristic, $\chi(\mathcal{R}) = a_2 - a_1 + a_0$. The theorem follows. ☐

The global version of the Gauss-Bonnet Theorem directly generalizes the local version so when one refers to *the* Gauss-Bonnet Theorem, one means the global version. Even at a first glance, the Gauss-Bonnet Theorem is profound because it connects local properties of curves on a surface (the geodesic curvature κ_g, the Gaussian curvature K, and angles associated to vertices) with global properties (the Euler characteristic of a region of a surface). In fact, since the Euler characteristic is a topological invariant, the Gauss-Bonnet Theorem connects local geometric properties to a topological property.

Let S be a compact regular surface (without boundary). An interesting particular case of the global Gauss-Bonnet Theorem occurs when we consider $\mathcal{R} = S$, which implies that $\partial \mathcal{R}$ is empty. This situation leads to the following strikingly simple and profound corollary.

Corollary 8.3.5 *Let S be an orientable, compact, regular surface of class C^3 without boundary. Then*

$$\iint_S K \, dS = 2\pi \chi(S).$$

Example 8.3.6 Consider the torus T parametrized by $\vec{X} : [0, 2\pi]^2 \rightarrow \mathbb{R}^3$, with

$$\vec{X}(u, v) = ((a + b\cos v)\cos u, (a + b\cos v)\sin u, b\sin v),$$

where $a > b$. We note that $\vec{X}((0, 2\pi)^2)$ covers all of T except for two curves on the torus, so $\vec{X}((0, 2\pi)^2)$ is dense in T. By the comment after Definition 8.3.3, we can use the usual surface integral over this one coordinate neighborhood to calculate $\iint_S K \, dS$ directly.

It is not hard to calculate that over the coordinate neighborhood described above, we have

$$K(u, v) = \frac{\cos v}{b(a + b\cos v)} \quad \text{and} \quad \|\vec{X}_u \times \vec{X}_v\| = b(a + b\cos v).$$

Thus,

$$\iint_S K \, dS = \int_0^{2\pi} \int_0^{2\pi} \frac{\cos v}{b(a + b\cos v)} b(a + b\cos v) \, du \, dv$$
$$= \int_0^{2\pi} \int_0^{2\pi} \cos v \, du \, dv = 0,$$

which proves that $\chi(T) = 0$. This agrees with the calculation provided by the triangulation in Figure 8.5.

We shall now present an application of the Gauss-Bonnet Theorem, as well as leave a few as exercises for the reader.

Proposition 8.3.7 *Let S be a compact, regular, orientable surface of class C^3 without boundary and with positive curvature everywhere. Then S is homeomorphic to a sphere.*

Proof: Since $K > 0$ over the whole surface S, then for the total Gaussian curvature we have

$$\iint_S K \, dS > 0.$$

By Corollary 8.3.5, this integral is $2\pi\chi(S)$. However, by the Classification Theorem of Surfaces and Equation (8.14), since $\chi(S) > 0$, we have $\chi(S) = 2$, and therefore S is homeomorphic to a sphere. $\square$

PROBLEMS

1. Provide all the details that establish Equation (8.10).

2. Provide the details for the proof of Lemma 8.3.2.

3. The following annular regions (in the plane) can be deformed continuously one into another, so have the same Euler characteristic.

Give a triangulation of the square annulus and use it to show that the Euler characteristic of the region is 0.

4. Find a triangulation of a polyhedral two-holed torus. Use it to show that the Euler characteristic of a two-holed torus is -2.

5. In our discussion of the Euler characteristic, we implied that "adding a handle" to a surface decreases the Euler characteristic by 2. This exercise explores that result. Let S be a polyhedral surface with V vertices, E edges and F faces. Without loss of generality, suppose that among its faces, it has 2 squares as depicted to the left below. Suppose we attached the "handle as depicted on the right by first removing the square and then adjoining the polyhedron shown. Let S' be the new polyhedron with the handle added. By counting vertices edges and faces of S', prove that $\chi(S') = \chi(S) - 2$.

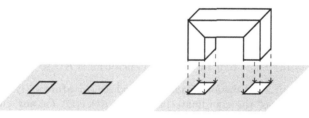

6. Verify directly the Gauss-Bonnet Theorem for the rectangular region $\mathcal{R}$ on a torus specified by $u_1 \leq u \leq u_2$ and $v_1 \leq v \leq v_2$, where the u-coordinate lines are the parallels and the v-coordinate lines are the meridians of the torus, as a surface of revolution.

7. Consider the surface of revolution parametrized by

$$\vec{X}(u, v) = ((2 + \sin v) \cos u, (2 + \sin v) \sin u, v)$$

with $(u, v) \in [0, 2\pi] \times \mathbb{R}$.

(a) Show that the geodesic curvature is constant along the coordinate line $v = v_0$.

(b) Determine the geodesic curvature along a coordinate line $u = u_0$.

(c) Use the above results and the Gauss-Bonnet Theorem to determine $\iint_{\mathcal{R}} K \, dS$ over a region $\mathcal{R}$ defined by $0 \leq u \leq 2\pi$ and $v_1 \leq v \leq v_2$, for constants v_1 and v_2.

8. Consider the surface parametrized by $\vec{X}(r, \theta) = (r\cos\theta, r\sin\theta, e^{-r^2})$ with $r \in \mathbb{R}$ and $\theta \in [0, 2\pi]$. Let $\mathcal{R}$ be the region of this surface that corresponds to $0 \le r \le c$ and $0 \le \theta \le \alpha$, where c and α are positive constants. Calculate all parts of the Global Gauss-Bonnet formula (Theorem 8.3.4) and confirm that the integral relation it leads to is correct. [Also explain why $\chi(\mathcal{R}) = 1$.]

9. Use theorems of this section applied to an ellipsoid to show that for all $a, b, c > 0$, the following integral holds

$$\int_0^{2\pi} \int_0^\pi \frac{a^2 b^2 c^2 \sin^2 v}{(b^2 c^2 \cos^2 u \sin^2 v + a^2 c^2 \sin^2 \sin^2 v + a^2 b^2 \cos^2 v)^2} \, dv \, du = 4\pi.$$

10. Let S be a regular, orientable, compact surface without boundary that has positive Gaussian curvature. Prove that the surface area of S is less than $4\pi/K_{\min}$, where $K_{\min} > 0$ is the minimum Gaussian curvature.

11. *Jacobi's Theorem.* Let $\vec{\alpha} : I \to \mathbb{R}^3$ be a closed, regular, parametrized curve. Suppose also that $\vec{\alpha}(t)$ has a curvature function $\kappa(t)$ that is never 0. Suppose also that the principal normal indicatrix, i.e., the curve $\vec{P} : I \to \mathbb{S}^2$, is simple, that is, that it cuts the sphere into only two regions. By viewing $\vec{P}(I)$ as a curve on the sphere $\mathbb{S}^2$, use the Gauss-Bonnet Theorem to prove that $\vec{P}(I)$ separates the sphere into two regions of equal area.

12. Consider the surface parametrized by $\vec{X}(u, v) = (u, v, uv)$ with $(u, v) \in \mathbb{R}^2$.

    (a) Show that the Gaussian curvature is $K = -1/(1 + u^2 + v^2)^2$.

    (b) Find the formula for the geodesic curvature $\kappa_g(t)s'(t)^3$ in terms of the coordinate functions $(u(t), v(t))$ for any curve on the surface.

    (c) Show that $\kappa_g(t) = 0$ along the coordinate lines.

    (d) Use the Gauss-Bonnet Theorem for this surface and a region $\mathcal{R}$ defined as $u_1 \le u \le u_2$ and $v_1 \le v \le v_2$ to prove the double integral formula

$$\int_{u_1}^{u_2} \int_{v_1}^{v_2} \frac{du \, dv}{(1 + u^2 + v^2)^{3/2}} =$$
$$\cos^{-1}\left(\frac{-u_1 v_1}{\sqrt{(1 + u_1^2)(1 + v_1^2)}}\right) + \cos^{-1}\left(\frac{u_2 v_1}{\sqrt{(1 + u_2^2)(1 + v_1^2)}}\right)$$
$$+ \cos^{-1}\left(\frac{-u_2 v_2}{\sqrt{(1 + u_2^2)(1 + v_2^2)}}\right) + \cos^{-1}\left(\frac{u_1 v_2}{\sqrt{(1 + u_1^2)(1 + v_2^2)}}\right) - 2\pi.$$

8.4 Geodesics

Classical geometry in the plane studies in great detail relationships between points, straight lines, and circles. One could characterize our theory of surfaces until now as a local theory in that we have concentrated our attention on the behavior of curves on surfaces at a point. Consequently, the notion of straightness (a global notion) and the concept of a circle (also a curve defined by a global property) do not yet make sense in our theory of curves on surfaces.

Euclid defines a line as a "breadthless width" and a straight line as "a line which lies evenly with the points on itself." These definitions do not particularly help us generalize the concept of a straight line to a general surface. However, it is commonly known that given any two points P and Q in $\mathbb{R}^n$, a line segment connecting P and Q provides the path of the shortest distance between these two points.

On a regular surface $S \subset \mathbb{R}^3$ that is not planar, even the notion of distance *in* S between two points P and Q poses some difficulty since we cannot assume a straight line in $\mathbb{R}^3$ connecting P and Q lies in S. However, since it is possible to talk about the arc length of curves, we can define the *distance on S* between P and Q as

inf{arc length of $C \mid C$ is a curve on S connecting P and Q}.

Therefore, one might wish to take as a first intuitive formulation of straightness on a regular surface S the following definition: A curve C on S is "straight" if for all pairs of points on the curve, the arc length between those two points P and Q is equal to the distance PQ between them. For general regular surfaces, this proposed definition turns out to be unsatisfactory, but it does lead to a more sophisticated way of generalizing straightness to surfaces.

Let S be a regular surface, and let P and Q be points of S. Suppose that P and Q are in a coordinate patch that is parametrized by $\vec{X} : U \to \mathbb{R}^3$, where $U \subset \mathbb{R}^2$. Consider curves on S parametrized by $\vec{\gamma}(t) = \vec{X}(u(t), v(t))$ such that $\vec{\gamma}(0) = P$ and $\vec{\gamma}(1) = Q$. According to Equation (6.2), the arc length of such a curve is

$$s = \int_0^1 \sqrt{g_{11}(u'(t))^2 + 2g_{12}u'(t)v'(t) + g_{22}(v'(t))^2}\, dt, \qquad (8.20)$$

where we understand that the g_{ij} coefficients are functions of u and v, which are in turn functions of t. To find a curve that connects P and Q with the shortest arc length, one must find parametric equations $(u(t), v(t))$ that minimize the integral in Equation (8.20). Such problems are studied in calculus of variations, a brief introduction to which is presented in Appendix B of [24].

According to the Euler-Lagrange Theorem in calculus of variations, if we set

$$f = \sqrt{g_{11}(u'(t))^2 + 2g_{12}u'(t)v'(t) + g_{22}(v'(t))^2},$$

then the parametric equations $(u(t), v(t))$ that optimize the arc length s in Equation (8.20) must satisfy

$$\mathcal{L}_u(f) \stackrel{\text{def}}{=} \frac{\partial f}{\partial u} - \frac{d}{dt}\left(\frac{\partial f}{\partial u'}\right) = 0 \qquad \text{and}$$

$$\mathcal{L}_v(f) \stackrel{\text{def}}{=} \frac{\partial f}{\partial v} - \frac{d}{dt}\left(\frac{\partial f}{\partial v'}\right) = 0.$$

We call $\mathcal{L}_u$ the Euler-Lagrange operator with respect to u and similarly for v. For a function with two intermediate variables u and v, as we have in this instance, let us also define the operator

$$\mathcal{L}(f) \stackrel{\text{def}}{=} \left(\mathcal{L}_u(f), \mathcal{L}_v(f)\right).$$

Proposition 8.4.1 *The Euler-Lagrange operator of*

$$f = \sqrt{g_{11}(u'(t))^2 + 2g_{12}u'(t)v'(t) + g_{22}(v'(t))^2}$$

satisfies

$$\mathcal{L}(f) = \sqrt{\det(g)}\kappa_g(t)\left(v'(t), -u'(t)\right). \tag{8.21}$$

Proof: (The proof is left as an exercise for the persistent reader and relies on the careful application of Euler-Lagrange equations and Equation (8.7).)   □

Corollary 8.4.2 *The parametric equations $(u(t), v(t))$ optimize the integral in (8.20) if and only if $\kappa_g(t) = 0$ or $(v'(t), u'(t)) = \vec{0}$.*

Obviously, $(v'(t), -u'(t)) = \vec{0}$ integrates to $(u, v) = (c_1, c_2)$, where c_1 and c_2 are constants, so the curve degenerates to a point. The other part of Corollary 8.4.2 motivates the following definition, which generalizes to a regular surface the notion of a straight line.

 Definition 8.4.3 A *geodesic* is a curve on a surface with geodesic curvature $\kappa_g(t)$ identically 0.

As a first remark, we notice that on a geodesic C, the curvature vector is $\vec{T}' = s'\kappa\vec{P} = s'\kappa_n\vec{N}$, so at each point on the surface, $\kappa = \pm\kappa_n$ and the curve's principal normal vector is equal to the surface unit normal vector up to a possible change of sign. The transition matrix between the Frenet frame and the $\{\vec{N}, \vec{T}, \vec{U}\}$ frame becomes

$$(\vec{T} \quad \vec{U} \quad \vec{N}) = (\vec{T} \quad \vec{P} \quad \vec{B}) \begin{pmatrix} 1 & 0 & 0 \\ 0 & 0 & \varepsilon \\ 0 & \varepsilon & 0 \end{pmatrix},$$

where $\varepsilon = \pm 1$. Consequently, along a geodesic, the osculating plane of the curve and the tangent plane to the surface are normal to each other.

As a second remark, the defining property that the geodesic curvature κ_g is identically 0 along a geodesic simplifies applications involving the Gauss-Bonnet Theorem. For example, suppose that S is a surface as in the conditions of the Gauss-Bonnet Theorem (Theorem 8.3.4) and that $\mathcal{R}$ is a region on S whose boundary consists of arcs of geodesics. Then the integration of κ_g along the regular arcs vanishes and the Gauss-Bonnet Theorem becomes

$$\iint_{\mathcal{R}} K \, dS + \sum_{i=1}^{k} \theta_i = 2\pi \chi(\mathcal{R}),$$

where θ_i are the exterior angles where the geodesic arcs meet. The following proposition gives another example of how the Gauss-Bonnet Theorem can provide profound geometric properties about geodesics just from this simplification.

Proposition 8.4.4 *Let S be a compact, connected, orientable, regular surface without boundary and of positive Gaussian curvature. If there exist two simple, closed geodesics γ_1 and γ_2 on S then they intersect.*

Proof: By Proposition 8.3.7, S is homeomorphic to a sphere. Suppose that γ_1 and γ_2 do not intersect. Then they form the boundary of region $\mathcal{R}$ that is homeomorphic to a cylinder with boundary. It is not hard to verify by supplying $\mathcal{R}$ with a triangulation that $\chi(\mathcal{R}) = 0$. However, applying the Gauss-Bonnet Theorem to this situation, we obtain

$$\iint_{\mathcal{R}} K \, dS = 0,$$

which is a contradiction since $K > 0$. □

We now address the problem of finding geodesics. From Equation (8.7), a curve $\vec{\gamma} = \vec{X} \circ \vec{\alpha}$, with $\vec{\alpha}(t) = (u(t), v(t))$, is a geodesic if and only if

$$\Gamma_{11}^2 (u')^3 + (2\Gamma_{12}^2 - \Gamma_{11}^1)(u')^2 v' + (\Gamma_{22}^2 - 2\Gamma_{12}^1) u'(v')^2$$
$$- \Gamma_{22}^1 (v')^3 + u'v'' - u''v' = 0. \quad (8.22)$$

This formula holds for any parameter t and not just when $\vec{\gamma}$ is parametrized by arc length.

We can approach the task of finding equations for geodesics in an alternative way. Suppose now that we consider curves on the surface parametrized by arc length so that $s' = 1$ and $s'' = 0$. Since $\vec{T}'$ is parallel to $\vec{N}$, we conclude that

$$\vec{T}' \cdot \vec{X}_u = 0 \qquad \text{and} \qquad \vec{T}' \cdot \vec{X}_v = 0.$$

Then using Equation (8.5) we deduce that

$$\vec{X}_{uu} \cdot \vec{X}_u (u')^2 + 2\vec{X}_{uv} \cdot \vec{X}_u u'v' + \vec{X}_{vv} \cdot \vec{X}_u (v')^2 + \vec{X}_u \cdot \vec{X}_u u'' + \vec{X}_v \cdot \vec{X}_u v'' = 0,$$
$$\vec{X}_{uu} \cdot \vec{X}_v (u')^2 + 2\vec{X}_{uv} \cdot \vec{X}_v u'v' + \vec{X}_{vv} \cdot \vec{X}_v (v')^2 + \vec{X}_u \cdot \vec{X}_v u'' + \vec{X}_v \cdot \vec{X}_v v'' = 0.$$

Solving algebraically for u'' and v'' in the above two equations, we obtain the following classical equations for a geodesic curve which we will express using tensor notation for simplicity with variables $x^1 = u$ and $x^2 = v$:

$$\frac{d^2 x^i}{ds^2} + \sum_{j,k=1}^{2} \Gamma^i_{jk} \frac{dx^j}{ds} \frac{dx^k}{ds} = 0 \qquad \text{for } i = 1, 2. \qquad (8.23)$$

Since i can be 1 or 2, this equation represents a system of two differential equations.

At first sight, we might see a discrepancy between (8.23), which involves two equations and (8.22), which involves only one. However, it is essential to point out that the system in (8.23) holds for geodesics parametrized by arc length. Furthermore, (8.23) is equivalent to the system of equations

$$\kappa_g(t) = 0 \qquad \text{and} \qquad (s')^2 = g_{11}(u')^2 + 2g_{12}u'v' + g_{22}(v')^2.$$

Finding explicit parametric equations of geodesic curves on a surface is often difficult since it involves solving a system of nonlinear second-order differential equations. Interestingly enough, there is a common strategy in differential equations that allows us to transform this system of second-order equations in two functions $x^1(s)$ and $x^2(s)$ into a system of first-order differential equations in four functions. (See the proof of Theorem 8.4.10.) There are common computational techniques to solve systems of first-order differential equations numerically, even if they are nonlinear. Consequently, there are algorithms to solve Equation (8.23) numerically.

In specific situations, there sometimes exist simplifications for Equation (8.23) that allow one to explicitly compute the geodesics on a particular surface.

Example 8.4.5 (The Cartesian Plane) Consider the xy-plane parametrized with the usual Cartesian coordinates. The Christoffel symbols are $\Gamma^i_{jk} = 0$, identically zero. Using Equation (8.23), the equations for a geodesic in the plane are simply

$$\frac{d^2u}{ds^2} = 0 \quad \text{and} \quad \frac{d^2v}{ds^2} = 0.$$

Integrating both equations twice, we obtain $u(s) = as + c$ and $v(s) = bs + d$. Furthermore, since $(u'(s))^2 + (v'(s))^2 = 1$, these constants must satisfy $a^2 + b^2 = 1$. Therefore, geodesics parametrized by arc length in the plane are given as $\vec{\gamma}(s) = \vec{p} + s\vec{u}$, where $\vec{p}$ is a point and $\vec{u}$ is a unit vector.

Example 8.4.6 (The Plane: Polar Coordinates) In contrast to the previous example, consider the xy-plane parametrized with polar coordinates so that as a surface in $\mathbb{R}^3$, the xy-plane is given by

$$\vec{X}(r, \theta) = (r \cos \theta, r \sin \theta, 0).$$

A short calculation gives

$$\Gamma^2_{12} = \Gamma^2_{21} = \frac{1}{r} \quad \text{and} \quad \Gamma^1_{22} = -r$$

and the remaining five other symbols are 0. Equations (8.23) become

$$\frac{d^2r}{ds^2} - r\left(\frac{d\theta}{ds}\right)^2 = 0, \tag{8.24}$$

$$\frac{d^2\theta}{ds^2} + \frac{2}{r}\frac{dr}{ds}\frac{d\theta}{ds} = 0. \tag{8.25}$$

We transform this system to obtain a differential equation relating r and θ as follows. Note that by repeatedly using the chain rule, we get

$$\frac{dr}{ds} = \frac{dr}{d\theta}\frac{d\theta}{ds} \quad \text{and} \quad \frac{d^2r}{ds^2} = \frac{d^2r}{d\theta^2}\left(\frac{d\theta}{ds}\right)^2 + \frac{dr}{d\theta}\frac{d^2\theta}{ds^2}.$$

Putting these two into Equations (8.24) and (8.25) leads to

$$\left(\frac{d\theta}{ds}\right)^2\left(\frac{d^2r}{d\theta^2} - \frac{2}{r}\left(\frac{dr}{d\theta}\right)^2 - r\right) = 0,$$

which breaks into the pair of equations

$$\frac{d\theta}{ds} = 0 \quad \text{or} \quad \frac{d^2r}{d\theta^2} - \frac{2}{r}\left(\frac{dr}{d\theta}\right)^2 - r = 0. \tag{8.26}$$

One notices that the first equation in Equation (8.26) is satisfied when θ is a constant, which corresponds to a line through the origin.

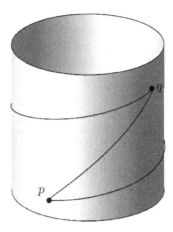

Figure 8.9: Two geodesics on a cylinder.

The other equation in Equation (8.26) does not appear particularly tractable but a substitution simplifies it greatly. Using the new variable $u = \frac{1}{r}$, one can check that

$$\frac{d^2 r}{d\theta^2} = \frac{2}{u^3}\left(\frac{du}{d\theta}\right)^2 - \frac{1}{u^2}\frac{d^2 u}{d\theta^2},$$

and hence the second equation in Equation (8.26) reduces to

$$\frac{d^2 u}{d\theta^2} + u = 0. \tag{8.27}$$

Using standard techniques in ordinary differential equations, the general solution to Equation (8.27) can be written as $u(\theta) = C\cos(\theta - \theta_0)$ where C and θ_0 are constants. Therefore, in polar coordinates, equations for geodesics in the plane (i.e., lines) are given by

$$\theta = C \qquad \text{or} \qquad r(\theta) = C\sec(\theta - \theta_0).$$

 Example 8.4.7 (Cylinder) Consider a right circular cylinder. A parametrization for this cylinder is $\vec{X}(u, v) = (\cos(u), \sin(u), v)$, with (u, v) in $[0, 2\pi] \times \mathbb{R}$. An easy calculation shows that $\Gamma^i_{jk} = 0$ for all i, j, k, and hence, that geodesics on a cylinder are curves of the form $\gamma = \vec{X} \circ \vec{\alpha}$ where

$$\vec{\alpha}(t) = (at + b, ct + d)$$

for a, b, c, d that are constant. As a result, the geodesics on a cylinder are either straight lines parallel to the axis of the cylinder, circles in planes perpendicular to the axis of the cylinder, or helices around that axis.

Figure 8.9 illustrates how on a surface, two different geodesics may connect two distinct points, a situation that does not occur in the plane. In fact, on a cylinder, there is an infinite number of geodesics that connect any two points. The difference between each geodesic connecting p and q is how many times the geodesic wraps around the cylinder and in which direction it wraps.

Example 8.4.8 (Sphere) Consider the sphere parametrization

$$\vec{X}(x^1, x^2) = (R\cos x^1 \sin x^2, R\sin x^1 \sin x^2, R\cos x^2),$$

where $x^1 = u$ is the longitude θ in spherical coordinates, and $x^2 = v$ is the colatitude angle φ down from the positive z-axis. In Example 7.1.6, we determined the Christoffel symbols for this parametrization. Equations (8.23) for geodesics on the sphere become

$$
\begin{aligned}
\frac{d^2 x^1}{ds^2} + 2\cot(x^2)\frac{dx^1}{ds}\frac{dx^2}{ds} &= 0, \\
\frac{d^2 x^2}{ds^2} - \sin(x^2)\cos(x^2)\left(\frac{dx^1}{ds}\right)^2 &= 0.
\end{aligned}
\tag{8.28}
$$

A geodesic on the sphere is now just a curve of the form $\vec{\gamma}(s) = \vec{X}(x^1(s), x^2(s))$, where $x^1(s)$ and $x^2(s)$ satisfy the system of differential equations in Equation (8.28). Taking a first derivative of $\vec{\gamma}(s)$ gives

$$
\begin{aligned}
\vec{\gamma}'(s) = R\Bigg(&-\sin x^1 \sin x^2 \frac{dx^1}{ds} + \cos x^1 \cos x^2 \frac{dx^2}{ds}, \\
&\cos x^1 \sin x^2 \frac{dx^1}{ds} + \sin x^1 \cos x^2 \frac{dx^2}{ds}, -\sin x^2 \frac{dx^2}{ds}\Bigg),
\end{aligned}
$$

and the second derivative, after simplification using (8.28), is

$$
\frac{d^2\vec{\gamma}}{ds^2} = -\left[\sin^2(x^2)\left(\frac{dx^1}{ds}\right)^2 + \left(\frac{dx^2}{ds}\right)^2\right]\vec{\gamma}(s).
$$

However, the term $R^2\left[\sin^2(x^2)\left(\frac{dx^1}{ds}\right)^2 + \left(\frac{dx^2}{ds}\right)^2\right]$ is the first fundamental form applied to

$$((x^1)'(s), (x^2)'(s)),$$

which is precisely the square of the speed of $\vec{\gamma}(s)$. However, since the geodesic is parametrized by arc length, its speed is identically 1. Thus, Equations (8.28) lead to the differential equation

$$\vec{\gamma}''(s) + \frac{1}{R^2}\vec{\gamma}(s) = 0.$$

Standard techniques with differential equations allow one to show that all solutions to this differential equation are of the form

$$\vec{\gamma}(s) = \vec{a}\cos\left(\frac{s}{R}\right) + \vec{b}\sin\left(\frac{s}{R}\right),$$

where $\vec{a}$ and $\vec{b}$ are constant vectors. Note that $\vec{\gamma}(0) = \vec{a}$ and that $\vec{\gamma}'(0) = \frac{1}{R}\vec{b}$. Furthermore, to satisfy the conditions that $\vec{\gamma}(s)$ lie on the sphere of radius R and be parametrized by arc length, we deduce that $\vec{a}$ and $\vec{b}$ satisfy

$$\|\vec{a}\| = R, \qquad \|\vec{b}\| = R, \qquad \text{and} \qquad \vec{a}\cdot\vec{b} = 0.$$

Therefore, we find that $\vec{\gamma}(s)$ traces out a great arc on the sphere that is the intersection of the sphere and the plane through the center of the sphere spanned by $\vec{\gamma}(0)$ and $\vec{\gamma}'(0)$.

Example 8.4.9 (Surfaces of Revolution) Consider a surface of revolution about the z-axis given by the parametric equations

$$\vec{X}(u, v) = \big(f(v)\cos u, f(v)\sin u, h(v)\big),$$

where f and h are functions defined over a common interval I such that over $[0, 2\pi] \times I$. We assume that over the open interval $(0, 2\pi) \times I$ the function $\vec{X}$ is a regular parametrization, which implies that $f(v) > 0$. A simple calculation reveals that the Christoffel symbols of the second kind are

$$\Gamma^1_{11} = 0, \qquad\qquad \Gamma^1_{12} = \frac{f'}{f}, \quad \Gamma^1_{22} = 0,$$

$$\Gamma^2_{11} = -\frac{ff'}{(f')^2 + (h')^2}, \quad \Gamma^2_{12} = 0, \quad \Gamma^2_{22} = \frac{f'f'' + h'h''}{(f')^2 + (h')^2}.$$

Equations (8.23) for geodesics parametrized by arc length on a surface of revolution become

$$\frac{d^2u}{ds^2} + \frac{2f'}{f}\frac{du}{ds}\frac{dv}{ds} = 0$$

$$\frac{d^2v}{ds^2} - \frac{ff'}{(f')^2 + (h')^2}\left(\frac{du}{ds}\right)^2 + \frac{f'f'' + h'h''}{(f')^2 + (h')^2}\left(\frac{dv}{ds}\right)^2 = 0. \tag{8.29}$$

As complicated as Equations (8.29) appear, it is possible to find a "solution" to this system of differential equations for u in terms of v. However, before establishing a general solution, we will determine which meridians ($u =$const.) and which parallels or latitude lines ($v =$const.) are geodesics.

Consider first the meridian lines that are defined by $u = C$, where C is a constant. Notice that the first equation in the system (8.29) is trivially satisfied for meridians and that the second equation becomes

$$\frac{d^2v}{ds^2} + \frac{f'f'' + h'h''}{(f')^2 + (h')^2}\left(\frac{dv}{ds}\right)^2 = 0. \tag{8.30}$$

It is easy to check that the first fundamental form on $(u'(t), v'(t))$ on the surface of revolution is

$$f(v)^2 u'(t)^2 + (f'(v)^2 + h'(v)^2)v'(t)^2. \tag{8.31}$$

However, assuming that we have a meridian that is parametrized by arc length, then

$$(f'(v)^2 + h'(v)^2)v'(s)^2 = 1$$

since the speed function of a curve parametrized by arc length is 1. Consequently,

$$v'(s)^2 = \frac{1}{f'(v)^2 + h'(v)^2},$$

and taking a derivative of this equation with respect to s, one obtains

$$2v'v'' = -\frac{2f'(v)f''(v)v' + 2h'(v)h''(v)v'}{(f'(v)^2 + h'(v)^2)^2}$$

$$= -2\frac{f'(v)f''(v) + h'(v)h''(v)}{f'(v)^2 + h'(v)^2}(v')^3.$$

Since $v'(s) \neq 0$ on the meridian parametrized by arc length, then

$$\frac{d^2v}{ds^2} = -\frac{f'(v)f''(v) + h'(v)h''(v)}{f'(v)^2 + h'(v)^2}\left(\frac{dv}{ds}\right)^2,$$

which shows that Equation (8.30) is satisfied on all meridians.

Now consider the parallel curves on a surface of revolution, which are defined by $v = v_0$, a real constant. In Equation (8.29), the first equation leads to $u'(s) = C$, a constant, and the second equation becomes

$$\frac{ff'}{(f')^2 + (h')^2}\left(\frac{du}{ds}\right)^2 = 0.$$

Since the parallel curves, parametrized by arc length, are regular curves, then $C \neq 0$, and the condition that the surface of revolution be a regular surface implies that $f(v) \neq 0$. Thus, the second equation is satisfied for parallels $v = v_0$ if and only if $f'(v_0) = 0$. (See Figure 8.10.)

Even if a geodesic is neither a meridian nor a parallel, the first equation in (8.29) is simple enough that we can nonetheless deduce

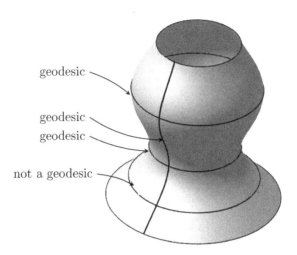

geodesic

geodesic
geodesic

not a geodesic

Figure 8.10: Geodesics on a surface of revolution.

some interesting consequences. Note that by taking a derivative with respect to the arc length parameter s, we have

$$\frac{d}{ds}(f^2 u') = 2ff'v'u' + f^2 u''.$$

Multiplying the first equation of (8.29) by f^2, we see that it can be written as

$$f(v)^2 u'(s) = C, \qquad (8.32)$$

where C is a constant. Note that when a curve on a surface is parametrized by arc length with coordinate functions $(u'(s), v'(s))$, the angle θ it makes with any given parallel curve satisfies

$$\cos\theta = \frac{I_p((u', v'), (1, 0))}{\sqrt{I_p((u', v'), (u', v'))}\sqrt{I_p((1, 0), (1, 0))}} = u'f.$$

As a geometric interpretation, since f is the radius r of surface of revolution at a given point, (8.32) leads to the relation

$$r\cos\theta = C \qquad (8.33)$$

for all nonparallel geodesics on a surface of revolution, where θ is the angle between the geodesic and the parallels. Equation (8.33), which is equivalent to (8.32) for nonparallel curves, is often called the *Clairaut relation*. Note that a curve satisfying Clairaut's relation is a meridian if and only if $C = 0$.

With the relation $u' = C/f^2$, since the speed is equal to 1, (8.31) leads to

$$\left(\frac{dv}{ds}\right)^2 ((f')^2 + (h')^2) = 1 - \frac{C^2}{f^2}. \tag{8.34}$$

Taking the derivative of this equation with respect to s, we obtain

$$2\frac{dv}{ds}\frac{d^2v}{ds^2}((f')^2 + (h')^2) + \left(\frac{dv}{ds}\right)^3 2(f'f'' + h'h'') = 2\frac{C^2 f'}{f^3}\frac{dv}{ds}$$

which is equivalent to

$$((f')^2 + (h')^2)\frac{dv}{ds}\left[\frac{d^2v}{ds^2} - \frac{ff'}{(f')^2 + (h')^2}\left(\frac{du}{ds}\right)^2 + \frac{f'f'' + h'h''}{(f')^2 + (h')^2}\left(\frac{dv}{ds}\right)^2\right] = 0.$$

Therefore, if a geodesic is not a parallel, the first equation of (8.29), which is equivalent to Clairaut's relation, implies the second equation.

Assuming that a geodesic is not a meridian, then from Clairaut's relation, we know that $u'(s)$ is never 0 so we can define an inverse function to $u(s)$, namely $s(u)$, and then v can be given as a function of u by $v = v(s(u))$. Then in (8.34), replacing $\frac{dv}{ds}$ with $\frac{dv}{du}\frac{du}{ds} = \frac{dv}{du}\frac{C}{f^2}$, we obtain

$$\left(\frac{dv}{du}\right)^2 \frac{C^2}{f^4}((f')^2 + (h')^2) = 1 - \frac{C^2}{f^2},$$

and hence,

$$\frac{dv}{du} = \frac{f}{C}\sqrt{\frac{f^2 - C^2}{(f')^2 + (h')^2}}.$$

Over an interval where this derivative is not 0, it is possible to take an inverse function of u with respect to v, and, by integration, this function satisfies

$$u = C\int \frac{1}{f(v)}\sqrt{\frac{f'(v)^2 + h'(v)^2}{f(v)^2 - C^2}}\, dv + D$$

for some constant of integration D. The constants C and D lead to a two-parameter family of solutions that parametrize segments of geodesics.

Theorem 8.4.10 *Let S be a regular surface of class C^3. For every point p on S and every unit vector $\vec{w} \in T_pS$, there exists a unique geodesic on S through p in the direction of $\vec{w}$.*

Proof: Let $\vec{X} : U \to \mathbb{R}^3$ be a regular parametrization of a neighborhood of p on S, and let g_{ij} and Γ^i_{jk} be the components of the metric tensor and Christoffel symbols of the second kind, respectively, in relation to $\vec{X}$. Note that since S is of class C^3, all of the functions Γ^i_{jk} are of class C^1 over their domain.

The proof of this proposition is an application of the existence and uniqueness theorem for first-order systems of differential equations (see [2, Section 31.8]), which states that if $F : \mathbb{R}^n \to \mathbb{R}^n$ is of class C^1, then there exists a unique function $\vec{x} : I \to \mathbb{R}^n$ that satisfies

$$\vec{x}' = F(\vec{x}) \qquad \text{and} \qquad \vec{x}(t_0) = \vec{C},$$

where I is an open interval containing t_0, and $\vec{C}$ is a constant vector.

Consider now the system of differential equations given in Equation (8.23), where we do not assume that a solution is parametrized by arc length. Setting the dependent variables $v^1 = (x^1)'$ and $v^2 = (x^2)'$, then Equation (8.23) is equivalent to the system of first-order differential equations

$$\begin{cases} (x^1)' & = v^1, \\ (x^2)' & = v^2, \\ (v^1)' & = -\Gamma^1_{11}(v^1)^2 - 2\Gamma^1_{12}v^1v^2 - \Gamma^1_{22}(v^2)^2, \\ (v^2)' & = -\Gamma^2_{11}(v^1)^2 - 2\Gamma^2_{12}v^1v^2 - \Gamma^2_{22}(v^2)^2, \end{cases} \tag{8.35}$$

where the functions Γ^i_{jk} are functions of x^1 and x^2. Therefore, according to the existence and uniqueness theorem, there exists a unique solution to Equation (8.23) with specific values given for $x^1(s_0)$, $x^2(s_0)$, $(x^1)'(s_0)$, and $(x^2)'(s_0)$. Set $\vec{\alpha}(s) = \vec{X}(x^1(s), x^2(s))$. Now Equation (8.23) is the formula for arc length parametrizations of geodesics. However, it is not hard to prove that if

$$f(s) = g_{11}(x^1(s), x^2(s))(x^1)'(s)^2 + 2g_{12}(x^1(s), x^2(s))(x^1)'(s)(x^2)'(s)$$
$$+ g_{22}(x^1(s), x^2(s))(x^2)'(s)^2,$$

where x^1 and x^2 satisfy Equation (8.23), then $f'(s) = 0$, so $f(s)$ is a constant function over its domain. Consequently, if $(x^1)'(s_0)$ and $(x^2)'(s_0)$ are given so that $\vec{\alpha}'(s_0) = \vec{w}$, then $\|\vec{\alpha}(s)\| = 1$ for all s. If, in addition, $x^1(s_0)$ and $x^2(s_0)$ are chosen so that $\vec{\alpha}(s_0) = p$, then $\vec{\alpha}(s)$ is an arc length parametrization of a geodesic passing through p with direction $\vec{w}$ and is unique. $\qquad\qquad \square$

The proof of the above theorem establishes an additional fact concerning the equation for a geodesic curve.

Proposition 8.4.11 *Let $\vec{X}$ be the parametrization of a neighborhood of a regular surface of class C^3. Any solution $(x^1(s), x^2(s))$ to Equation (8.23), where one makes no prior assumptions on the parameter* *s*, *is such that the parametric curve* $\vec{\gamma}(s) = \vec{X}(x^1(s), x^2(s))$, *whose locus is a geodesic, has constant speed.*

Example 8.4.12 For a last example of the section, let us revisit the Poincaré upper-half plane from Example 7.1.7. Since the geodesic curvature only depending on the metric coefficients $g_{ij}(u, v)$ (i.e., is intrinsic), we can find the expression for the geodesic curvature of a curve in the Poincaré upper half-plane with the metric $g_{11}(u, v) = g_{22}(u, v) = 1/v^2$ and $g_{12}(u, v) = 0$, using the computations from Example 7.1.7. From (8.7), we find that the geodesic curvature of a curve in the Poincaré upper half plane satisfies

$$s'(t)^3 \kappa_g(t) = \frac{1}{v(t)^2} \left(\frac{1}{v(t)} u'(t)^3 + \frac{1}{v(t)} u'(t) v'(t)^2 \right.$$

$$\left. + u'(t) v''(t) - u''(t) v'(t) \right). \quad (8.36)$$

Since $s'(t)^2 = (1/v(t)^2) u'(t)^2 + (1/v(t)^2) v'(t)^2$, we get

$$\kappa_g = \frac{(u')^3 + u'(v')^2 + v(u'v'' - u''v')}{((u')^2 + (v')^2)^{3/2}}.$$

PROBLEMS

1. Let S be a regular surface, and let $\vec{\gamma}$ be a geodesic on S. Prove that the geodesic torsion of γ is equal to $\pm\tau$, where τ is the torsion function of $\vec{\gamma}(t)$ as a space curve. [Hint: The possible difference in sign stems from the possible change in sign of κ_n which may come from a reparametrization of the surface. This property of τ_g justifies its name as geodesic torsion.]

2. **(ODE)** Consider a right circular cone with opening angle α, where $0 < \alpha < \pi/2$. Consider the coordinate patch parametrized by $\vec{X}(u, v) = (v \sin \alpha \cos u, v \sin \alpha \sin u, v \cos \alpha)$, where we assume $v > 0$. Determine equations for the geodesics on this cone. [Hint: Find a differential equation that expresses $\frac{dv}{du}$ in terms of u and v and then solve this to find an equation for u in terms of v.]

3. Consider the torus parametrized by $\vec{X}(u, v) = ((a + b \cos v) \cos u, (a + b \cos v) \sin u, b \sin v)$ where $a > b$. Show that the geodesics on a torus satisfy the differential equation

$$\frac{dr}{du} = \frac{1}{Cb} r \sqrt{r^2 - C^2} \sqrt{b^2 - (r - a)^2},$$

where C is a constant and $r = a + b \cos v$.

4. Find the differential equations that determine geodesics on a function graph $z = f(x, y)$.

5. If $\vec{X} : U \to \mathbb{R}^3$ is a parametrization of a coordinate patch on a regular surface S such that $g_{11} = E(u)$, $g_{12} = 0$ and $g_{22} = G(u)$ show that

(a) the u-parameter curves (i.e., over which v is a constant) are geodesics;

(b) the v-parameter curve $u = u_0$ is a geodesic if and only if $G_u(u_0) = 0$;

(c) the curve $\vec{x}(u, v(u))$ is a geodesic if and only if

$$v = \pm \int \frac{C\sqrt{E(u)}}{\sqrt{G(u)}\sqrt{G(u) - C^2}} \, du,$$

where C is a constant.

6. Fill in the details in the proof of Proposition 8.4.10, namely, prove that if $x^1(s)$ and $x^2(s)$ satisfy Equation (8.23), then the function $f(s)$, which is the square of the speed of $\vec{X}(x^1(s), x^2(s))$, is constant.

7. *Liouville Surface.* A regular surface is called a *Liouville surface* if it can be covered by coordinate patches in such a way that each patch can be parametrized by $\vec{X}(u, v)$ such that

$$g_{11} = g_{22} = U(u) + V(v) \qquad \text{and} \qquad g_{12} = 0,$$

where U is a function of u alone and V is a function of v alone. (Note that Liouville surfaces generalize surfaces of revolution.) Prove the following facts:

(a) Show that the geodesics on a Liouville surface can be given as solutions to an equation of the form

$$\int \frac{du}{\sqrt{U(u) - c_1}} = \pm \int \frac{dv}{\sqrt{V(v) + c_1}} + c_2,$$

where c_1 and c_2 are constants.

(b) Show that if ω is the angle a geodesic makes with the curve $v = $const., then

$$U \sin^2 \omega - V \cos^2 \omega = C$$

for some constant C.

8. Let $\vec{\alpha} : I \to \mathbb{R}^3$ be a regular space curve parametrized by arc length with nowhere 0 curvature. Consider the ruled surface parametrized by

$$\vec{X}(s, t) = \vec{\alpha}(s) + t\vec{B}(s),$$

where $\vec{B}$ is the binormal vector of $\vec{\alpha}$. Suppose that $\vec{X}$ is defined over $I \times (-\varepsilon, \varepsilon)$, with $\varepsilon > 0$.

(a) Prove that if ε is small enough, the image S of $\vec{X}$ is a regular surface.

(b) Prove that if S is a regular surface, $\vec{\alpha}(I)$ is a geodesic on S.

(This shows that every regular space curve that has nonzero curvature is the geodesic of some surface.)

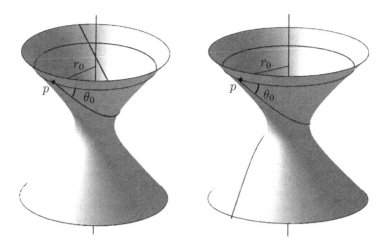

Figure 8.11: Geodesics on the hyperboloid of one sheet.

9. Consider the elliptic paraboloid given by $z = x^2 + y^2$. Note that this is a surface of revolution. Consider the geodesics on this surface that are not meridians.

   (a) Suppose that the geodesic intersects the parallel at $z = z_0$ with an angle of θ_0. Find the lowest parallel that the geodesic reaches.

   (b) Prove that any geodesic that is not a meridian intersects itself an infinite number of times.

10. Consider the hyperboloid of one sheet given by the equation $x^2 + y^2 - z^2 = 1$ and let p be a point in the upper half-space defined by $z > 0$. Consider now geodesic curves that go through p and make an acute angle of θ_0 with the meridian of the hyperboloid passing through p. Call r_0 the distance from p to the z-axis.

    (a) Prove that if $\cos\theta_0 > 1/r_0$, then the geodesic remains in the upper half-space $z > 0$.

    (b) Prove that if $\cos\theta_0 < 1/r_0$, then the geodesic crosses the $z = 0$ plane, and descends indefinitely in the negative z-direction.

    (c) Prove that if $\cos\theta_0 = 1/r_0$, then the geodesic, as it descends from p, asymptotically approaches the meridian given by $x^2 + y^2 = 1$ at $z = 0$.

    (d) Using the parametrization

$$\vec{X}(u, v) = \left(\cosh v \cos u, \cosh v \sin u, \sinh v\right),$$

    suppose that the initial conditions for the geodesic are $u(0) = 0$ and $v(0) = v_0$, so that $p = \vec{X}(0, v_0)$. Find the initial conditions $u'(0)$ and $v'(0)$ so that the geodesic is parametrized by arc length and satisfies the condition in (c).

(See Figure 8.11 for examples of nonasymptotic behavior.)

11. We revisit the Poincaré upper half plane of Example 8.4.12. Consider the curve parametrized by $(u(t), v(t)) = (a + R \cos t, b + R \sin t)$, where $R > 0$ and a, b are constants. We only use t such that $v(t) > 0$.

    (a) Show that this curve has constant geodesic curvature b/R.

    (b) Deduce that if $b = 0$, then the resulting curve is a geodesic in the Poincaré upper half-plane.

    (c) Also show that a curve parametrized by (a, t), where a is a constant, is a geodesic for this metric.

    (d) Assume that $b > R > 0$. Find a point (a, c) such that the distances from (a, c) to $(a, b + R)$ and from (a, c) to $(a, b - R)$ are equal. [Ans: $c = \sqrt{b^2 - R^2}$. Note:Though more work is necessary to show this, (a, c) is equidistant from every point on every point on the curve $(u(t), v(t))$, showing that this curve is a "circle" in the Poincaré metric of the upper half-plane.]

12. Let S be a regular, orientable surface of class C^3 in $\mathbb{R}^3$ that is homeomorphic to the sphere. Let γ be a simple closed geodesic in S. The curve γ separates S into two regions A and B that share γ as their boundary. Let $n : S \to \mathbb{S}^2$ be the Gauss map induced from a given orientation of S. Prove that $n(A)$ and $n(B)$ have the same area. [Hint: Use the Gauss-Bonnet Theorem.]

13. Let S be an orientable surface with Gaussian curvature $K \geq 0$, and let $p \in S$.

    (a) Let γ_1 and γ_2 be two geodesics that intersect at p. Prove that γ_1 and γ_2 do not intersect at another point q in such a way that γ_1 and γ_2 form the boundary of a simple region $\mathcal{R}$.

    (b) Prove also that a geodesic on S cannot intersect itself in such a way as to enclose a simple region.

8.5 Geodesic Coordinates

8.5.1 General Geodesic Coordinates

Definition 8.5.1 Let S be a regular surface of class C^3. A system of *geodesic coordinates* is an orthogonal regular parametrization $\vec{X}$ of S such that, for one of the coordinates, all the coordinate lines are geodesics.

There are many ways to define geodesic coordinates on a regular surface of class C^3. We introduce one general way first.

Suppose that $\vec{\alpha}(t)$ for $t \in [a, b]$ is a regular curve with image C on S of class C^2. According to Theorem 8.4.10, for each $t_0 \in [a, b]$, there exists a unique geodesic on S through $\vec{\alpha}(t_0)$ perpendicular to C. (See Figure 8.12.) Furthermore, one can parametrize the geodesics $\vec{\gamma}_{t_0}(s)$

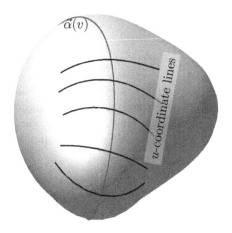

Figure 8.12: Geodesic coordinate system generated by $\vec{\alpha}(v)$.

by arc length such that $\vec{\gamma}_{t_0}(0) = \vec{\alpha}(t_0)$ and $t \mapsto \vec{\gamma}'_t(0)$ is continuous in t. Define the function

$$\vec{X}(s,t) = \vec{\gamma}_t(s). \qquad (8.37)$$

We wish to show that over an open set containing C, the function $\vec{X}(s,t)$ is a regular parametrization of class C^2. (One desires class C^2 since this is required for the first and second fundamental forms to exist.) However, we must assume that S is of class C^5.

Proposition 8.5.2 *If S is of class C^5, there exists an $\varepsilon > 0$ such that $\vec{X}(s,t)$, as defined in Equation (8.37), is a regular parametrization of class C^2 over $(-\varepsilon, \varepsilon) \times (a, b)$.*

Proof: The proof relies on standard theorems of existence and uniqueness for differential equations as well as on some basic topology.

Let p be a point on C, and let $\vec{X}(x^1, x^2)$ be a regular parametrization of class C^5 of a neighborhood $V_p = \vec{X}(U_p)$ of S containing p. Let $\alpha^1(t)$ and $\alpha^2(t)$ be functions such that $\vec{\alpha}(t) = \vec{X}(\alpha^1(t), \alpha^2(t))$, the given parametrization of C on S in the neighborhood V_p. Call I_p the domain of $\vec{\alpha}$ such that $\vec{\alpha}(I_p) \subset V_p$, and suppose that $p = \vec{\alpha}(t_0)$.

Let $x^1(s,t)$ and $x^2(s,t)$ be functions that map onto the geodesics $\vec{\gamma}_t(s)$ via

$$\vec{X}(x^1(s,t), x^2(s,t)) = \vec{\gamma}_t(s).$$

By Theorem 8.4.10, for all parameters t, over an interval $s \in (-\varepsilon_t, \varepsilon_t)$, there exists a unique solution for $x^1(s,t)$ and for $x^2(s,t)$ to Equation

(8.23) given the initial conditions

$$x^1(0,t) = \alpha^1(t), \qquad x^2(0,t) = \alpha^2(t),$$

$$\frac{\partial x^1}{\partial s}(0,t) = u(t), \qquad \frac{\partial x^2}{\partial s}(0,t) = v(t),$$

where u and v are any functions of class C^1. Furthermore, these solutions are of class C^2 in the variable s. Now since for each t we want $\vec{\gamma}_t(s)$ to be an arc length parametrization of a geodesic perpendicular to C, we impose the following two conditions on u and v for all t:

(i) $g_{11}u^2 + 2g_{12}uv + g_{22}v^2 = 1$,

(ii) $g_{11}u\dfrac{d\alpha^1}{dt} + g_{12}u\dfrac{d\alpha^2}{dt} + g_{12}v\dfrac{d\alpha^1}{dt} + g_{22}v\dfrac{d\alpha^2}{dt} = 0$,

where (i), for example, means that

$$g_{11}(\alpha^1(t), \alpha^2(t))u(t)^2 + 2g_{12}(\alpha^1(t), \alpha^2(t))u(t)v(t)$$
$$+ g_{22}(\alpha^1(t), \alpha^2(t))v(t)^2 = 1 \quad (8.38)$$

for all t. The requirement to parametrize each $\vec{\gamma}_t(s)$ so that $\vec{\gamma}_t'(0)$ is continuous in t is equivalent to requiring that u and v be continuous. These conditions completely specify $u(t)$ and $v(t)$ for all t. Notice that for (i) and (ii) to both be satisfied, $((\alpha^1)'(t), (\alpha^2)'(t))$ and $(u(t), v(t))$ cannot be linear multiples of each other, and hence,

$$\begin{vmatrix} (\alpha^1)'(t) & u(t) \\ (\alpha^2)'(t) & v(t) \end{vmatrix} \neq 0 \quad \text{for all } t.$$

Now write the equations for geodesics as a first-order system, as in Equation (8.35). Since $\vec{X}$ is of class C^5, then all the functions Γ^i_{jk} are of class C^3. Then according to theorems of dependency of solutions of differential equations on initial conditions (see [2, Theorem 32.4]), solutions to the system from Equation (8.35) are of class C^2 in terms of initial conditions, which implies that

$$\frac{\partial^2 x^1}{\partial s \partial t}, \quad \frac{\partial^2 x^2}{\partial s \partial t}, \quad \frac{\partial^2 x^1}{\partial t^2}, \quad \text{and} \quad \frac{\partial^2 x^2}{\partial t^2}$$

are continuous over a possibly smaller open neighborhood U'_p of the point $(0, t_0)$. Hence, $x^1(s,t)$ and $x^2(s,t)$ are of class C^2 over U'_p.

To prove regularity of $\vec{X}(s,t)$, note that

$$\begin{vmatrix} (\alpha^1)'(t_0) & u(t_0) \\ (\alpha^2)'(t_0) & v(t_0) \end{vmatrix} = \frac{\partial(x^1, x^2)}{\partial(s,t)}\bigg|_{(0,t_0)},$$

the Jacobian of the change of variables $x^1(s,t)$ and $x^2(s,t)$ at the point $(0, t_0)$. Since this Jacobian is a continuous function and is nonzero at $(0, t_0)$, there is an open neighborhood U_p'' of $(0, t_0)$ such that $\frac{\partial(x^1, x^2)}{\partial(s, t)} \neq 0$. By Proposition 5.4.9,

$$\frac{\partial \vec{X}}{\partial s} \times \frac{\partial \vec{X}}{\partial t} = \frac{\partial(x^1, x^2)}{\partial(s, t)} \left(\frac{\partial \vec{X}}{\partial x^1} \times \frac{\partial \vec{X}}{\partial x^2} \right),$$

so over U_p'' the parametrization $\vec{X}(s, t)$ is regular. Consequently, there exists an $\varepsilon_p > 0$ such that

$$\bar{U}_p \stackrel{\text{def}}{=} (-\varepsilon_p, \varepsilon_p) \times (t_0 - \varepsilon_p, t_0 + \varepsilon_p) \subset U_p' \cap U_p'',$$

and over $\bar{U}_p$, the parametrization $\vec{X}(s, t)$ is regular and of class C^2. Call $\bar{V}_p = \vec{X}(\bar{U}_p)$.

Finally, consider the whole curve C on S. The curve C can be covered by open sets of the form $\bar{V}_p$ for various $p \in C$. Let p and q be two points on C, and let us write $\vec{X}_p : \bar{U}_p \to \bar{V}_p$ and $\vec{X}_q : \bar{U}_q \to \bar{V}_q$ for associated parametrizations of $\bar{V}_p$ and $\bar{V}_q$. If

$$\bar{U}_p = (-\varepsilon_p, \varepsilon_p) \times I_p \quad \text{and} \quad \bar{U}_q = (-\varepsilon_q, \varepsilon_q) \times I_q$$

overlap, since for all $t \in I_p \cap I_q$ we have $\vec{X}_p(s, t) = \vec{X}_q(s, t)$ by the uniqueness of geodesics, then we can extend $\vec{X}_p$ and $\vec{X}_q$ to a function $\vec{X}$ over

$$(-\min(\varepsilon_p, \varepsilon_q), \min(\varepsilon_p, \varepsilon_q)) \times (I_p \cup I_q).$$

Since $[a, b]$ is compact, $[a, b]$ can be covered by a finite number of sets $\bar{U}_p$. Therefore, there exists $\varepsilon > 0$ and a single function $\vec{X}$ defined over $(-\varepsilon, \varepsilon) \times (a, b)$ such that $\vec{X}(0, t) = \vec{\alpha}(t)$, and for $t \in (a, b)$, $\vec{X}(s, t)$ parametrizes a geodesic by arc length. $\square$

Proposition 8.5.2 leads to the following general theorem.

Theorem 8.5.3 *Let S be a regular surface of class C^5, and let $\vec{\alpha} : [a, b] \to S$ be a regular parametrization of a simple curve of class C^2. Then there exists a system of geodesic coordinates $\vec{X}(u, v)$ of class C^2 defined over $-\varepsilon < u < \varepsilon$ and $a < v < b$ such that $\vec{X}(0, v) = \vec{\alpha}(v)$, and the u-coordinate lines parametrize geodesics by arc length.*

Proof: Proposition 8.5.2 constructs the function $\vec{X}(u, v)$ as desired. However, it remains to be proven that $\vec{X}(u, v)$ is an orthogonal parametrization.

Consider the derivative of g_{12} with respect to u

$$g_{12,1} = \frac{\partial g_{12}}{\partial u} = \frac{\partial}{\partial u}(\vec{X}_1 \cdot \vec{X}_2) = \vec{X}_{11} \cdot \vec{X}_2 + \vec{X}_1 \cdot \vec{X}_{12}.$$

Since each u-coordinate line is parametrized by arc length, $\vec{X}_1 \cdot \vec{X}_1 = 1$, so by differentiating with respect to v, we find that $g_{11,2} = 2\vec{X}_1 \cdot \vec{X}_{12} = 0$. By the same reasoning, over the u-coordinate lines, in the $\{\vec{N}, \vec{T}, \vec{U}\}$ frame one has

$$\frac{\partial \vec{X}}{\partial u} = \vec{T} \quad \text{and} \quad \frac{\partial^2 \vec{X}}{\partial u^2} = \kappa \vec{P} = \kappa_g \vec{U} + \kappa_n \vec{N}.$$

But since $\vec{X}(u,v)$ with v fixed is a geodesic, $\kappa_g = 0$. Thus, $\vec{X}_{11} = \kappa_n \vec{N}$ and, in particular, $\vec{X}_{11} \cdot \vec{X}_2 = 0$.

Consequently, $g_{12,1} = 0$, and therefore, g_{12} is a function of v only. We can write $g_{12}(u,v) = g_{12}(0,v)$. However, by construction of $\vec{X}(u,v)$ in Proposition 8.5.2, $g_{12}(0,v) = 0$ for all v. Thus, g_{12} is identically 0, and hence, $\vec{X}(u,v)$ is an orthogonal parametrization. $\square$

The class of geodesic coordinate systems described in Theorem 8.5.3 is of a particular type. Not every geodesic coordinate system needs to be defined in reference to a curve C on S as done above. Let $\vec{X}(u,v)$ be any system of geodesic coordinates where the u-coordinate lines are geodesics. A priori, we know only that $g_{12} = 0$. However, much more can be said.

Proposition 8.5.4 *Let S be a surface of class C^3, and let $\vec{X} : U \to V$ be a regular parametrization of class C^3 of a neighborhood V of S. The parametrization $\vec{X}(u,v)$ is a system of geodesic coordinates in which the u-coordinate lines are geodesics if and only if the metric tensor is of the form*

$$g = \begin{pmatrix} E(u) & 0 \\ 0 & G(u,v) \end{pmatrix}.$$

Proof: First, the parametrization $\vec{X}$ is an orthogonal if and only if $g_{12} = 0$.

By Equation (8.7), along the u-coordinate lines, the geodesic curvature is

$$\kappa_g = \left(\frac{du}{ds}\right)^3 \Gamma_{11}^2 \sqrt{g_{11}g_{22}}.$$

Since the metric tensor must be positive definite everywhere, $g_{11}g_{22}$ is never 0. Since $du/ds \neq 0$, along a u-coordinate line, $\kappa_g = 0$ if and only if $\Gamma_{11}^2 = -\frac{1}{2}g^{22}\frac{\partial g_{11}}{\partial v} = 0$. The result follows. $\square$

Along u-coordinate lines, since $v' = 0$, the speed function $\dfrac{ds}{du} = \sqrt{E(u)}$ is independent of v. Regardless of v, the arc length formula between $u = u_0$ and u is

$$s(u) = \int_{u_0}^{u} \sqrt{E(u)}\, du.$$

Therefore, it is possible to reparametrize u along the u-coordinate lines by arc length, with $u(s) = s^{-1}(u)$. This leads to the following proposition.

Proposition 8.5.5 *Let S be a regular surface of class C^3, and let $\vec{X}(\bar{u}, v)$ be a system of geodesic coordinates in a neighborhood V of S. Over the same neighborhood V, there exists a system of geodesic coordinates $\vec{X}(u, v)$ such that the u-coordinate lines are geodesics parametrized by arc length. Furthermore, if this is the case, then the coefficients of the metric tensor are of the form*

$$g_{11}(u, v) = 1, \qquad g_{12}(u, v) = 0, \qquad g_{22}(u, v) = G(u, v),$$

and the Gaussian curvature of S is given by

$$K = -\frac{1}{\sqrt{g_{22}}} \frac{\partial^2 \sqrt{g_{22}}}{\partial u^2}. \tag{8.39}$$

Proof: The fact that $g_{11}(u, v) = 1$ follows from the u-coordinate lines being parametrized by arc length and Equation (8.39) is a direct application of Equation (7.23). $\square$

8.5.2   Geodesic Polar Coordinates

Let p be a point on a regular surface of class C^3. By Theorem 8.4.10, for every $\vec{v} \in T_p S$, there exists a unique geodesic $\vec{\gamma}_{p,\vec{v}} : (-\varepsilon, \varepsilon) \to S$, with $\vec{\gamma}_{p,\vec{v}}(0) = p$. Furthermore, by Proposition 8.4.11, we know that $\|\vec{\gamma}'_{p,\vec{v}}(t)\| = \|\vec{v}\|$. We remark then that for any constant scalar λ,

$$\vec{\gamma}_{p,\vec{v}}(\lambda t) = \vec{\gamma}_{p,\lambda\vec{v}}(t)$$

for all $t \in (-\varepsilon/\lambda, \varepsilon/\lambda)$.

Definition 8.5.6 Let S be a regular surface of class C^3, and let $p \in S$ be a point. For any tangent $\vec{v} \in T_p S$, we define the *exponential map* at p as

$$\exp_p(\vec{v}) = \begin{cases} p, & \text{if } \vec{v} = \vec{0}, \\ \vec{\gamma}_{p,\vec{v}}(1), & \text{if } \vec{v} \neq \vec{0}, \end{cases}$$

whenever $\vec{\gamma}_{p,\vec{v}}(1)$ is well defined.

The map $\exp_p : U \to S$, where U is a neighborhood of $\vec{0}$ in T_pS corresponds to traveling along the geodesic through p with direction $\vec{v}$ over the distance $\|\vec{v}\|$. We will show that the exponential map can lead to some nice parametrizations of a neighborhood of p on S, but we first need to prove the following two propositions.

Proposition 8.5.7 *For all $p \in S$, the exponential map is defined over an open neighborhood U of $\vec{0}$. Furthermore, if S is of class C^4, then $\exp_p$ is differentiable over U as a function from T_pS into $\mathbb{R}^3$.*

Proof: We first show that $\exp_p$ is defined over some open disk centered at $\vec{0}$.

Let $C_1 \subset T_pS$ be the set of all vectors of unit length. Set R to be a large positive real number. For each $\vec{v} \in C_1$, let $\varepsilon(\vec{v})$ be the largest positive real $\varepsilon \leq R$ such that $\vec{\gamma}_{p,\vec{v}} : (-\varepsilon, \varepsilon) \to S$ parametrizes a geodesic, or in other words, solves the system of equations in Equation (8.23). The theorems of existence and uniqueness of solutions to differential equations tell us that the solutions to Equation (8.23) depend continuously on the initial conditions, so $\varepsilon(\vec{v})$ is continuous over C_1. (Setting R as an upper bound ensures that $\varepsilon(\vec{v})$ is defined for all $\vec{v} \in C_1$.)

Since C_1 is a compact set, as a function to $\mathbb{R}$, $\varepsilon(\vec{v})$ attains a minimum ε_0 on C_1. However, since $\varepsilon(\vec{v}) > 0$ for all $\vec{v} \in C_1$, then $\varepsilon_0 > 0$. Consequently, $\exp_p$ is defined over the open ball $B_{\varepsilon_0}(\vec{0})$ and is continuous.

Let $\vec{X} : U' \to S$ be a regular parametrization of a neighborhood of p on S. According to the theorem of dependence of solutions to differential equations on initial conditions, the function

$$\vec{\gamma}_p(\vec{v}, t) : U \times (-\varepsilon, \varepsilon) \longrightarrow \mathbb{R}^3$$

is of class C^1 in the initial conditions $\vec{v}$ if the Christoffel symbol functions are of class C^2, which means that $\vec{X}$ needs to be of class C^4. The result follows. $\square$

Proposition 8.5.8 *Let S be a regular surface of class C^4. There is a neighborhood U of $\vec{0}$ in T_pS such that $\exp_p$ is a homeomorphism onto $V = \exp_p(U)$, which is an open neighborhood of p on S.*

Proof: The tangent space T_pS is isomorphic as a vector space to $\mathbb{R}^2$, and one can therefore view $\exp_p$ as a function from $U' \subset \mathbb{R}^2$ into $\mathbb{R}^3$. Let $\vec{v} \in T_pS$ be a nonzero vector, and let $\vec{\alpha}(t) = t\vec{v}$ be defined for $t \in (-\varepsilon, \varepsilon)$ for some $\varepsilon > 0$. Consider the curve on S defined by

$$\exp_p(\vec{\alpha}(t)) = \exp_p(t\vec{v}) = \vec{\gamma}_{p,t\vec{v}}(1) = \vec{\gamma}_{p,\vec{v}}(t).$$

According to the chain rule,

$$\frac{d}{dt}\left(\exp_p(t\vec{v})\right)\Big|_0 = d(\exp_p)_0 \frac{d\vec{\alpha}}{dt}\Big|_0 = d(\exp_p)_0(\vec{v}),$$

where $\vec{v}$ in this expression is viewed as an element in T_pS, and hence, by isomorphism, in $\mathbb{R}^2$. However, by construction of the geodesics, $\vec{\gamma}'_{p,\vec{v}}(0) = \vec{v}$, where we view $\vec{v}$ as an element of $\mathbb{R}^3$. This result proves, in particular, that $d(\exp_p)_0$ is nonsingular.

Using the Implicit Function Theorem from analysis (see Theorem 8.27 in [7]), the fact that $d(\exp_p)_0$ is invertible implies that there exists an open neighborhood U of $\vec{0}$ such that $\exp_p : U \to \exp_p(U)$ is a bijection. Furthermore, setting $V = \exp_p(U)$, by the Implicit Function Theorem, the inverse function $\exp_p^{-1} : V \to U$ is at least of class C^1, and hence, it is continuous.

Hence, $\exp_p$ is a homeomorphism between U and $\exp_p(U)$. □

The identification of T_pS with $\mathbb{R}^2$ allows one to define parametrizations of neighborhoods of p with some nice properties.

Definition 8.5.9 Let S be a surface of class C^4. A system of *Riemann normal coordinates* of a neighborhood of p is a parametrization defined by

$$\vec{X}(u,v) = \exp_p(u\vec{w}_1 + v\vec{w}_2),$$

where $\{\vec{w}_1, \vec{w}_2\}$ is an orthonormal basis of T_pS.

Propositions 8.5.7 and 8.5.8 imply that Riemann normal coordinates provide a regular parametrization of a neighborhood V of p. Furthermore, the theorem of the dependence of solutions to differential equations on initial conditions shows that Riemann normal coordinates form a parametrization of class C^r if S is a surface of class C^{r+3}.

Definition 8.5.10 Let S be a surface of class C^4. The *geodesic polar coordinates* of a neighborhood of p give a parametrization defined by

$$\vec{X}(r,\theta) = \exp_p\left((r\cos\theta)\vec{w}_1 + (r\sin\theta)\vec{w}_2\right),$$

where $\{\vec{w}_1, \vec{w}_2\}$ is an orthonormal basis of T_pS. A curve on S that can be parametrized by

$$\vec{\gamma}(t) = \exp_p\left((R\cos t)\vec{w}_1 + (R\sin t)\vec{w}_2\right), \qquad \text{for } t \in [0, 2\pi]$$

is called a *geodesic circle* of center p and radius R.

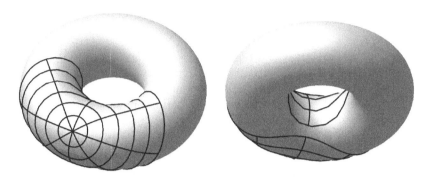

Figure 8.13: Geodesic polar coordinate lines on a torus.

Figure 8.13 gives an example of coordinate lines of a geodesic polar coordinate system on a torus. It is important to note that a geodesic circle is neither a geodesic curve on the surface nor a circle in $\mathbb{R}^3$.

We present the following propositions about the above coordinate systems but leave the proofs as exercises for the reader.

Theorem 8.5.11 *Let S be a regular surface of class C^5, and let p be a point on S. There exists an open neighborhood U of $(0,0)$ in $\mathbb{R}^2$ such that the Riemann normal coordinates defined in Definition 8.5.9 form a system of geodesic coordinates.*

Proof: This theorem follows immediately from Theorem 8.5.3, but using the curve
$$\vec{\alpha}(t) = \exp_p(t\vec{w}_1),$$
where $\vec{w}_1$ is a unit vector in T_pS.   □

Proposition 8.5.12 *Let S be a regular surface of class C^5, and let p be a point on S. Let $\vec{X}$ be a parametrization of a Riemann normal coordinate system in a neighborhood of p so that $\vec{X}(0,0) = p$, and let g_{ij} be the coefficients of the metric tensor associated to $\vec{X}$. The coefficients satisfy $g_{ij}(0,0) = \delta_i^j$, and all the first partial derivatives of all the g_{ij} functions vanish at $(0,0)$.*

Proof: (Left as an exercise for the reader. See Problem 8.5.3.)   □

The next theorem discusses the existence of geodesic polar coordinate systems in a neighborhood of a point $p \in S$.

Theorem 8.5.13 *Let S be a regular surface of class C^5, and let p be a point on S. There exists an $\varepsilon > 0$ such that the parametrization*

$\vec{X}(r, \theta)$ in Definition 8.5.10, with $0 < r < \varepsilon$ and $0 < \theta < 2\pi$, is a regular parametrization and defines a geodesic coordinate system in a neighborhood V whose closure contains p in its interior.

The parametrization $\vec{X}(r, \theta)$ is defined as a function for $-r_0 < r < r_0$ and $\theta \in \mathbb{R}$ for some $r_0 > 0$. However, similar to usual polar coordinates in $\mathbb{R}^2$, one must restrict one's attention to $r > 0$ and $0 < \theta < 2\pi$ to ensure that $\vec{X}$ is a homeomorphism. The image $\vec{X}((0, \varepsilon) \times (0, 2\pi))$ is then a "geodesic disk" on S centered at p with a "geodesic radius" removed. By definition, the parametrization $\vec{X}(r, \theta)$ is such that all the coordinate lines for one of the variables are geodesics, but Theorem 8.5.13 asserts that, within a small enough radius, the parametrization $\vec{X}(r, \theta)$ is regular and orthogonal.

The existence of geodesic polar coordinates at any point p on a surface S of high enough class leads to interesting characterizations of the Gaussian curvature K of S at p. First, we remind the reader that if $\vec{X}(r, \theta)$ is a system of geodesic polar coordinates, then the coefficients of the associated metric tensor are of the form

$$g_{11}(r, \theta) = 1, \quad g_{12}(r, \theta) = 0, \quad g_{22}(r, \theta) = G(r, \theta) \qquad (8.40)$$

for some function $G(r, \theta)$ defined over the domain of $\vec{X}$.

Proposition 8.5.14 *Let S be a surface of class C^5, and let $\vec{X}(r, \theta)$ be a system of geodesic polar coordinates at p on S. Then the function $G(r, \theta)$ in Equation (8.40) satisfies*

$$\sqrt{G(r, \theta)} = r - \frac{1}{6}K(p)r^3 + R(r, \theta),$$

where $K(p)$ is the Gaussian curvature of S at p, and $R(r, \theta)$ is a function that satisfies

$$\lim_{r \to 0} \frac{R(r, \theta)}{r^3} = 0.$$

Proof: (Left as an exercise for the reader. See Problem 8.5.5.) $\square$

Proposition 8.5.14 shows that the perimeter of a geodesic circle at p of radius R is

$$C = \int_0^{2\pi} \sqrt{G(R, \theta)} \, d\theta = 2\pi R - \frac{1}{3}K(p)\pi R^3 + F(R)$$

where $F(r)$ is a function of the radial variable r and satisfies

$$\lim_{r \to 0} F(r)/r^3 = 0.$$

This, and a similar consideration of the area of geodesic disks around p, leads to the following geometric characterization of the Gaussian curvature at p.

Theorem 8.5.15 *Let S be a surface of class C^5, and let $p \in S$ be a point. Define $C(r)$ (resp. $A(r)$) as the perimeter (resp. the area) of the geodesic circle (resp. disk) centered at p and of radius r. The Gaussian curvature $K(p)$ of S at p satisfies*

$$K(p) = \lim_{r \to 0} \frac{3}{\pi}\left(\frac{2\pi r - C(r)}{r^3}\right) \text{ and } K(p) = \lim_{r \to 0} \frac{12}{\pi}\left(\frac{\pi r^2 - A(r)}{r^4}\right).$$

Proof: (Left as an exercise for the reader. See Problem 8.5.6.)   □

Theorem 8.5.15 is particularly interesting because, like the Theorema Egregium, it establishes the Gaussian curvature to a surface as a point as an intrinsic property of the surface, but in an original way. Geodesics and geodesic coordinate systems are intrinsic properties of the surface, and hence geodesic circles around a point are intrinsically defined. Some authors go so far as to use the second formula in Theorem 8.5.15 as a definition of the Gaussian curvature.

PROBLEMS

1. Let S be the unit sphere and suppose that p is a point on S.

   (a) Find a formula (using vectors) for the exponential map $\exp_p : T_pS \to S$.

   (b) Use this to give a formula for the geodesic polar coordinates (Definition 8.5.10) of a patch of S around p in terms of some $\vec{w}_1$ and $\vec{w}_2$.

   (c) Show that if p is the north pole, then with a proper choice of $\vec{w}_1$ and $\vec{w}_2$ we recover our usual colatitude-longitude parametrization of the sphere.

2. Prove that if $\vec{X} : U \to V$ is a parametrization in which both families of coordinate lines are families of geodesics, then the Gaussian curvature satisfies $K(u, v) = 0$.

3. Prove Proposition 8.5.12. [Hint: First use Equation (8.23) to prove that all Γ^i_{jk} vanish at $(0, 0)$, i.e., at p.]

4. (*) Prove Theorem 8.5.13.

5. Prove Proposition 8.5.14. [Hint: Use a Taylor series expansion of $G(r, \theta)$ and Equation (8.39).]

6. Prove Theorem 8.5.15.

7. Consider the usual parametrization of the sphere S of radius R

$$\vec{X}(u, v) = (R \cos u \sin v, R \sin u \sin v, R \cos v).$$

   Define the new parametrization $Y(r, \theta) = \vec{X}(\theta, r/R)$.

   (a) Prove that $Y(r, \theta)$ is a geodesic polar coordinate system of S at $p = (R, 0, 0)$.

   (b) Prove the result of Proposition 8.5.14 directly and determine the corresponding remainder function $R(r, \theta)$.

8.6 Applications to Plane, Spherical, and Elliptic Geometry

8.6.1 Plane Geometry

In plane geometry, we consider the plane to be a surface with Gaussian curvature identically 0 and curves in the plane we consider to be curves on a surface. Consider now a region $\mathcal{R}$ in the plane such that the boundary $\partial\mathcal{R}$ is a piecewise regular, simple, closed curve. It is easy to calculate that the Euler characteristic of a region that is homeomorphic to a disk is $\chi(\mathcal{R}) = 1$.

As a first case, let $\partial\mathcal{R}$ be a polygon. Since the regular arcs are straight lines, the geodesic curvature of each regular arc of $\partial\mathcal{R}$ is identically 0. The Gauss-Bonnet formula then reduces to the well-known fact that the sum of the exterior angles $\{\theta_1, \theta_2, \ldots, \theta_k\}$ around a polygon is 2π. Also, since the exterior angle θ_i at any corner is $\pi - \alpha_i$, where α_i is the interior angle, we deduce that the sum of the interior angles of an n-sided polygon is

$$\sum_{i=1}^{n} \alpha_i = (n-2)\pi.$$

This statement is in fact equivalent to a theorem that occurs in every high school geometry curriculum, namely, that the sum of the interior angles of a triangle is π radians.

As a second case of the Gauss-Bonnet Theorem in plane geometry, suppose that $\partial\mathcal{R}$ is a simple, closed, regular curve of class C^2. In this case, $\partial\mathcal{R}$ has no corners or cusps and hence no exterior angles. Assuming the boundary is oriented so that the interior $\mathcal{R}$ is in the positive $\vec{U}$ direction, then the Gauss-Bonnet Theorem reduces to the formula

$$\oint_{\partial\mathcal{R}} \kappa_g \, ds = 2\pi.$$

Therefore, the Gauss-Bonnet Theorem subsumes the proposition that the rotation index of a simple, closed, regular curve is 1 (see Propositions 2.2.1 and 2.2.11).

8.6.2 Spherical Geometry

Ever since navigators confirmed that the Earth is (approximately) spherical, geometry of the sphere became an important area of study. Improper calculations could send explorers and navigators in drastically wrong directions. A key result particularly valuable to navigators is that the sum of the interior angles of a triangle is not equal to two right angles.

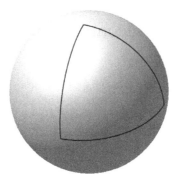

Figure 8.14: Triangle on a sphere.

Figure 8.14 gives an example of a triangle on a sphere in which each vertex has a right angle, and hence, the sum of the interior angles is $3\pi/2$.

Example 8.4.8 showed that a geodesic on a sphere is a great arc of the sphere, i.e., an arc on the circle of intersection between the sphere and a plane through the center of the sphere. Therefore, any two geodesic lines intersect in at least two points: the common intersection of the sphere and the two planes supporting the geodesics. This remark, along with the fact that geometry on the sphere satisfies the first four postulates of Euclid's *Elements*, justifies the claim that geometry of the sphere is an elliptic geometry.

Consider now a triangle $\mathcal{R}$ on a sphere of radius R. By definition of a geometric triangle, $\partial\mathcal{R}$ has three vertices, and its regular arcs are geodesic curves. The Gaussian curvature of a sphere of radius R is $\frac{1}{R^2}$. In this situation, the Gauss-Bonnet Theorem gives

$$(\theta_1 + \theta_2 + \theta_3) + \frac{A}{R^2} = 2\pi, \qquad (8.41)$$

where A is the surface area of $\mathcal{R}$, and θ_i are the exterior angles of the vertices. With the interior angles $\alpha_i = \pi - \theta_i$, then Equation (8.41) reduces to

$$\alpha_1 + \alpha_2 + \alpha_3 = \pi + \frac{A}{R^2} = \pi + 4\pi\left(\frac{A}{4\pi R^2}\right).$$

This shows that in a spherical triangle, the excess sum of the angles, i.e., $\alpha_1 + \alpha_2 + \alpha_3 - \pi$, is 4π times the ratio of the surface area of the triangle to the surface area of the whole sphere. As an example, the spherical triangle in Figure 8.14 has an excess of $\pi/2$ and covers $1/8$

of the sphere. The quantity $\alpha_1 + \alpha_2 + \alpha_3 - \pi$ is often referred to as the *angle excess* of a triangle on sphere.

For navigators, the Gauss-Bonnet Theorem matters significantly because if they traveled in a circuit, generally following straight lines and only making sharp turns at discrete specific locations, the sum of the exterior angles of the circuit does not add up to 2π but to a quantity that is strictly less than 2π.

We propose to study (geodesic) circles on a sphere of radius R. Recall that a circle on a sphere is the set of points that are at a fixed distance r from a point, where distance means geodesic distance. Call $\mathcal{D}$ the closed disk inside this circle. An example of a geodesic circle is any latitude line, since points on a latitude line are equidistant from the North Pole. Using the parametric equations for the sphere,

$$\vec{X}(u, v) = (R \cos u \sin v, R \sin u \sin v, R \cos v),$$

a circle of radius r around the north pole $(0, 0, R)$ is given by $\vec{\gamma}(t) = \vec{X}(t, v_0)$. Note that the geodesic radius is the length of an arc of radius R and angle equal to the colatitude, so $r = R v_0$. It is not hard to tell by symmetry that the geodesic curvature κ_g of a circle on a sphere is a constant. The Gauss-Bonnet Theorem then gives

$$\kappa_g(\text{perimeter of } \partial\mathcal{D}) + \frac{1}{R^2}(\text{surface area of } \mathcal{D}) = 2\pi.$$

Using the surface area formula for a surface of revolution from single-variable calculus, it is possible to prove (Problem 8.6.1) that the area of $\mathcal{D}$ is

$$\text{surface area of } \mathcal{D} = 2\pi R^2 (1 - \cos v_0) = 2\pi R^2 \left(1 - \cos\left(\frac{r}{R}\right)\right). \quad (8.42)$$

As a circle in $\mathbb{R}^3$, the radius is $R \sin v_0$, so the perimeter of $\mathcal{D}$ is

$$\text{perimeter of } \mathcal{D} = 2\pi R \sin v_0 = 2\pi R \sin\left(\frac{r}{R}\right).$$

Consequently, the Gauss-Bonnet Theorem gives the geodesic curvature of a circle on a sphere as

$$\kappa_g = \frac{1}{R} \cot\left(\frac{r}{R}\right).$$

Using a Maclaurin series for the even function $x \cot x$, we can show that this geodesic curvature is approximately

$$\kappa_g \approx \frac{1}{R}\left(\left(\frac{r}{R}\right)^{-1} - \frac{1}{3}\left(\frac{r}{R}\right) - \frac{1}{45}\left(\frac{r}{R}\right)^3 - \cdots\right)$$

$$\approx \frac{1}{r} - \frac{1}{3R^2}r - \frac{1}{45R^4}r^3 - \cdots$$

In this expression, as $r \to 0$, then κ_g behaves asymptotically like $\frac{1}{r}$, where $\frac{1}{r}$ is the usual curvature of a circle of radius r. Also, as $R \to \infty$, which corresponds to a sphere with a radius that grows so large that the sphere approaches planar, we have

$$\lim_{R \to \infty} \kappa_g = \lim_{R \to \infty} \frac{1}{R} \cot\left(\frac{r}{R}\right) = \frac{1}{r},$$

which again recovers the familiar curvature of a circle of radius r.

8.6.3 Elliptic Geometry

As mentioned in the introduction to this chapter, in his *Elements*, Euclid proves all his propositions from 23 definitions and five postulates. The fifth postulate reads as follows:

> If a straight line crossing two straight lines makes the interior angles on the same side less than two right angles, the two straight lines, if extended indefinitely, meet on that side on which are the angles less than the two right angles.

This postulate is much wordier than the other four, and many mathematicians over the centuries attempted to prove the fifth postulate from the others, but never succeeded. Studying Euclid's *Elements*, it appears that Euclid himself uses the fifth postulate sparingly. Proposition I.29 is the first proposition to cite Postulate 5. Furthermore, some proofs could be simplified if Euclid cited Postulate 5. One could speculate that Euclid avoided a liberal use of Postulate 5 so that, were someone to prove it as a consequence of the other four, then the proofs of theorems would still be given in a minimal form. Indeed, many commonly known properties about plane geometry hold without reference to the fifth postulate. However, examples of a few commonly known theorems that rely on it are:

1. Playfair Axiom: Given a line L and a point P not on L, there exists a unique line through P that does not intersect L.

2. The sum of the interior angles of a triangle is π.

3. Pythagorean Theorem: the squared length of the hypotenuse of a right triangle is the sum of the squares of the lengths of the sides adjacent to the right angle.

In fact, in the above list, (1) is equivalent to the fifth postulate under the assumption that the first four postulates hold. That is why this theorem became known as the Playfair Axiom. In the 19th century, mathematicians Lobachevsky and Bolyai independently considered geometries that retained the first four postulates of Euclid's *Elements* but assuming alternatives to the Playfair Axiom. The two logical alternatives to the Playfair Axiom are:

Elliptic Given a line L and a point P not on L, there does not exist a line through P that does not intersect L.

Hyperbolic Given a line L and a point P not on L, there exists more than one line through P that does not intersect L.

Depending on whether we use one or the other of the above two alternatives we obtain alternate geometries that are respectively called *elliptic geometry* and *hyperbolic geometry*. Collectively, elliptic and hyperbolic geometries were first called *non-Euclidean geometry* but then this label soon encompassed all types of geometry where the metric differs from the flat Euclidean metric.

It is possible to prove many theorems in these different geometries using synthetic techniques, that is involving proofs that avoid the use of coordinates. (See [10], [30], or [8] for a comprehensive synthetic treatment of elliptic or hyperbolic geometry.) One well-known result in these geometries is that the sum of the interior angles of a triangle is greater than π in the case of elliptic geometry and less than π in the case of hyperbolic geometry.

When Lobachevsky first investigated alternatives to Euclidean he considered hyperbolic geometry but rejected the elliptic hypothesis. Spherical geometry satisfies the elliptic hypothesis. However, in geometry on the sphere, any two great circles intersect at two points, and this does not satisfy the first axiom of Euclid. Consequently, it is possible that he believed that since spherical geometry has the elliptic hypothesis but fails Euclid's first axiom, the elliptic geometry was inconsistent. However, in the late 19th century, it was realized that on a real projective plane obtained by identifying opposite points on the sphere, all of the first four Euclidean axioms remain valid and that the elliptic hypothesis holds as well. A number of the theorems that were known for spheres were valid in this elliptic geometry on a non-orientable surface. (Because a proper topological introduction to the real projective plane would take us too far afield, we do not describe it explicitly in this textbook but encourage the reader to discover it in subsequent studies in geometry.)

We would like to now consider applications of the Gauss-Bonnet Theorem to elliptic and hyperbolic geometry. We first need to clarify the use of the terms "line," "right angle," and "parallel." Let us assume that we have a metric. In the geometry on a surface of class C^3, the word "line" (or straight line) means a geodesic. Two lines meet at a right angle at a point p if the direction vectors of the lines at p have a first fundamental form that vanishes. The word "parallel" is more problematic. In Problem 1.3.21, we defined a parallel curve to a given curve $\vec{\gamma}$ as a curve whose locus is always a fixed orthogonal distance r from the locus of $\vec{\gamma}$. This is true of parallel lines in the plane. However, in this sense of parallelism, a parallel curve to a line (geodesic) need not be another line. Consequently, we do not directly use the word parallel but prefer to discuss whether two lines intersect or not.

Elliptic and hyperbolic geometry involve the notion of congruence. Recall that in any geometry that has a notion of congruence, the concept of area must satisfy the following three axioms:

Axiom 1: The area of any set must be nonnegative.

Axiom 2: The area of congruent sets must be the same.

Axiom 3: The area of the union of disjoint sets must equal the sum of the areas of these sets.

From Axiom 2 and Theorem 8.5.15, we can deduce that the Gaussian curvature is constant in elliptic and hyperbolic geometries.

Consider a triangle $\mathcal{R}$ in a space of constant Gaussian curvature K. Because the edges of the triangle are geodesics, by the Gauss-Bonnet Theorem,

$$\iint_{\mathcal{R}} K \, dS + \theta_1 + \theta_2 + \theta_2 = 2\pi\chi(\mathcal{R}),$$

where θ_i is the exterior angle at the ith vertex. Note that $\theta_i = \pi - \alpha_i$, where α_i are the interior angles of the triangle. Furthermore, the Euler characteristic of any triangle is $\chi(\mathcal{R}) = 3 - 3 + 1 = 1$. Thus

$$K \cdot \text{Area}(\mathcal{R}) + (\pi - \alpha_1) + (\pi - \alpha_2) + (\pi - \alpha_3) = 2\pi$$
$$\implies \alpha_1 + \alpha_2 + \alpha_3 = \pi + K \cdot \text{Area}(\mathcal{R}). \tag{8.43}$$

We mentioned earlier that in elliptic (resp. hyperbolic) geometry, the interior sum of angles of a triangle is greater (resp. less) than pi. We deduce that in elliptic geometry the Gaussian curvature is positive and constant, whereas in hyperbolic geometry, the Gaussian curvature is negative and constant. It is possible to prove from synthetic geometry in either elliptic or hyperbolic geometry) that the sum of the interior angles of a triangle satisfies

$$\alpha_1 + \alpha_2 + \alpha_3 = \pi + c \cdot \text{Area}(\mathcal{R}),$$

where c is a positive constant for elliptic geometry and a negative constant for hyperbolic geometry. However, the Gauss-Bonnet Theorem shows that this constant c is precisely the Gaussian curvature.

PROBLEMS

1. Calculate the surface area of a disk (Equation (8.42)) on a sphere using the surface area formula for a surface of revolution. Calculate directly the geodesic curvature function of a circle on a sphere.

2. Using Equation (8.42) for the surface area of a disk on a sphere and Theorem 8.5.15 recover the familiar result that the Gaussian curvature of a sphere of radius R is $1/R^2$.

3. Suppose that a sphere of radius R has the longitude-colatitude (u, v) system of coordinates. Suppose that a simply connected region $\mathcal{R}$ of this sphere does not include the north or south pole and suppose that the boundary $\partial\mathcal{R}$ is a single regular, closed curve of class C^2 parametrized by $(u(t), v(t))$ for $t \in [a, b]$ in its coordinate plane. Prove that the area of the region is

$$2\pi R^2 + R^2 \int_a^b (\cos v)u' + \frac{(\cos v)u'(v')^2 + (\sin v)(u''v' - u'v'')}{(\sin^2 v)(u')^2 + (v')^2}\, dt.$$

4. Using Euclidean style constructions, describe a procedure for finding the midpoint of a given great circle arc on the unit sphere with endpoints A and B.

8.7 Hyperbolic Geometry

8.7.1 Synthetic Hyperbolic Geometry

Recall from the previous section that hyperbolic geometry is a geometry that involves the undefined terms of points, lines, and circles and that satisfies the first four postulates of Euclid's *Elements* along with the fifth hyperbolic axiom: (H) for any line ℓ and any point P not on ℓ, there pass more than one line that does intersect ℓ.

Synthetic hyperbolic geometry is like Euclidean geometry in its style of proof, in that it does not use coordinate systems. Various texts (see Chapter 2 in [8] for example) explore this geometry in detail. Some theorems are identical to Euclidean geometry if they only involve the first four postulates and some theorems differ considerably from Euclidean geometry. We mention a few results without proof in order to motivate the applications of the Gauss-Bonnet Theorem to hyperbolic geometry.

Let ℓ be a line and let P be a point not on ℓ. We can invoke Proposition I.12 in the *Elements* because its proof does not rely on the fifth postulate. Hence, we can construct a line through P that is perpendicular to ℓ. In Figure 8.15, this is the line $\overleftrightarrow{PQ}$. (Figure 8.15 only shows the segment $\overline{PQ}$.)

Obviously, the diagram in Figure 8.15 is sketched in the Euclidean plane as a "local approximation" and we must understand that as ℓ_1 and ℓ_2 are extended indefinitely, they do not intersect ℓ.

Consider the lines through P as they sweep out angles away from $\overleftrightarrow{PQ}$. There must exist a first line on the right side of $\overleftrightarrow{PQ}$ that does not intersect ℓ and similarly on the left side. These lines are called the

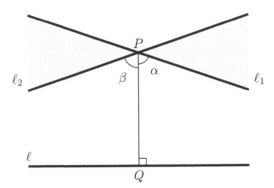

Figure 8.15: Left and right sensed-parallels.

right (resp. left) *sensed-parallel* of ℓ through P; they are indicated
in Figure 8.15 by ℓ_1 (resp. ℓ_2). The angle α (resp. β) is called the
angle of right (resp. left) parallelism. It is not hard to prove that
$\alpha = \beta$ and that this common angle is acute. All the lines through P
that lie in the shaded region in Figure 8.15 are called *ultraparallel* to
ℓ through P.

In addition to usual triangles, in hyperbolic geometry, one also
considers asymptotic triangles. If ℓ_1 and ℓ_2 are right (resp. left)
sensed-parallels, then we say that they meet at an *ideal point* Ω. If
A is on ℓ_1 and B is on ℓ_2, then the figure consisting of the ray $\overrightarrow{A\Omega}$,
the segment $\overline{AB}$, and the ray $\overrightarrow{B\Omega}$ is called an asymptotic triangle and
denoted $\triangle AB\Omega$. It is also possible to have asymptotic triangles with
two or three ideal points as we shall see in models below.

It is natural to try to prove theorems in hyperbolic geometry that
mirror the development of Euclidean geometry as presented in *The
Elements*. A number of surprising results arise. For example, in any
triangle, the sum of the interior angles is less than two right angles.
For a triangle $\triangle ABC$, the *defect* of $\triangle ABC$ is the difference between
π and the sum of the interior angles. Furthermore, with a few basic
axioms of how area works, it is possible to prove that there exists a
positive constant c such that for all triangles $\triangle ABC$, the area-defect
formula holds

$$\text{Area}(\triangle ABC) = c^2 \, \text{defect}(\triangle ABC).$$

In the previous section, we offered a reason why the Gaussian
curvature in elliptic and in hyperbolic geometry should be constant.
Under this assumption, we saw that because the sum of the interior
angles of the triangle is less than π, then the Gaussian curvature in
hyperbolic geometry must be negative. Let us suppose that $K = -\frac{1}{c^2}$

in imitation of what happens for a sphere. Then using the Gauss-Bonnet Theorem applied to triangle $\mathcal{R}$ in hyperbolic geometry, we get

$$-\frac{1}{c^2}\text{Area}(\mathcal{R}) + (\pi - \alpha_1) + (\pi - \alpha_2) + (\pi - \alpha_3) = 2\pi,$$

where α_i are the interior angles of the triangle. Thus

$$\text{Area}(\mathcal{R}) = c^2(\pi - (\alpha_1 + \alpha_2 + \alpha_3)),$$

which recovers the area-defect formula. The Gauss-Bonnet approach is not a synthetic approach but it leads to the same relationship.

As mentioned earlier, the diagram offered in Figure 8.15 is only of limited value because if we extend ℓ_1 and ℓ for example, they intersect. The integrity of a synthetic proof cannot rely on a diagram but it is unfortunate that such a diagram fails to capture key properties, such as the angle defect of a triangle.

This made it difficult for some mathematicians to initially accept hyperbolic geometry since it deviated from sense perception. However, it was soon discovered that by using different metrics, it is possible to depict the hyperbolic plane more effectively.

8.7.2 The Poincaré Upper Half-Plane

As early as Section 6.1, we introduced the alternate metric

$$g_{ij} = \begin{pmatrix} \dfrac{1}{v^2} & 0 \\ 0 & \dfrac{1}{v^2} \end{pmatrix}$$

on the upper half-plane $H = \{(u,v) \in \mathbb{R}^2 \mid v > 0\}$. We called this the Poincaré upper half-plane. In Example 6.1.9, we calculated certain lengths of curves and also showed the area of the region $a \leq u \leq b$ and $c \leq v \leq d$ is

$$(b-a)\left(\frac{1}{c} - \frac{1}{d}\right).$$

The area of any region $\mathcal{R}$ in the plane is

$$\text{Area}(\mathcal{R}) = \iint_{\mathcal{R}} \frac{1}{v^2}\, du\, dv.$$

It is possible to prove that, given this metric, the Gaussian curvature is constant with $K = -1$. Building on other examples in the text, in Problem 8.4.11 we proved that geodesics in this space are either vertical lines or semicircles of radius R and with origin on the u-axis. Consequently, the Poincaré upper half-plane offers a model of hyperbolic geometry in which "lines" consist of these geodesics.

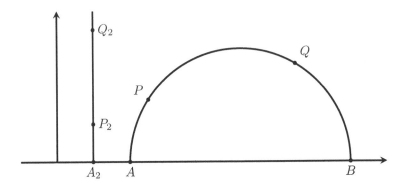

Figure 8.16: Geodesics in the Poincaré half-plane.

Interestingly enough, unlike for the cylinder in which there exist points such that there are many geodesics connecting them, given any two points in the Poincaré, there exists only one geodesic connecting them. In Problem 8.7.3, the reader is asked to prove a distance formula between points by calculating the length between the points along the unique geodesic that connects them.

Finally, consider two vectors $\vec{a}$ and $\vec{b}$ based at a point p with coordinates (u_0, v_0). We can understand $\vec{a}$ and $\vec{b}$ as vectors in the tangent plane to p. The angle θ between $\vec{a}$ and $\vec{b}$ satisfies

$$
\begin{aligned}
\cos \theta &= \frac{I_p(\vec{a}, \vec{b})}{\sqrt{I_p(\vec{a}, \vec{a})}\sqrt{I_p(\vec{b}, \vec{b})}} \\
&= \frac{\frac{1}{v_0^2} a_1 b_1 + \frac{1}{v_0^2} a_2 b_2}{\sqrt{\frac{1}{v_0^2} a_1^2 + \frac{1}{v_0^2} a_2^2}\sqrt{\frac{1}{v_0^2} a_1^2 + \frac{1}{v_0^2} a_2^2}} \\
&= \frac{a_1 b_1 + a_2 b_2}{\sqrt{a_1^2 + a_2^2}\sqrt{b_1^2 + b_2^2}}.
\end{aligned}
$$

This is the usual formula for the cosine of an angle between two vectors in the plane. Consequently, the Poincaré upper half-plane model of hyperbolic geometry accurately reflects angles as the angles measured in the usual sense.

The diagram below depicts three lines (geodesics) in the Poincaré upper half-plane model of hyperbolic geometry that meet to form a triangle. This particular triangle is a usual and not an asymptotic one.

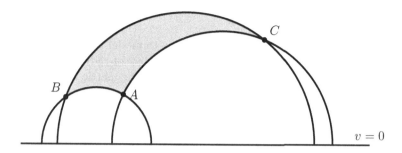

The reader might have guessed that there is a right angle at A and that would be correct. The circles in the above diagram were chosen specifically so that the two smaller circles meet at a right angle. Because of the above comment, since they meet at right angles as circles in the model shown in the Euclidean plane, the corresponding lines in the hyperbolic geometry meet at right angles as well. (It is easy to prove in Euclidean geometry that two circles of radius r and R respectively intersect orthogonally if the distance between them is $\sqrt{r^2 + R^2}$.)

On the other hand, in the diagrams below, the first figure depicts an asymptotic triangle with one ideal point, the second figure depicts an asymptotic triangle with two ideal points for vertices, and the third figure depicts an asymptotic triangle with three ideal points.

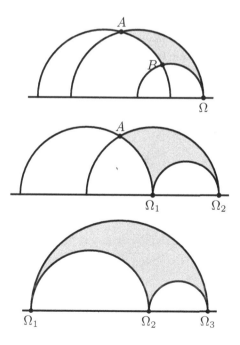

8.7.3   The Poincaré Disk

Another model for hyperbolic geometry is to model the points in the hyperbolic plane as the points in the open disk of radius R, namely $\mathcal{D}_R = \{(u,v) \in \mathbb{R}^2 \,|\, u^2 + v^2 < R^2\}$ but to use the metric tensor

$$
g_{ij} = \begin{pmatrix} \dfrac{4R^4}{(R^2 - u^2 - v^2)^2} & 0 \\[2ex] 0 & \dfrac{4R^4}{(R^2 - u^2 - v^2)^2} \end{pmatrix}.
$$

The disk $\mathcal{D}_R$ equipped with this metric is called the *Poincaré disk*. If $(u(t), v(t))$ parametrizes a curve in $\mathcal{D}_R$, then the length of the curve with $t_1 \le t \le t_2$ is

$$
\int_{t_1}^{t_2} \frac{2R^2}{(R^2 - u(t)^2 - v(t)^2)} \sqrt{(u'(t))^2 + (v'(t))^2} \, dt.
$$

Consequently, as the curve approaches the boundary of the disk $\mathcal{D}_R$, the length becomes arbitrarily large. Hence, we can intuitively think of the boundary $\partial \mathcal{D}_R$ as heading far away, becoming unbounded. We leave it as an exercise for the reader to compute the Christoffel symbols and to prove that the Gaussian curvature of the Poincaré disk is $-1/R^2$. Consequently, the Poincaré disk has constant negative Gaussian curvature. (If R is unspecified, we assume that $R = 1$ and that $\mathcal{D}$ is the unit disk.)

Using the so called Möbius transformations from complex analysis (a technique that is just beyond the scope of this book) from the Poincaré upper half-plane to the Poincaré disk, it is possible to prove that the geodesics on the Poincaré are either diameters of the disk or arcs of circles that meet the boundary $\partial \mathcal{D}$ at right angles.

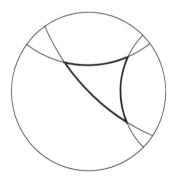

The diagram above depicts a (non-asymptotic) triangle in the Poincaré disk. An asymptotic triangle in the Poincaré disk involves geodesic edges that meet at the boundary of $\mathcal{D}$.

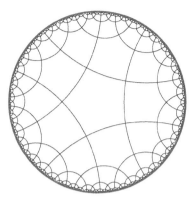

Figure 8.17: Tiling of the Poincaré disk with regular pentagons.

Because of the relationship between the defect of triangles and by extension the defect of other polygons, it is possible to tessellate (tile) the hyperbolic plane with regular polygons in ways that are impossible in the Euclidean plane. For example, it is possible to tessellate the hyperbolic plane with pentagons in such a way that at each vertex of the tessellation, four regular pentagons meet. (See Figure 8.17.) Since four pentagons meet at a vertex, each interior angle of the pentagon is $\frac{\pi}{2}$. Each pentagon is the union of five isosceles triangles, each congruent to each other, meeting at the center of the pentagon. Each of these triangles has for interior angles $\frac{\pi}{4}$, $\frac{\pi}{4}$, and $\frac{2\pi}{5}$. Hence the defect of each triangle is $\frac{\pi}{10}$ and its area is $c^2\frac{\pi}{10}$, when the Gaussian curvature is a constant $K = -\frac{1}{c^2}$. Thus, in order to tessellate the hyperbolic plane when $K = -1$ with pentagons such that four meet at each vertex, we simply must choose regular pentagons of area $\frac{\pi}{2}$.

Artist M. C. Escher (1891–1972) was known for artwork that explores mathematical symmetries in creative ways. Among a few of his more esoteric works are symmetry patterns based on tessellations of the hyperbolic plane.

8.7.4 The Pseudosphere Revisited

The Poincaré upper half-plane and Poincaré disk offer effective models of the hyperbolic plane. One may feel still somewhat dissatisfied because these models do not correspond to regular surfaces in $\mathbb{R}^3$. The pseudosphere introduced in Example 6.6.6 is a regular surface in $\mathbb{R}^3$ that has constant Gaussian curvature equal to $K = -1$. Consequently, the pseudosphere offers another model of hyperbolic space.

Figure 8.18 depicts a triangle (the edges are geodesics) on the pseudosphere. Problem 8.7.8 guides the reader to explore some of the

Figure 8.18: A geodesic triangle on a pseudosphere.

properties of the pseudosphere starting from its metric.

PROBLEMS

1. In the Poincaré upper half-plane, consider the isosceles triangle with extended edges given by the three geodesics: $x^2 + y^2 = 36$, $(x-3)^2 + y^2 = 16$, and $(x+3)^2 + y^2 = 16$ (with $y > 0$). Calculate the area of this triangle.

2. Consider the Poincaré upper half-plane.

   (a) Find the geodesic curvature of the line $(u(t), v(t)) = (t, mt)$, with $t > 0$, where m is some constant.

   (b) Explicitly evaluate the line integral $\displaystyle\int_{\partial \mathcal{R}} \kappa_g \, ds$, where $\mathcal{R}$ is the region bounded by three Euclidean straight lines from $(1,1)$ to $(2,2)$ to $(1,2)$ and back to $(1,1)$.

   (c) Without additional calculations, deduce $\displaystyle\iint_{\mathcal{R}} K \, dS$.

   (d) Directly evaluate the double integral in the previous part to confirm your result. [Recall that $K = -1$.]

3. Consider the Poincaré half-plane. Let P and Q be two points on H. We determine the distance $d(P, Q)$ between these two points in the Poincaré metric on H (see Figure 8.16).

   (a) If P and Q lie on a geodesic that is a half-circle, prove that the distance between them is

   $$d(P, Q) = \left| \ln \frac{|PA|/|PB|}{|QA|/|QB|} \right|,$$

   where A and B are points as shown in Figure 8.16 and $|PA|$ is the usual Euclidean distance between P and A and similarly for all the others.

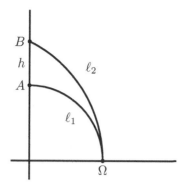

Figure 8.19: Angle of parallelism.

(b) If P and Q lie on a geodesic that is a vertical line, prove that the distance between them is

$$d(P,Q) = \left| \ln \frac{|P_2 A_2|}{|Q_2 A_2|} \right|,$$

where A_2 is again as shown in Figure 8.16.

4. Describe a procedure for finding the midpoint of a given hyperbolic segment on the Poincaré upper half-plane. In particular, find the midpoint of the hyperbolic segment from $(0,1)$ to $\left(\frac{1}{\sqrt{2}}, \frac{1}{\sqrt{2}}\right)$. [Hint: See Problem 8.7.3.]

5. Show that the set of points at a fixed distance d from the vertical line $u = 0$ in the hyperbolic plane consists of the points on two Euclidean lines of the form $y = mx$.

6. Using the Poincaré upper half-plane, Figure 8.19 depicts two lines ℓ_1 and ℓ_2 that are parallel (right-sensed). The line $\overleftrightarrow{AB}$ is perpendicular to ℓ_1 so the acute angle between $\overleftrightarrow{AB}$ and ℓ_2 is the angle of parallelism. Suppose that in this diagram ℓ_1 is a circle of radius 1 and let h be the distance between A and B. Calculate the angle of parallelism in terms of the distance h. Show that as $h \to 0$, the angle of parallelism goes to $\pi/2$, and as $h \to \infty$, the angle of parallelism goes to 0.

7. Prove that in the Poincaré disk, the angle at which curves meet as observed in the model in $\mathbb{R}^2$ is precisely the angle at which the curves meet in the Poincaré disk metric. [We say that the Poincaré disk is a *conformal* model of the hyperbolic plane.]

8. Consider the upper half-plane $\mathcal{H} = \{(x,y) \in \mathbb{R}^2 \,|\, y > 0\}$ with the metric tensor

$$(g_{ij}) = \begin{pmatrix} e^{-2y} & 0 \\ 0 & 1 \end{pmatrix}.$$

(a) Prove that the Gaussian curvature of such a surface has $K = -1$ everywhere.

(b) Show that the parametrization $\vec{X} : \mathcal{H} \to \mathbb{R}^3$ defined by

$$\vec{X}(x,y) = \left(e^{-y}\cos x, e^{-y}\sin x, \ln(1 + \sqrt{1 - e^{-2y}}) + y - \sqrt{1 - e^{-2y}}\right)$$

has the above metric coefficients.

(c) Prove that $\vec{X}(x,y)$ is a regular reparametrization of the pseudosphere as described in Example 6.6.6.

(d) (**ODE**) Find parametric equations $(u(t), v(t))$ for the geodesic curves on $\mathcal{H}$ with this given metric.

9. Consider the Poincaré half-plane. We can consider $\mathbb{R}^2$ as the set of complex numbers $\mathbb{C}$. In this context, $\mathcal{H} = \{z \in \mathbb{C} \,|\, \mathrm{Im}(z) > 0\}$. Consider a fractional linear transformation of $\mathbb{C}$ of the form

$$w = \frac{az + b}{cz + d},$$

where $a, b, c, d \in \mathbb{R}$ and $ad - bc = 1$. Write $z = u + iv$ and $w = x + iy$.

(a) Prove that the function $w = f(z)$ sends $\mathcal{H}$ into $\mathcal{H}$ bijectively.

(b) Prove that the metric tensor for $\mathcal{H}$ is unchanged under this transformation, more precisely, that the metric coefficients of $\mathcal{H}$ in the w-coordinate are

$$(\bar{g}_{kl}) = \begin{pmatrix} y^{-2} & 0 \\ 0 & y^{-2} \end{pmatrix}.$$

[We say that the Poincaré metric on the upper half-plane is invariant under the action of $\mathrm{SL}(2, \mathbb{R})$.]

(c) Show explicitly that this fractional linear transformation sends geodesics to geodesics.

CHAPTER 9

Curves and Surfaces in n-Dimensional Space

Up to this point, this text emphasized curves and surfaces in $\mathbb{R}^3$. However, our only reason to do so was based on a discrimination by dimensionality: As beings living in three-dimensional space, we are accustomed to visualizing in $\mathbb{R}^3$. In this chapter, we generalize a number of definitions and theorems about curves and surfaces to Euclidean n-dimensional space.

This brief chapter is not meant as a complete treatment of the subject for two reasons. After our classical approach to curves and surfaces, the topic of differential geometry generalizes to manifolds (see [25]). Manifolds provide a much broader and more general framework through which to study geometric objects in any number of dimensions. Secondly, especially in regards to curves, much of our presentation simply illustrates the logical progression started in our treatment of local properties of plane curves and then local properties of space curves.

9.1 Curves in n-Dimensional Euclidean Space

In Chapters 1 and 3, we studied local properties of curves with an emphasis on notions of velocity, acceleration, speed, curvature, and torsion. Along the way, we introduced the natural orthonormal frames attached to curves at a point, namely the $(\vec{T}, \vec{U})$ for a regular plane curve and the Frenet frame $(\vec{T}, \vec{P}, \vec{B})$ for a regular space curve of class C^2 in $\mathbb{R}^3$.

Many local properties and definitions have immediate generalizations to curves in n-dimensional Euclidean space, $\mathbb{R}^n$. Let I be an interval of $\mathbb{R}$ and let $\vec{X} : I \to \mathbb{R}^n$ be a parametrized curve. Then

- the velocity is the vector function $\vec{X}'(t)$;

- the speed is real-valued function $s'(t) = \|\vec{X}'(t)\|$;

DOI: 10.1201/9781003295341-9

- the arc length function with origin $t = t_0$, is the antiderivative of $s'(t)$ with $s(t_0) = 0$;
- the acceleration is the vector function $\vec{X}''(t)$;
- the parametrization $\vec{X}$ is regular at $t = t_0$ if $\vec{X}'(t_0)$ exists and is nonzero;
- the parametrization $\vec{X}$ is regular if it is regular at all t_0 in its domain;
- a curve C in $\mathbb{R}^n$ is regular if it is the locus (image) of a regular parametrized curve;
- at any regular point $t = t_0$, the unit tangent vector is $\vec{T}(t_0) = \vec{X}'(t_0)/\|\vec{X}'(t_0)\|$.

One hitch that occurs in generalizing our theory from curves in space to curves in $\mathbb{R}^n$ comes from the fact that for $n \neq 3$ there does not exist a cross product in $\mathbb{R}^n$. So any formula or reasoning must avoid that nice property about $\mathbb{R}^3$.

9.1.1   Curvatures and the Frenet Frame

Recall that any unit vector function $\vec{Y} : I \to \mathbb{R}^n$ satisfies $\vec{Y}(t) \cdot \vec{Y}(t) = 1$ so after differentiating this expression, for all $t \in I$,

$$2\vec{Y}'(t) \cdot \vec{Y}(t) = 0 \implies \vec{Y}'(t) \perp \vec{Y}(t). \tag{9.1}$$

Thus, if the derivative exists, $\vec{T}'(t)$ is perpendicular to $\vec{T}(t)$, though $\vec{T}'(t)$ is not necessarily of unit length.

Suppose that $\vec{X}(t)$ is a regular parametrized curve of class C^2. We define the *first curvature* $\kappa_1(t)$ of $\vec{X}(t)$ implicitly as the unique nonnegative function such that

$$\vec{T}'(t) = s'(t)\kappa_1(t)\vec{P}_1(t),$$

and where $\vec{P}_1(t)$ is a unit vector function. This implicit definition might not produce a well-defined vector at $\vec{P}_1(t_0)$ if $\kappa_1(t_0) = 0$, i.e., if $\vec{T}'(t_0) = \vec{0}$. This problem can be remedied if $\vec{T}'(t) \neq \vec{0}$ for $0 < |t - t_0| < \varepsilon$, for some positive ε. In this case, $\vec{P}_1(t)$ is a well-defined and continuous vector function on the unit sphere in $\mathbb{R}^n$ for $0 < |t - t_0| < \varepsilon$. Then a value for $\vec{P}_1(t_0)$ can be assigned by completing $\vec{P}_1(t)$ by continuity. On the other hand, if $\kappa_1(t) = 0$ over an interval of t, then for the points in this interval, the vector function $\vec{P}_1(t)$ is undefined.

As with curves in $\mathbb{R}^3$, assuming the derivatives exist, we have

$$\vec{X}'(t) = s'(t)\vec{T}(t)$$
$$\vec{X}''(t) = s''(t)\vec{T}(t) + s'(t)^2\kappa_1(t)\vec{P}_1(t).$$

The vector function $\vec{P}_1(t)$ is called the *first normal* or first principal normal vector function. As we consider higher derivatives of $\vec{X}(t)$ in reference to this type of decomposition, we need to make sense of the derivative of $\vec{P}_1(t)$ and the derivatives of subsequent vector functions that arise in a similar way.

Now if $\vec{Y}(t)$ and $\vec{Z}(t)$ are unit vector functions with the same domain interval I and are everywhere perpendicular to each other, then for all $t \in I$,

$$\vec{Y}(t) \cdot \vec{Z}(t) = 0 \Longrightarrow \vec{Y}'(t) \cdot \vec{Z}(t) + \vec{Y}(t) \cdot \vec{Z}'(t) = 0$$
$$\Longrightarrow \vec{Y}'(t) \cdot \vec{Z}(t) = -\vec{Y}(t) \cdot \vec{Z}'(t). \qquad (9.2)$$

Now consider the vector function $\vec{P}_1'(t)$. From Equation (9.1), we know that $\vec{P}_1'(t) \cdot \vec{P}_1(t) = 0$ and from (9.2), we have $\vec{P}_1'(t) \cdot \vec{T}(t) = -s'(t)\kappa_1(t)$. If $n > 3$, we now define the *second curvature* of $\vec{X}(t)$ as the unique nonnegative function $\kappa_2 : I \to \mathbb{R}^{\geq 0}$ such that

$$\vec{P}_1'(t) = -s'(t)\kappa_1(t)\vec{T}(t) + s'(t)\kappa_2(t)\vec{P}_2(t),$$

for some unit vector function $\vec{P}_2(t)$ that is perpendicular to both $\vec{T}$ and $\vec{P}_1$. As long as $\kappa_2(t)$ is not zero, $\vec{P}_2(t)$ is well-defined. It is called the second normal or second principal normal vector function.

By repeating this process, we recursively define the ith curvature function and the associated ith unit normal vector function, for $1 \leq i \leq n-1$. However, an exception is made for the $(n-1)$th "curvature" and normal vector. In defining $\vec{T}$, $\vec{P}_1$, up to $\vec{P}_{n-1}$ as above, we establish an orthonormal set of vectors (unit length and mutually perpendicular). By construction, the n-tuple of vectors organized into a matrix as

$$A(t) = \begin{pmatrix} \vec{T}(t) & \vec{P}_1(t) & \cdots & \vec{P}_{n-1}(t) \end{pmatrix} \qquad (9.3)$$

defines an orthogonal matrix. Orthogonal matrices can have a determinant that is 1 or -1. Knowing the vectors $\vec{T}$ through $\vec{P}_{n-2}$ along with the condition that $A(t)$ be orthogonal leaves exactly two possibilities for $\vec{P}_{n-1}$. For calculations and other reasons, it is desirable that $A(t)$ be a positive orthogonal matrix, i.e., satisfy $\det A(t) = 1$. Thus, with the requirement that $\det A(t) = 1$, the $(n-1)$th normal vector $\vec{P}_{n-1}$ is uniquely determined from $\vec{T}$ through $\vec{P}_{n-2}$. However, because we use $\vec{T}$ through $\vec{P}_{n-2}$ to determine $\vec{P}_{n-1}(t)$, then we cannot define the $(n-1)$th curvature function $\kappa_{n-1}(t)$ to be a nonnegative function.

Definition 9.1.1 Let $\vec{X} : I \to \mathbb{R}^n$ be a parametrized curve. The n-tuple of vector functions $(\vec{T}(t), \vec{P}_1(t), \cdots, \vec{P}_{n-1}(t))$ as constructed above is called the *Frenet frame* of $\vec{X}$.

As with Equation (3.5), we can summarize the derivatives of the Frenet frame vectors by

$$\frac{d}{dt}A(t) = A(t)\begin{pmatrix} 0 & -s'\kappa_1 & 0 & \cdots & 0 & 0 \\ s'\kappa_1 & 0 & -s'\kappa_2 & \cdots & 0 & 0 \\ 0 & s'\kappa_2 & 0 & \cdots & 0 & 0 \\ \vdots & \vdots & \vdots & \ddots & \vdots & \vdots \\ 0 & 0 & 0 & \cdots & 0 & -s'\kappa_{n-1} \\ 0 & 0 & 0 & \cdots & s'\kappa_{n-1} & 0 \end{pmatrix}.$$

where $A(t)$ is the positive orthogonal matrix defined in Equation (9.3). See [17] for the original introduction to the notion of higher curvatures of curves in $\mathbb{R}^n$.

Equation (3.9) for a formula of the curvature of a space curve relied on taking the cross product of two vectors. However, this is not a valid operation in $\mathbb{R}^n$ for $n > 3$. Consequently, we need another formula for $\kappa_1(t)$. Note that as before, we have

$$\vec{X}' = s'\vec{T}$$
$$\vec{X}'' = s''\vec{T} + (s')^2\kappa_1\vec{P}_1.$$

The vector formula of Problem 3.1.6 shows

$$\|\vec{a} \times \vec{b}\|^2 = \begin{vmatrix} \vec{a}\cdot\vec{a} & \vec{a}\cdot\vec{b} \\ \vec{b}\cdot\vec{a} & \vec{b}\cdot\vec{b} \end{vmatrix}.$$

Since Formula (3.9) involves $\|\vec{X}' \times \vec{X}''\|$, we are inspired to consider the following calculation:

$$\begin{vmatrix} \vec{X}'\cdot\vec{X}' & \vec{X}'\cdot\vec{X}'' \\ \vec{X}''\cdot\vec{X}' & \vec{X}''\cdot\vec{X}'' \end{vmatrix} = \begin{vmatrix} (s')^2 & s's'' \\ s's'' & (s'')^2 + (s')^4\kappa_1^2 \end{vmatrix}$$
$$= (s')^2(s'')^2 + (s')^6\kappa_1^2 - (s')^2(s'')^2$$
$$= (s')^6\kappa_1^2.$$

This leads to the following formula for first curvature:

$$\kappa_1(t) = \frac{1}{\|\vec{X}'\|^3}\sqrt{\begin{vmatrix} \vec{X}'\cdot\vec{X}' & \vec{X}'\cdot\vec{X}'' \\ \vec{X}''\cdot\vec{X}' & \vec{X}''\cdot\vec{X}'' \end{vmatrix}} = \frac{1}{\|\vec{X}'\|^3}\sqrt{\det(B(t)^\top B(t))},$$

where $B(t)$ is the $n \times 2$ matrix $\begin{pmatrix} \vec{X}'(t) & \vec{X}''(t) \end{pmatrix}$. In order to prove the more general formula, we must use the Cauchy-Binet formula.

Theorem 9.1.2 (Cauchy-Binet) *Define $[m] = \{1, 2, \ldots, m\}$. If M is an $n \times m$ matrix and $S \subseteq [n]$ and $S' \subseteq [m]$, use the notation $M_{S,S'}$ to mean the submatrix consisting of the entries from the* *rows taken from the set S with columns taken from the set S'. If A is an $m \times n$ matrix and B an $n \times m$ matrix with $m \leq n$, then*

$$\det(AB) = \sum_{\substack{S \subseteq [n] \\ |S|=m}} \det(A_{[m],S}) \det(B_{S,[m]}).$$

Proof: (A proof can be found in the appendices of [24].) □

Proposition 9.1.3 *Let $\vec{X} : I \to \mathbb{R}^n$ be a regular parametrization of class C^n of a curve in $\mathbb{R}^n$ for $n \geq 2$. Define the matrix $B_m(t) = \begin{pmatrix} \vec{X}' & \vec{X}'' & \cdots & \vec{X}^{(m)} \end{pmatrix}$ for any integer m with $1 \leq m \leq n$. Then if $1 \leq m \leq n-2$,*

$$s'(t)^{(m+1)(m+2)/2} \kappa_1(t)^m \kappa_2(t)^{m-1} \cdots \kappa_m(t)$$
$$= \sqrt{\det(B_{m+1}(t)^\top B_{m+1}(t))}.$$

and for the $n-1$ curvature function $\kappa_{n-1}(t)$, we have

$$s'(t)^{n(n+1)/2} \kappa_1(t)^{n-1} \kappa_2(t)^{n-2} \cdots \kappa_{n-1}(t) = \det(B_n(t)).$$

Proof: Since the standard frame and the Frenet frame are orthonormal bases, there exists an orthogonal matrix $M(t)$ that gives the transition matrix from coordinates in the Frenet frame to standard coordinates. By the definition of the Frenet frame, $B_m(t) = M(t)D_m(t)$ where $D_m(t)$ is an $n \times m$ upper triangular matrix (in the sense that entries are 0 below the main diagonal). For example, if $n = 4$ then, using Equations (3.6), (3.7), and (3.10) as generalized to $\mathbb{R}^4$, we have

$$D_2 = \begin{pmatrix} s' & s'' \\ 0 & (s')^2 \kappa_1 \\ 0 & 0 \\ 0 & 0 \end{pmatrix} \quad D_3 = \begin{pmatrix} s' & s'' & s''' - (s')^3 \kappa_1^2 \\ 0 & (s')^2 \kappa_1 & 3s''(s')^2 \kappa_1 + (s')^2 \kappa_1' \\ 0 & 0 & (s')^3 \kappa_1 \kappa_2 \\ 0 & 0 & 0 \end{pmatrix}.$$

By the manner in which we defined the curvatures, we can see that the $(1,1)$ entry of $D_m(t)$ is s' and that the jth diagonal entry of $D_m(t)$ is $\vec{X}^{(j)} \cdot \vec{P}_{j-1}$. By repeatedly using the orthonormality property of the Frenet frame, we obtain

$$\vec{X}^{(j)} \cdot \vec{P}_{j-1} = s' \vec{T}^{(j-1)} \cdot \vec{P}_{j-1}$$
$$= (s')^2 \kappa_1 \vec{P}_1^{(j-2)} \cdot \vec{P}_{j-1}$$
$$= (s')^3 \kappa_1 \kappa_2 \vec{P}_2^{(j-3)} \cdot \vec{P}_{j-1}$$
$$= \vdots$$
$$= (s')^j \kappa_1 \kappa_2 \cdots \kappa_{j-1}.$$

Since $M(t)$ is an orthogonal matrix

$$B_m(t)^\top B_m(t) = D_m(t)^\top M(t)^\top M(t) D_m(t) = D_m(t)^\top D_m(t).$$

Writing simply D for $D_m(t)$, by the Cauchy-Binet Theorem,

$$\det(D^\top D) = \sum_{\substack{S \subseteq [n] \\ |S|=m}} \det((D_{S,[m]})^\top) \det(D_{S,[m]}) = \sum_{\substack{S \subseteq [n] \\ |S|=m}} \det(D_{S,[m]})^2.$$

However, since D is upper triangular, if $S \neq [m]$, then $D_{S,[m]}$ contains a row of 0s and hence its determinant vanishes. Thus $\det(D^\top D) = \det(D_{[m],[m]})^2$, which is the square of the product of its diagonal elements. Having determined the diagonal elements above, and since all the functions involved are nonnegative,

$$\sqrt{\det(B_m(t)^\top B_m(t))} = (s')^{\sum_{j=1}^m j} \kappa_1^{m-1} \kappa_2^{m-2} \cdots \kappa_{m-1}$$
$$= (s')^{m(m+1)/2} \kappa_1^{m-1} \kappa_2^{m-2} \cdots \kappa_{m-1}.$$

The first part of the proposition follows.

We cannot use the same recursive formula for κ_{n-1} simply because $\kappa_{n-1}(t)$ is not necessarily a nonnegative function. However, in this case $B_n(t)$ is a square matrix. Also, $B_n(t) = M(t) D_n(t)$ and since $M(t)$ is orthogonal $\det(B_n(t)) = \det(D_n(t))$. However, $D_n(t)$ is upper triangular so its determinant is the product of its diagonal elements. Thus

$$\det(B_n(t)) = (s')^{n(n+1)/2} \kappa_1^{n-1} \kappa_2^{n-2} \cdots \kappa_{n-1}.$$

The theorem follows.   □

Proposition 9.1.3 gives a recursive formula for the higher curvatures and consequently it is a straightforward exercise, though sometimes tedious to calculate the curvature functions of a parametrized curve in $\mathbb{R}^n$. The reader should also note that Proposition 9.1.3 directly generalizes the formula for curvature of a plane curve (Equation 1.12), the formula for the curvature of a space curve (Equation 3.9), and the formula for the torsion of a space curve (Equation 3.11).

 Example 9.1.4 We illustrate Proposition 9.1.3 with a curve that generalizes the twisted cubic to $\mathbb{R}^4$: the curve parametrized by

$$\vec{X}(t) = (t, t^2, t^3, t^4) \qquad \text{for } t \in \mathbb{R}.$$

Some of the calculations are onerous so we do not show every step. (In fact, for some of these calculations we would encourage the reader to use the assistance of a computer algebra system.) Note that

$$B_4(t) = \begin{pmatrix} 1 & 0 & 0 & 0 \\ 2t & 2 & 0 & 0 \\ 3t^2 & 6t & 6 & 0 \\ 4t^3 & 12t^2 & 24t & 24 \end{pmatrix}$$

and that for $m = 1, 2, 3$, the matrix of functions $B_m(t)$ consists of the first m columns of $B_4(t)$. It is easy to see that we have

$$s'(t) = \sqrt{B_1(t)^\top B_1(t)} = \sqrt{1 + 4t^2 + 9t^4 + 16t^6}.$$

For the first curvature, we calculate that $\det(B_2(t)^\top B_2(t)) = 144t^8 + 256t^6 + 180t^4 + 36t^2 + 4$, so

$$\kappa_1(t) = \frac{\sqrt{\det(B_2(t)^\top B_2(t))}}{s'(t)^3} = \frac{\sqrt{144t^8 + 256t^6 + 180t^4 + 36t^2 + 4}}{(1 + 4t^2 + 9t^4 + 16t^6)^{3/2}}.$$

For the second curvature, we calculate $\det(B_3(t)^\top B_3(t)) = 2304t^6 + 5184t^4 + 2304t^2 + 144$. Proposition 9.1.3 states that

$$(s')^6 \kappa_1^2 \kappa_2 = ((s')^3 \kappa_1)^2 \kappa_2 = \sqrt{\det(B_3(t)^\top B_3(t))}.$$

After simplification, we get

$$\kappa_2(t) = \frac{\sqrt{2304t^6 + 5184t^4 + 2304t^2 + 144}}{144t^8 + 256t^6 + 180t^4 + 36t^2 + 4}.$$

For the third curvature, it is easy to see that $\det(B_4(t)) = 288$. The recursive formula of Proposition 9.1.3 states that $(s')^{10} \kappa_1^3 \kappa_2^2 \kappa_3 = \det(B_4(t))$. After simplification, we obtain

$$\kappa_3(t) = \frac{288\sqrt{144t^8 + 256t^6 + 180t^4 + 36t^2 + 4}}{(2304t^6 + 5184t^4 + 2304t^2 + 144)\sqrt{1 + 4t^2 + 9t^4 + 16t^6}}.$$

Proposition 9.1.3 gives a recursive formula to find the higher curvatures of a curve in $\mathbb{R}^n$. After doing a few calculations, we may notice how many common terms cancel. This is not a coincidence. The following proposition deduces from the recursive formula a non-recursive formula for higher curvatures.

Proposition 9.1.5 *Let $\vec{X} : I \to \mathbb{R}^n$ be a regular parametrization of class C^n of a curve in $\mathbb{R}^n$ for $n \geq 2$. Define the functions $c_m(t)$ as*

$$c_m(t) = \begin{cases} 1 & \text{if } m = 0 \\ \sqrt{\det(B_m(t)^\top B_m(t))} & \text{if } 1 \leq m \leq n - 1 \\ \det(B_n(t)) & \text{if } m = n. \end{cases}$$

Then $s'(t)\kappa_m(t) = \dfrac{c_{m+1}(t)c_{m-1}(t)}{c_m(t)^2}$ for all m with $1 \leq m \leq n-1$.

Proof: (Left as an exercise for the reader. See Problem 9.1.6.) □

9.1.2 Osculating Planes, Circles, and Spheres

Following the local theory of plane curves and space curves as developed earlier in the text, for curves in $\mathbb{R}^n$ we can consider osculating k-planes or osculating k-spheres, where k is any integer with $1 \leq k \leq n-1$.

Definition 9.1.6 Let $\vec{X} : I \to \mathbb{R}^n$ be a regular parametrized curve of class C^n. The *osculating k-plane* to $\vec{X}$ at $t = t_0$ is the k-plane through the point $\vec{X}(t_0)$ and spanned by $\{\vec{T}(t_0), \vec{P}_1(t_0), \ldots, \vec{P}_{k-1}(t_0)\}$.

Note that the osculating line to a curve at a point is simply the tangent line.

We call a k-sphere in $\mathbb{R}^n$, any k-dimensional sphere in a $(k+1)$ dimensional plane in $\mathbb{R}^n$. Note that in this indexing, a circle is a 1-sphere (in a 2-plane), a usual sphere is a 2-sphere (in a 3-plane), and so forth. We leave as exercises to the reader formulas for osculating k-spheres.

9.1.3 The Fundamental Theorem of Curves in $\mathbb{R}^n$

The Fundamental Theorem of Space Curves immediately generalizes to an n dimensional case. The proof of this generalization involves no new strategy so we omit the proof.

Theorem 9.1.7 (Fundamental Theorem of Curves)   *Given curvature functions $\kappa_i(s) \geq 0$, for $1 \leq i \leq n-2$ and $\kappa_{n-1}(s)$ continuously differentiable over some interval $J \subseteq \mathbb{R}$ containing 0, there exists an open interval I containing 0 and a regular vector function $\vec{X} : I \to \mathbb{R}^n$ that parametrizes its locus by arc length, with $\kappa_i(s)$ as the ith curvature functions. Furthermore, any two curves C_1 and C_2 with the same $(n-1)$-tuple of curvature functions can be mapped onto one another by a rigid motion of $\mathbb{R}^n$.*

For the same reason as in the case of space curves, we call the $(n-1)$-tuple of curvature functions $(\kappa_1(s), \kappa_2(s), \ldots, \kappa_{n-1}(s))$ the *natural equations* of the curve in $\mathbb{R}^n$.

PROBLEMS

1. Clearly show all the simplifications in Example 9.1.4.

2. Let m and n be positive integers and let r_1, r_2 be positive real numbers. Show by direct computation that all three curvature functions $\kappa_1(t)$, $\kappa_2(t)$, and $\kappa_3(t)$ associated to the regular curve $\vec{X}(t) = (r_1 \cos(mt), r_1 \sin(mt), r_2 \cos(nt), r_2 \sin(nt))$ are constant.

3. Find the curvature functions $\kappa_1, \kappa_2, \kappa_3$ for the curve parametrized by $\vec{X}(t) = (t, t^2, \cos t, \sin t)$.

4. Find the curvature functions $\kappa_1, \kappa_2, \kappa_3$ for the curve parametrized by $\vec{X}(t) = (t, \cos t, \sin t, \sin 2t)$. [Hint: Use a computer algebra system if necessary. Write all your answers as functions of $\sin(2t)$.]

5. Consider regular curves in $\mathbb{R}^n$. Use Proposition 9.1.3 to show that the curvature functions are invariant under a positively oriented reparametrization. Prove that under a negatively oriented reparametrization the curvatures $\kappa_1(t), \ldots, \kappa_{n-2}(t)$ remain invariant and that the $n-1$ curvature changes according to $\bar{\kappa}_{n-1}(t) = (-1)^n \kappa_{n-1}(t)$.

6. Prove Proposition 9.1.5. [Hint: Use strong induction.]

7. Let $\vec{X}(t) = (p_1(t), p_2(t), \ldots, p_n(t))$ be the parametrization of a curve in $\mathbb{R}^n$, where $p_i(t)$ are polynomials in t. Let d be the maximum degree of all the polynomials $p_i(t)$. Prove that the higher curvature functions $\kappa_m(t)$ are identically 0 for all $m \geq d$.

8. Proposition 9.1.3 defines the curvatures in a recursive way but does not show any simplifications if such exist. Show an explicit formula for $\kappa_3(t)$ for curves in $\mathbb{R}^n$, first when $n = 4$ and then for $n \geq 5$.

9. Calculate the osculating 2-sphere to $\vec{X}(t) = (t, t^2, t^3, t^4)$ in $\mathbb{R}^4$ at $t = 0$. How is this similar or different from the osculating 2-sphere to $\vec{X}(t) = (t, t^2, t^3)$ in $\mathbb{R}^3$? [See Problem 3.3.4.]

10. Follow the ideas in Section 3.3 to find the center and the radius of the osculating 3-sphere to a curve in $\mathbb{R}^4$. [Note that the osculating 3-sphere to a curve C at a point P will be a 3-sphere with contact of order 4 to C at P.]

11. Let $A(t)$ and $B(t)$ be an $n \times n$ matrix of differentiable real-valued functions defined in an interval $I \subseteq \mathbb{R}$. If $A(t) = (a_{ij}(t))$, then denote by $A'(t)$ the $n \times n$ matrix $(a'_{ij}(t))$. Prove that:

    (a) $\frac{d}{dt}(A(t)^\top) = (A'(t))^\top$;

    (b) $\frac{d}{dt}(A(t) + B(t)) = A'(t) + B'(t)$;

    (c) $\frac{d}{dt}(A(t)B(t)) = A'(t)B(t) + A(t)B'(t)$;

    (d) if $A(t)$ is invertible for all $t \in I$, then $\frac{d}{dt}(A(t)^{-1}) = -A(t)^{-1}A'(t)A(t)^{-1}$.

12. Suppose that $A(t)$ is an $n \times n$ matrix of differentiable real-valued functions defined in an interval $I \subseteq \mathbb{R}$ that is everywhere orthogonal, i.e. $A(t)^\top A(t) = I$. Use the previous exercise to prove that $A'(t) = A(t)M(t)$, where $M(t)$ is an antisymmetric matrix $(M(t)^\top = -M(t))$.

13. Prove that a curve in $\mathbb{R}^n$ lies in a k-dimensional "plane" (a subspace of $\mathbb{R}^n$ of the form $\vec{p} + W$, where W is a k-dimensional subspace) if and only if $\kappa_k(s) = \cdots = \kappa_{n-1}(s) = 0$.

9.2   Surfaces in Euclidean n-Space

9.2.1   Regular Surfaces in $\mathbb{R}^n$

In Chapter 5, we introduced the concept of a regular surface in $\mathbb{R}^3$. As an overarching intuition, a regular surface in $\mathbb{R}^3$ is a set of points S obtained from a parametrization such that at each point $p \in S$ the set of tangent vectors forms a plane. In order to make all the notions in the previous sentence precise we needed to discuss parametrizations of surfaces, which in turn made sense of tangent vectors to surfaces at a point, and address the conditions under which the set of tangent vectors forms a tangent plane. This culminated in Definition 5.3.1.

Subsequent to establishing a workable definition for a regular surface, we proceeded to define orientability, the metric tensor, and then a variety of intrinsic and extrinsic properties of surfaces. Our investigations culminated in the Gauss-Bonnet Theorem, from which we gave a variety of applications.

Much of the presentation given for regular surfaces in $\mathbb{R}^3$ easily generalizes to define regular surfaces in $\mathbb{R}^n$ with $n > 3$. There is one main exception: the relevance of normality. In particular, given any two-dimensional plane of vectors in $\mathbb{R}^n$, the set of vectors perpendicular to the plane has dimension $n - 2$. So if $n > 3$, the vector space of perpendicular (normal) vectors to a plane is not one-dimensional. Consequently, though we have a dot product in $\mathbb{R}^n$, we do not have a cross-product with the usual algebraic properties in $\mathbb{R}^n$ if $n \geq 4$.

In Proposition 5.2.4, we proved that a parametrized surface $\vec{X} : U \rightarrow \mathbb{R}^3$, where U is an open set in $\mathbb{R}^2$, has a tangent plane at p if $\vec{X}^{-1}(p)$ is a single point $q = (u_0, v_0) \in U$ and $\vec{X}_u(u_0, v_0) \times \vec{X}_v(u_0, v_0) \neq \vec{0}$. Subsequently, we gave an alternate characterization which led to the definition of a regular surface, Definition 5.3.1. Happily, that alternate characterization generalizes immediately.

Definition 9.2.1 A subset $S \subseteq \mathbb{R}^n$ is a *regular surface* if for each $p \in S$, there is an open set $U \subseteq \mathbb{R}^2$, an open neighborhood V of p in $\mathbb{R}^n$, and a surjective continuous function $\vec{X} : U \rightarrow V \cap S$ such that

1) $\vec{X}$ is continuously differentiable: if we write the parametrization as $\vec{X}(u,v) = (x_1(u,v), x_2(u,v), , \ldots, x_n(u,v))$, then for $i = 1, 2, \ldots, n$, the functions $x_i(u,v)$ have continuous partial derivatives with respect to u and v;

2) $\vec{X}$ is a homeomorphism: $\vec{X}$ is continuous and has an inverse $\vec{X}^{-1} : V \cap S \rightarrow U$ such that $\vec{X}^{-1}$ is continuous;

3) $\vec{X}$ satisfies the regularity condition: for each $(u,v) \in U$, the differential $d\vec{X}_{(u,v)} : \mathbb{R}^2 \rightarrow \mathbb{R}^n$ is a one-to-one linear transformation.

Furthermore, a regular surface is said to be of class C^n (resp. C^∞) if each function $\vec{X} : U \to \mathbb{R}^n$ has continuous n derivatives (resp. continuous derivatives of all orders).

The regularity condition can be restated to say that $\vec{X}_u(u,v)$ and $\vec{X}_u(u,v)$ are linearly dependent over the domain.

The concepts of a regular, positively oriented, or negatively oriented reparametrization of a surface carry over in an identical fashion from surfaces in $\mathbb{R}^3$ to surfaces in $\mathbb{R}^n$.

9.2.2 Intrinsic Geometry for Surfaces

Recall that intrinsic geometry is a property of surfaces that depend entirely on the metric coefficients (and the partial derivatives thereof) of the regular. Since intrinsic properties do not depend on a normal vector to a surface at a point, then such properties apply to any surface as long as we have a metric tensor.

For a regular surface in $\mathbb{R}^n$, we define the first fundamental form in precisely the same way as we did for surfaces in $\mathbb{R}^3$. The first fundamental form $I_p(\_,\_)$ to a regular surface S at p is the inner product on T_pS obtained as the restriction of the dot product in $\mathbb{R}^n$ to the tangent plane T_pS.

Suppose that a coordinate patch of S has a regular parametrization $\vec{X} : U \to \mathbb{R}^n$. Suppose also that $p = \vec{X}(u_0, v_0)$ with $(u_0, v_0) \in U$ is a point on the surface. Consider the standard ordered basis $\mathcal{B} = (\vec{X}_u(u_0, v_0), \vec{X}_v(u_0, v_0))$ of the tangent space T_pS. Suppose also that two vectors $\vec{a}, \vec{b} \in T_p(S)$ are expressed with the following components with respect to $\mathcal{B}$,

$$[\vec{a}]_{\mathcal{B}} = \begin{pmatrix} a_1 \\ a_2 \end{pmatrix} \quad \text{and} \quad [\vec{b}]_{\mathcal{B}} = \begin{pmatrix} b_1 \\ b_2 \end{pmatrix}.$$

Then the first fundamental form has

$$\begin{aligned} I_p(\vec{a}, \vec{b}) &= \vec{a} \cdot \vec{b} \\ &= (a_1 \vec{X}_u(u_0, v_0) + a_2 \vec{X}(u_0, v_0)) \cdot (b_1 \vec{X}_u(u_0, v_0) + b_2 \vec{X}(u_0, v_0)) \\ &= \begin{pmatrix} a_1 & a_2 \end{pmatrix} \begin{pmatrix} \vec{X}_u(u_0, v_0) \cdot \vec{X}_u(u_0, v_0) & \vec{X}_u(u_0, v_0) \cdot \vec{X}_v(u_0, v_0) \\ \vec{X}_v(u_0, v_0) \cdot \vec{X}_u(u_0, v_0) & \vec{X}_v(u_0, v_0) \cdot \vec{X}_v(u_0, v_0) \end{pmatrix} \begin{pmatrix} b_1 \\ b_2 \end{pmatrix} \\ &= \begin{pmatrix} a_1 & a_2 \end{pmatrix} g(u_0, v_0) \begin{pmatrix} b_1 \\ b_2 \end{pmatrix}, \end{aligned}$$

where we define the metric tensor $g = (g_{ij})$ as the matrix of functions defined on $U \subseteq \mathbb{R}^2$

$$g(u, v) = \begin{pmatrix} \vec{X}_u(u, v) \cdot \vec{X}_u(u, v) & \vec{X}_u(u, v) \cdot \vec{X}_v(u, v) \\ \vec{X}_v(u, v) \cdot \vec{X}_u(u, v) & \vec{X}_v(u, v) \cdot \vec{X}_v(u, v) \end{pmatrix}.$$

Since the metric tensor depends on dot products, which exist in any $\mathbb{R}^n$, the usual definition of the metric tensor remains unchanged.

Equations (6.1) and (6.2) for the arc length of a curve on a regular surface remain identical for surfaces in $\mathbb{R}^n$ with exactly the same proof. This is also true for Equation (6.3) to calculate angles between tangent vectors. However, Proposition 6.1.7 concerning the area is still true but its original proof referred to the cross product of $\vec{X}_u \times \vec{X}_v$ in order to find the area element. We need to prove the proposition concerning area but without reference to the cross product or its magnitude.

Proposition 9.2.2 *Let S be a regular surface in $\mathbb{R}^n$, with $n \geq 3$, and let $\vec{X} : U \to \mathbb{R}^n$ be a parametrization of a coordinate patch of S, where U is an open subset of $\mathbb{R}^2$. Let Q be a compact (closed and bounded) subset of U and call $\mathcal{R} = \vec{X}(Q)$ the corresponding region on S. The area of $\mathcal{R}$ is given by*

$$A(\mathcal{R}) = \iint\limits_Q \sqrt{\det(g)}\, du\, dv. \tag{9.4}$$

Proof: The proof of Equation (6.4) relied on the fact that S is a regular surface in $\mathbb{R}^3$ and made reference to the cross product of $\vec{X}_u$ and $\vec{X}_v$, namely, using the formula

$$\iint\limits_Q \|\vec{X}_u \times \vec{X}_v\|\, du\, dv$$

from multivariable calculus. The Riemann sum behind this integral involves approximating the surface area traced out by $\vec{X}$ over $[u_i, u_{i+1}] \times [v_i, v_{i+1}]$ by a parallelogram spanned on two sides by $\vec{X}_u(u^*, v^*)$ and $\vec{X}_v(u^*, v^*)$, where (u^*, v^*) is a selection point in the subrectangle $[u_i, u_{i+1}] \times [v_i, v_{i+1}]$. The approach used by the Riemann sum is still correct but we must find a convenient formula for the area of the parallelogram spanned by $\vec{X}_u$ and $\vec{X}_v$ in $\mathbb{R}^n$.

Regardless of the dimension of the ambient Euclidean space, we can calculate the area of these parallelograms as $\|\vec{X}_u\| \|\vec{X}_v\| \sin\theta$, where θ is the angle between the two vectors. From properties of the dot product, we can obtain $\cos\theta$ from $\vec{X}_u \cdot \vec{X}_v = \|\vec{X}_u\| \|\vec{X}_v\| \cos\theta$. Thus,

$$(\vec{X}_u \cdot \vec{X}_v)^2 = (\vec{X}_u \cdot \vec{X}_u)(\vec{X}_v \cdot \vec{X}_v) \cos^2\theta$$
$$\Longrightarrow (\vec{X}_u \cdot \vec{X}_v)^2 = (\vec{X}_u \cdot \vec{X}_u)(\vec{X}_v \cdot \vec{X}_v)(1 - \sin^2\theta)$$
$$\Longrightarrow (\vec{X}_u \cdot \vec{X}_u)(\vec{X}_v \cdot \vec{X}_v) - (\vec{X}_u \cdot \vec{X}_v)^2 = (\vec{X}_u \cdot \vec{X}_u)(\vec{X}_v \cdot \vec{X}_v) \sin^2\theta$$
$$\Longleftrightarrow \sqrt{(\vec{X}_u \cdot \vec{X}_u)(\vec{X}_v \cdot \vec{X}_v) - (\vec{X}_u \cdot \vec{X}_v)^2} = \|\vec{X}_u\| \|\vec{X}_v\| \sin\theta.$$

Thus, at each point $(u, v) \in U$, we have $\sqrt{\det(g)} = \|\vec{X}_u\| \|\vec{X}_v\| \sin \theta$. Taking the limit of the Riemann sum as the norm of a mesh that goes to 0 gives Equation (9.4).   □

Another proof for the above result relies on the geometric fact that the area of a parallelogram spanned by $\vec{a}, \vec{b} \in \mathbb{R}^n$ is

$$\sqrt{\det(C^\top C)} = \sqrt{\begin{vmatrix} \vec{a} \cdot \vec{a} & \vec{a} \cdot \vec{b} \\ \vec{b} \cdot \vec{a} & \vec{b} \cdot \vec{b} \end{vmatrix}},$$

where C is the $n \times 2$ matrix $C = \begin{pmatrix} \vec{a} & \vec{b} \end{pmatrix}$ with the vectors $\vec{a}$ and $\vec{b}$ (expressed in standard coordinates) as columns.

In Chapter 7, we introduced the Christoffel symbols and used them to prove the Theorema Egregium, which affirms that the Gaussian curvature is an intrinsic property. In particular, Christoffel symbols and the Gaussian curvature can be computed for regular surfaces of class C^2 in $\mathbb{R}^n$ in precisely the same way that we computed them in Chapter 7.

Example 9.2.3 (Flat Torus) The topological definition of a torus is any set that is homeomorphic to $\mathbb{S}^1 \times \mathbb{S}^1$, where $\mathbb{S}^1$ is a circle. In other parts of this book, we discuss the torus in $\mathbb{R}^3$ given by the parametrization

$$\vec{X}(u, v) = ((b + a \cos v) \cos u, (b + a \cos v) \sin u, \sin v)$$

for $(u, v) \in [0, 2\pi] \times [0, 2\pi]$, and where $0 < a < b$. This torus has a non-identity metric tensor and in Example 6.4.5, we saw that this torus has elliptic points whenever $0 < v < \pi$, hyperbolic points when $\pi < v < 2\pi$, and parabolic points when v is a multiple of π. In particular, these are regions where the Gaussian curvature is respectively positive, negative, and zero.

Consider the following parametrization of the torus in $\mathbb{R}^4$:

$$\vec{X}(u, v) = (\cos u, \sin u, \cos v, \sin v). \tag{9.5}$$

The set traced out in $\mathbb{R}^4$ by the function $\vec{X}$ over $[0, 2\pi] \times [0, 2\pi]$ satisfies the topological definition of a torus. It is an easy exercise (Problem 9.2.3) to prove that the metric tensor is constant with

$$g_{ij}(u, v) = \begin{pmatrix} 1 & 0 \\ 0 & 1 \end{pmatrix}.$$

This is the same metric tensor as a plane equipped with an orthonormal basis. Furthermore, it is obvious that the Christoffel symbols of the first kind will be identically 0, since they involve partial derivatives of the g_{ij}, and hence the Christoffel symbols of the second kind and the Gaussian curvature will be 0.

This surface in $\mathbb{R}^n$ is called the *flat torus*. Speaking intuitively, we can say that there is "enough room" in $\mathbb{R}^4$ to embed a torus by bending but without stretching. A common way to imagine creating a torus in $\mathbb{R}^3$, is to take a piece of paper and first roll it into a cylinder as shown below.

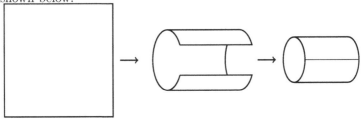

If the cylinder can be stretched, we can imagine bending it around so that the two circular ends meet up. This bending is required when we are in $\mathbb{R}^3$. In four dimensions, suppose that Π_1 and Π_2 are planes that meet orthogonally at one point P. Suppose also that there is a line segment L_1 in Π_1 and another line segment L_2 in Π_2 that are perpendicular to each other that also meet at P. Let Π be a plane parallel and unequal to the plane spanned by L_1 and L_2. We can first roll Π around L_1 along a circle that lies in Π_2 and the resulting cylinder can be rolled *without bending* around L_2 along a circle that lies in Π_2. The resulting torus never bends the plane Π.

We remind the reader that the formulas for the Gaussian curvature arise in Equations (7.16)-(7.18). In particular,

$$K = \frac{R_{1212}}{\det(g_{ij})} = \frac{1}{\det(g_{ij})} \left(R^1_{121}g_{12} + R^2_{121}g_{22} \right), \qquad (9.6)$$

where the symbols R^l_{ijk} are defined in Equation (7.16). The Riemann symbols satisfy certain symmetry relations which imply that $R_{1212} = R_{2121}$. Hence, we also have

$$K = \frac{R_{2121}}{\det(g_{ij})} = \frac{1}{\det(g_{ij})} \left(R^1_{121}g_{11} + R^2_{121}g_{21} \right).$$

In Section 8.1, we studied the Darboux frame associated to a curve on a surface. Because a surface in $\mathbb{R}^n$ with $n \geq 4$ does not have a one-dimensional normal space at a point, we do not have a natural $\vec{N}(t)$ function associated to a curve on a surface. Consequently, neither the normal curvature nor the geodesic torsion make sense. Surprisingly, the geodesic curvature does still make sense for curves on surfaces in $\mathbb{R}^n$. In Section 8.1.3, we arrived at Formula (8.7) by use of the normal vector $\vec{N}(t)$. Despite the intermediate steps, this formula only involves intrinsic quantities. When working in $\mathbb{R}^n$ we could take (8.7) as a definition for geodesic curvature and ignore how we arrived at it, but this is unsatisfactory. In Proposition 8.4.1, the same function $\kappa_g(t)$ arises in another (more complicated) way, one that makes no reference to a normal vector. Consequently, geodesic curvature has meaning for curves on surfaces in $\mathbb{R}^n$ and we can still use Formula (8.7).

Example 9.2.4 For an example illustrating the concepts we have presented so far, consider the surface S in $\mathbb{R}^4$ parametrized by

$$\vec{X}(u, v) = (\cos u \cosh v, \sin u \cosh v, \cos u \sinh v, \sin u \sinh v)$$

with $(u, v) \in [0, 2\pi] \times \mathbb{R}$. We point out that this surface is a subset of the quadratic hypersurface with equation $x^2 + y^2 - z^2 - w^2 = 1$. By way of an example, we will calculate the coefficients of the metric tensor, find the Christoffel symbols, and deduce the Gaussian curvature function. Taking derivatives, we find

$$\vec{X}_u(u, v) = (-\sin u \cosh v, \cos u \cosh v, -\sin u \sinh v, \cos u \sinh v)$$
$$\vec{X}_v(u, v) = (\cos u \sinh v, \sin u \sinh v, \cos u \cosh v, \sin u \cosh v).$$

By taking appropriate dot products (and using the identity $\cosh^2 v + \sinh^2 v = \cosh(2v)$), we have

$$(g_{ij}) = \begin{pmatrix} \cosh 2v & 0 \\ 0 & \cosh 2v \end{pmatrix}.$$

Using (7.2), the Christoffel symbols of the first kind are

$$[11, 1] = 0 \qquad [11, 2] = -\sinh 2v$$
$$[12, 1] = \sinh 2v \quad [12, 2] = 0$$
$$[22, 1] = 0 \qquad [22, 2] = \sinh 2v.$$

Using (7.3), the Christoffel symbols of the second kind are

$$\Gamma^1_{11} = 0 \qquad \Gamma^2_{11} = -\tanh 2v$$
$$\Gamma^1_{12} = \tanh 2v \quad \Gamma^2_{12} = 0$$
$$\Gamma^1_{22} = 0 \qquad \Gamma^2_{22} = \tanh 2v.$$

We can calculate the Gaussian curvature from (7.18), using (7.16) and (7.17). We have

$$R_{1212} = R^1_{121}g_{12} + R^2_{121}g_{22} = R^2_{121}g_{22},$$

where the last equality holds because $g_{12} = 0$ for this surface. We
have

$$R^2_{121} = \frac{\partial \Gamma^2_{11}}{\partial v} - \frac{\partial \Gamma^2_{21}}{\partial u} + \Gamma^1_{11}\Gamma^2_{12} + \Gamma^2_{11}\Gamma^2_{22} - \Gamma^1_{21}\Gamma^2_{11} - \Gamma^2_{21}\Gamma^2_{21}$$

$$= -2\,\text{sech}^2\,2v - 0 + 0 - \tanh^2 2v + \tanh^2 2v - 0$$

$$= -2\,\text{sech}^2\,2v.$$

Then

$$K = \frac{1}{\det(g)}R^2_{121}g_{22} = \frac{-2\,\text{sech}^2\,2v\cosh 2v}{\cosh^2 2v} = -2\,\text{sech}^3\,2v.$$

Using Formula (8.7), we can also determine the geodesic curvature of
a curve on this surface parametrized by $(u(t), v(t))$. We first have

$$s'(t) = \sqrt{\cosh(2v)((u')^2 + (v')^2)}$$

$$\text{and } \det(g) = \cosh^2(2v),$$

so the geodesic curvature is

$$\kappa_g(t) = \frac{-\tanh(2v)((u')^3 + u'(v')^2) + u'v'' - u''v'}{(\cosh(2v))^{1/2}((u')^2 + (v')^2)^{3/2}}.$$

From this calculation, we can easily see that u-coordinate lines, curves
with parametric functions (u_0, t) where u_0 is constant, have $\kappa_g = 0$, so
are geodesics. In contrast, v-coordinate lines, curves with parametric
functions (t, v_0) where v_0 is constant, have a geodesic curvature of

$$\kappa_g(t) = \frac{-\tanh(2v_0)}{\sqrt{\cosh(2v_0)}},$$

which is constant.

In the above example, as we calculated various functions, we did
not explicitly need the parametrization of S or its derivatives once we
had calculated the metric tensor components.

9.2.3   Orientability

The property of orientability becomes more challenging to define for
surfaces in $\mathbb{R}^n$ precisely because in $\mathbb{R}^n$ with $n > 3$, there does not
exist a one-dimensional normal line to a regular surface at a point.
Consequently, we cannot talk about a normal vector $\vec{N}(u, v)$. From
an intuitive perspective, we would like to say that a surface is orientable if it is possible to cover the surface continuously with a sense
of clockwise rotation.

Without a normal vector, we cannot refer to a positive rotation around a normal vector. Instead, we refer to the sign of a determinant. Recall that the definition of a regular surface implied that S is covered by a collection of coordinate patches. Recall that for any two coordinate patches U and $\bar{U}$ with coordinates (x^1, x^2) and $(\bar{x}^1, \bar{x}^2)$, the change of coordinates corresponds to a regular reparametrization of the surface. Consequently, the Jacobian

$$\frac{\partial(\bar{x}^1, \bar{x}^2)}{\partial(x^1, x^2)} = \begin{vmatrix} \dfrac{\partial \bar{x}^1}{\partial x^1} & \dfrac{\partial \bar{x}^1}{\partial x^2} \\ \dfrac{\partial \bar{x}^2}{\partial x^1} & \dfrac{\partial \bar{x}^2}{\partial x^2} \end{vmatrix}$$

is never zero. The sign of a 2×2 determinant is the sign of the angle between the first column and the second column vector. Suppose that surface is parametrized by $\vec{X} : U \to \mathbb{R}^n$ and also by $\vec{Y} : \bar{U} \to \mathbb{R}^n$, then over subsets of the domains that correspond to the intersection of $\vec{X}(U)$ and $\vec{Y}(\bar{U})$, the sign of the determinant corresponds to whether the angle from $\vec{X}_{x^1}$ to $\vec{X}_{x^2}$ has the same sign as the angle from $\vec{Y}_{\bar{x}^1}$ to $\vec{Y}_{\bar{x}^2}$. Following the criterion given in Proposition 5.4.11, we use this sign to define orientability.

Definition 9.2.5 A regular surface is called orientable if it is covered by a collection of coordinate patches such that for any two overlapping coordinate patches U and $\bar{U}$ with coordinates (x^1, x^2) and $(\bar{x}^1, \bar{x}^2)$, the Jacobian satisfies
$$\frac{\partial(\bar{x}^1, \bar{x}^2)}{\partial(x^1, x^2)} > 0.$$

Whenever this condition holds between two coordinate patches, we say they have a compatible orientation.

This condition is challenging to check for surfaces in general. However, for some surfaces that arise as the image of a single parametrized function, not necessarily a bijection, it may be easy to check if the surface is orientable. First, if a single parametrization $\vec{X} : U \to S$ provides the required homeomorphism between U and S, then the surface is automatically orientable.

Example 9.2.6 (Flat Torus) As a second example, consider the flat torus described in Example 9.2.3. The parametrization

$$\vec{X}(u, v) = (\cos u, \sin u, \cos v, \sin v)$$

with $(u, v) \in \mathbb{R}^2$ has the flat torus as its locus, though not bijectively. Now any point p on the flat torus has $p = \vec{X}(u_1, v_1) = \vec{X}(u_2, v_2)$, where $u_2 - u_1$ and $v_2 - v_1$ are integer multiples of 2π. However, we have

$$\vec{X}_u(u,v) = (-\sin u, \cos u, \sin v, \cos v) \quad \text{and}$$
$$\vec{X}_v(u,v) = (\cos u, \sin u, -\sin v, \cos v)$$

and for any two pairs (u_1, v_1) with (u_2, v_2) with $u_2 - u_1, v_2 - v_1 \in 2\pi\mathbb{Z}$, we have

$$\vec{X}_u(u_1, v_1) = \vec{X}_u(u_2, v_2) \quad \text{and} \quad \vec{X}_v(u_1, v_1) = \vec{X}_v(u_2, v_2).$$

We can cover the flat torus with two coordinate patches $U = (0, 2\pi) \times (0, 2\pi)$ and $\bar{U} = (\pi, 3\pi) \times (\pi, 3\pi)$, each using the same function $\vec{X}$. Over regions where these coordinate patches overlap, the Jacobian of the change of coordinates is constant. Hence the flat torus is orientable.

Example 9.2.7 (Klein Bottle)   An example of a non-orientable surface in $\mathbb{R}^4$ is the Klein bottle. This is similar to the torus but with a twist. Consider the parametrization $\vec{X} : [0, 2\pi] \times [0, 2\pi] \to \mathbb{R}^4$ defined by

$$\vec{X}(u,v) = \Big((b + a\cos v)\cos u, (b + a\cos v)\sin u,$$
$$b\sin v \cos\left(\frac{u}{2}\right), b\sin v \sin\left(\frac{u}{2}\right)\Big),$$

with $0 < a < b$. The first two components of this should remind the reader of the usual torus in $\mathbb{R}^3$. (See Problem 6.6.5.) However, the last two components are in the x^3 and x^4 direction and are both perpendicular to the internal ring of the torus $(b\cos u, b\sin u, 0, 0)$. In these components, we trace out a small ring as it goes around the larger internal ring, but making a half-turn twist in the process. It is not hard to see that the image of $\vec{X}$ over the domain $[0, 2\pi] \times [0, 2\pi]$ defines a regular surface without boundary. We leave it as an exercise to check that over the domain $[0, 2\pi) \times [0, 2\pi)$, the parametrized surface is a bijection.

Now consider the coordinate patch with this parametrization but over the domain $U = (0, 2\pi) \times (0, 2\pi)$. Consider the point

$$\vec{X}(0, \pi/2) = (b, 0, b, 0) = \vec{X}(2\pi, 3\pi/2),$$

which is not in $\vec{X}(U)$. We can calculate that

$$\vec{X}_u\left(0, \frac{\pi}{2}\right) = \left(0, b, 0, \frac{b}{2}\right) \quad \text{and} \quad \vec{X}_v\left(0, \frac{\pi}{2}\right) = (-a, 0, 0, 0)$$

and also that

$$\vec{X}_u\left(2\pi, \frac{3\pi}{2}\right) = \left(0, b, 0, \frac{b}{2}\right) \quad \text{and} \quad \vec{X}_v\left(2\pi, \frac{\pi}{2}\right) = (a, 0, 0, 0).$$

The pair of vectors $(\vec{X}_u(u, v), \vec{X}_v(u, v))$ vary continuously on the surface. Assume there does exist a cover of the Klein bottle using coordinate patches that make it an orientable surface. Then the Jacobian of the coordinate change with respect to the coordinates defined by $\vec{X}$ and that of any other patch must have the same sign. Consider a coordinate patch $\vec{Y} : \bar{U} \to S$ such that $p \in \vec{Y}(\bar{U})$ and that $\vec{Y}(\bar{U})$ is connected. The image $\vec{Y}(\bar{U}) \cap \vec{X}(U)$ consists of two connected components V_1 and V_2 with $V_1 \cap V_2 = \emptyset$. At p, the coordinate change matrix on T_pS over V_1 and the coordinate change matrix on T_pS over V_2 cannot have the same sign because $\vec{X}_u\left(0, \frac{\pi}{2}\right) = \vec{X}_u\left(2\pi, \frac{3\pi}{2}\right)$, whereas $\vec{X}_v\left(0, \frac{\pi}{2}\right) = -\vec{X}_v\left(2\pi, \frac{3\pi}{2}\right)$.

9.2.4 The Gauss-Bonnet Theorem

In Chapter 8, we approached the Gauss-Bonnet Theorem starting from the perspective of curves on surfaces. After introducing the concept of geodesic curvature, we were able to prove the Gauss-Bonnet Theorem. One key part of the proof relied on using an orthogonal coordinate system. We originally cited the orthogonal coordinate system involving lines of curvature; however, lines of curvature are extrinsic properties and hence cannot be naturally generalized to surfaces in $\mathbb{R}^n$. On the other hand, geodesics are intrinsic properties of surfaces and in Section 8.5 we proved the existence of geodesic coordinate systems that are orthogonal. More precisely, around every point p on the surface S there exists a neighborhood U of p that is parametrized by a geodesic coordinate system.

Such a coordinate system, which is an intrinsic property, leads to the proof of the global Gauss-Bonnet Theorem. Consequently, Theorem 8.3.4 is an intrinsic theorem and holds for regular surfaces in $\mathbb{R}^n$. The Gauss-Bonnet Theorem requires the surface to be orientable in order to put a compatible orientation on the boundary curve C. This orientation on C affects the sign of

$$\int_C \kappa_g \, ds,$$

which in turns affects the Gauss-Bonnet formula. If $\mathcal{R}$ is a region of the surface without boundary, then the orientation of the surface is irrelevant for the formula and we deduce the following corollary.

Corollary 9.2.8 *If $\mathcal{R}$ is a regular surface without boundary of class C^2 in $\mathbb{R}^n$, then*

$$\iint_{\mathcal{R}} K\,dS = 2\pi\chi(\mathcal{R}).$$

PROBLEMS

1. Consider the parametrized surface S in $\mathbb{R}^4$ given as the image of

$$\vec{X}(u,v) = (u\cos v \sin v, u\sin^2 v, u\cos v, u)$$

with $(u,v) \in \mathbb{R} \times [0, 2\pi]$.

  (a) Prove that S is regular everywhere except at $(0,0,0,0)$.

  (b) Find the tangent plane at $(u_0, v_0) = (1, \pi/2)$.

  (c) Find the normal plane to the surface at $(u_0, v_0) = (1, \pi/2)$.

2. Consider the parametrized surface in $\mathbb{R}^4$ given by

$$\vec{X}(u,v) = (a\cos u \sin v, a\sin u \sin v, a\cos v, bu)$$

for $(u,v) \in \mathbb{R} \times [0, \pi]$, where a and b are positive constants. We can view this as a generalization of a helix in the sense that, instead of parametrizing a circle that rises along a perpendicular axis, it parametrizes a sphere that rises along an axis perpendicular to the plane of the sphere.

  (a) Prove that this surface is a regular.

  (b) Find the tangent plane at $(u_0, v_0) = (\pi/2, \pi/2)$.

  (c) Calculate the metric tensor components.

3. Prove the claim concerning the metric tensor of the flat torus in Example 9.2.3.

4. Let a, b, c, d be positive real constants and consider the surface in $\mathbb{R}^4$ parametrized by

$$\vec{X}(u,v) = (a\cos u, b\sin u, c\cos v, d\sin v)$$

with $(u,v) \in [0, 2\pi] \times [0, 2\pi]$. Show that in general the coefficients of the metric tensor are non-constant functions but that the Gaussian curvature is identically 0.

5. A *Veronese surface* in $\mathbb{R}^5$ is a surface parametrized by $\vec{X}(u,v) = a(u, v, u^2, uv, v^2)$ for $(u,v) \in \mathbb{R}^2$ and a being a constant. Calculate the metric tensor, the Christoffel symbols, and the Gaussian curvature.

6. We revisit the surface in $\mathbb{R}^4$ introduced in Example 9.2.4.

  (a) Prove that the surface is not a subset of any 3-dimensional plane.

     (b) Apply the Gauss-Bonnet Theorem on a region $\mathcal{R}$ defined by $u_1 \leq u \leq u_2$ and $v_1 \leq v \leq v_2$ and confirm the resulting identity of integrals that ensues.

7. Calculate the metric tensor and the Gaussian curvature for the surface in $\mathbb{R}^4$ parametrized by

$$\vec{X}(u, v) = (u^3, u^2 v, uv^2, v^3) \qquad \text{for } (u, v) \in \mathbb{R}^2.$$

8. The graph of a change of coordinates in $F : \mathbb{R}^2 \to \mathbb{R}^2$ is a parametrized surface in $\mathbb{R}^4$. In particular, if $F(u, v) = (f(u, v), g(u, v))$ over a domain U, then the graph of f is parametrized by

$$\vec{X}(u, v) = (u, v, f(u, v), g(u, v)) \quad \text{for } (u, v) \in U.$$

     (a) Calculate the components of the metric tensor of this function graph.

     (b) Using Equation (9.6), calculate the Gaussian curvature.

     (c) As an application, give the metric tensor and the Gaussian curvature for polar coordinate transformation, i.e., when $u = r$, $v = \theta$, $f(r, \theta) = r \cos \theta$, and $g(r, \theta) = r \sin \theta$.

9. Let $f, h : I \to \mathbb{R}$ be differentiable functions over and interval I and consider the surface S in $\mathbb{R}^4$ that can be parametrized by

$$\vec{X}(u, v) = (f(v) \cos u, f(v) \sin u, h(v) \cos u, h(v) \sin u)$$

with $(u, v) \in [0, 2\pi] \times I$. Determine the metric coefficients and the Christoffel symbols of the second kind. Then write down the system of differential equations as in (8.23) that describe the differential equations governing geodesics.

10. Let $\vec{\alpha} : I \to \mathbb{R}^2$ and $\vec{\beta} : J \to \mathbb{R}^2$ be regular curves in the plane with component functions $\vec{\alpha}(t) = (\alpha_1(t), \alpha_2(t))$ and $\vec{\beta}(t) = (\beta_1(t), \beta_2(t))$. Calculate the metric tensor and the Gaussian curvature function of the surface parametrized by

$$\vec{X}(u, v) = (\alpha_1(u), \alpha_2(u), \beta_1(v), \beta_2(v))$$

for $(u, v) \in I \times J$.

11. Consider the flat torus with the parametrization given in Equation (9.5). Use Equation (8.7) to show that curves of the form $(u, v) = (at + c, bt + d)$ for a, b, c, d constants and $t \in \mathbb{R}$ are geodesics.

12. Justify the claim in Example 9.2.7 that $\vec{X} : [0, 2\pi] \times [0, 2\pi] \to \mathbb{R}^4$ is bijective with its image.

13. Calculate the Gaussian curvature of the Klein bottle using the parametrization given in Example 9.2.7.

14. There are many ways to define a two-dimensional tube around a curve $\vec{\gamma} : I \to \mathbb{R}^4$. Suppose that we define the tube as

$$\vec{X}(u, v) = \vec{\gamma}(u) + (r \cos v) \vec{P}_1(u) + (r \sin v) \vec{P}_2(u),$$

with $(u, v) \in I \times [0, 2\pi]$, and for some radius r chosen small enough so that the surface is regular.

(a) Determine the metric tensor of this tube.

(b) Calculate the Gaussian curvature function $K(u, v)$.

(c) Find all the points where $K(u, v) = 0$.

[Compare to Problem 6.6.10.]

APPENDIX A

Tensor Notation

A.1   Tensor Notation

In Chapter 6, we gave to the first fundamental form the alternate name of metric tensor and delayed the explanation of what a tensor is. Mathematicians and physicists often present tensors and the tensor product in very different ways, sometimes making it difficult for a reader to see that authors in different fields are talking about the same thing. In this section, we introduce tensor notation in what one might call the "physics style," which emphasizes how components of objects change under a coordinate transformation. Readers who are well acquainted with tensor algebras on vector spaces might find this approach unsatisfactory, but physicists should recognize it. (The reader who wishes to understand the full modern mathematical formulation of tensors and see how the physics style meshes with the mathematical style should consult Appendix C in [24].)

The description of tensors we introduce below relies heavily on transformations between coordinate systems. Though we discuss coordinates and transformations between them generally, one can keep in mind as running examples Cartesian or polar coordinates for regions of the plane or Cartesian, cylindrical, and spherical coordinates for regions of $\mathbb{R}^3$. Ultimately, our discussion will apply to changes of coordinates between overlapping coordinate patches on regular surfaces. (The reader should be aware that [24], which follows the present text, provides a rigorous, coordinate-free introduction to these concepts.)

A.1.1   Curvilinear Coordinate Systems

Let S be an open set in $\mathbb{R}^n$. A continuous surjective function $f : U \to S$, where U is an open set in $\mathbb{R}^n$, defines a coordinate system on S by associating to every point $P \in S$ an n-tuple $x(P) = (x^1(P), x^2(P), \ldots, x^n(P))$, such that $f(x(P)) = P$. In this notation, the superscripts do not indicate powers of a variable x but the ith coordinate for that point in the given coordinate system. Though a possible source of confusion at the beginning, differential geometry literature uses superscripts instead of the usual subscripts in order to mesh properly with subsequent tensor notation. One also sometimes says that $f : U \to S$ parametrizes S. As with polar coordinates, where (r_0, θ_0) and $(r_0, \theta_0 + 2\pi)$ correspond to the same point in the plane, the n-tuple need not be uniquely associated to the point P.

Let S be an open set in $\mathbb{R}^n$, and consider two coordinate systems on S relative to which the coordinates of a point P are denoted by $(x^1, x^2, \ldots, x^n)$ and $(\bar{x}^1, \bar{x}^2, \ldots, \bar{x}^n)$.

Suppose that the open set $U \subset \mathbb{R}^n$ parametrizes S using the n-tuple of coordinates $(x^1, x^2, \ldots, x^n)$ and that the open set $V \subset \mathbb{R}^n$ parametrizes S using the coordinates $(\bar{x}^1, \bar{x}^2, \ldots, \bar{x}^n)$. We assume that there exists a bijective change-of-coordinates function $F : U \to V$ so that we can write

$$(\bar{x}^1, \bar{x}^2, \ldots, \bar{x}^n) = F(x^1, x^2, \ldots, x^n). \tag{A.1}$$

Again using the superscript notation, we might write explicitly

$$\begin{cases} \bar{x}^1 &= F^1(x^1, x^2, \ldots, x^n), \\ &\vdots \\ \bar{x}^n &= F^n(x^1, x^2, \ldots, x^n), \end{cases}$$

where F^i are functions from U to $\mathbb{R}$. We will assume from now on that the change of variables function F is always of class C^2, i.e., that all the second partial derivatives are continuous. Unless it becomes necessary for clarity, one often abbreviates the notation and writes

$$\bar{x}^i = \bar{x}^i(x^j),$$

by which one understands that the coordinates $(\bar{x}^1, \bar{x}^2, \ldots, \bar{x}^n)$ functionally depend on the coordinates $(x^1, x^2, \ldots, x^n)$. Thus, we write

$$\frac{\partial \bar{x}^i}{\partial x^j} \quad \text{for} \quad \frac{\partial F^i}{\partial x^j},$$

and the matrix of the differential dF_P (see Equation (5.3)) is given by

$$[dF_P]_{ij} = \left(\frac{\partial \bar{x}^i}{\partial x^j} \right).$$

Just as the functions $\bar{x}^i = \bar{x}^i(x^j)$ represent the change of variables F, we write $x^j = x^j(\bar{x}^k)$ to indicate the component functions of the inverse $F^{-1} : V \to U$. In this notation, Proposition 5.4.3 states that, as matrices,

$$\left(\frac{\partial x^j}{\partial \bar{x}^i} \right) = \left(\frac{\partial \bar{x}^i}{\partial x^j} \right)^{-1}, \tag{A.2}$$

where we assume that the functions in the first matrix are evaluated at p in the $(\bar{x}^1, \ldots, \bar{x}^n)$-coordinates, while the functions in the second matrix are evaluated at p in the $(x^1, \ldots, x^n)$-coordinates. One can express the same relationship in an alternate way by writing $x^i = x^i(\bar{x}^j(x^k))$ and applying the chain rule when differentiating with respect to x^k as follows:

$$\frac{\partial x^i}{\partial x^k} = \frac{\partial x^i}{\partial \bar{x}^1} \frac{\partial \bar{x}^1}{\partial x^k} + \frac{\partial x^i}{\partial \bar{x}^2} \frac{\partial \bar{x}^2}{\partial x^k} + \cdots + \frac{\partial x^i}{\partial \bar{x}^n} \frac{\partial \bar{x}^n}{\partial x^k} = \sum_{j=1}^{n} \frac{\partial x^i}{\partial \bar{x}^j} \frac{\partial \bar{x}^j}{\partial x^k}.$$

However, by definition of a coordinate system in $\mathbb{R}^n$, there must be no function dependence of one variable on another, so

$$\frac{\partial x^i}{\partial x^k} = \delta^i_k,$$

where δ^i_k is the Kronecker delta symbol defined by

$$\delta^i_j = \begin{cases} 1, & \text{if } i = j, \\ 0, & \text{if } i \neq j. \end{cases} \tag{A.3}$$

Therefore, since δ^i_j are essentially the entries of the identity matrix, we conclude that

$$\sum_{j=1}^{n} \frac{\partial x^i}{\partial \bar{x}^j} \frac{\partial \bar{x}^j}{\partial x^k} = \delta^i_k \tag{A.4}$$

and hence recover Equation (A.2).

The space $\mathbb{R}^n$ is a vector space, so to each point P, we can associate the position vector $\vec{r} = \overrightarrow{OP}$. Let $(x^1, x^2, \ldots, x^n)$ be a coordinate system of an open set containing P. The natural basis of $\mathbb{R}^n$ at a point P associated to this coordinate system is the set of vectors

$$\left\{ \frac{\partial \vec{r}}{\partial x^1}, \frac{\partial \vec{r}}{\partial x^2}, \ldots, \frac{\partial \vec{r}}{\partial x^n} \right\},$$

where all the derivatives are evaluated at P. Note that one often expresses the vector $\vec{r}$ in terms of the Cartesian coordinate system, and this expression is precisely the transformation functions between the given coordinate system and the Cartesian coordinate system.

Example A.1.1 (Spherical Coordinates) Spherical coordinates for points in $\mathbb{R}^3$ consist of a triple (ρ, θ, φ), where ρ is the distance of

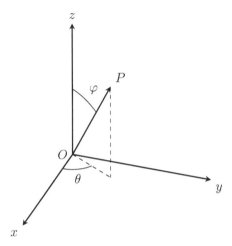

Figure A.1: Spherical coordinates.

P to the origin, θ is the angle between the xz-plane and the vertical plane containing P, and φ is the angle between the positive z-axis and the ray $[OP)$ (see Figure A.1).

In Cartesian coordinates, the position vector $\vec{r}$ for a point with spherical coordinates (ρ, θ, φ) is

$$\vec{r} = (\rho \cos \theta \sin \varphi, \rho \sin \theta \sin \varphi, \rho \cos \varphi). \qquad (A.5)$$

This corresponds to the coordinate transformation from spherical to Cartesian coordinates. Then

$$\frac{\partial \vec{r}}{\partial \rho} = (\cos \theta \sin \varphi, \sin \theta \sin \varphi, \cos \varphi),$$

$$\frac{\partial \vec{r}}{\partial \theta} = (-\rho \sin \theta \sin \varphi, \rho \cos \theta \sin \varphi, 0),$$

$$\frac{\partial \vec{r}}{\partial \varphi} = (\rho \cos \theta \cos \varphi, \rho \sin \theta \cos \varphi, -\rho \sin \varphi).$$

It is interesting to note that these three vectors are orthogonal for all triples (ρ, θ, φ). When this is the case, we say that the coordinate system is *orthogonal*. In this case, one obtains a natural orthonormal basis at P associated to this coordinate system by simply dividing by the length of each vector (or its negative). For spherical coordinates, the associated orthonormal basis at any point consists of the three vectors

$$\vec{e}_\rho = (\cos\theta \sin\varphi, \sin\theta \sin\varphi, \cos\varphi),$$
$$\vec{e}_\theta = (-\sin\theta, \cos\theta, 0),$$
$$\vec{e}_\varphi = (\cos\theta \cos\varphi, \sin\theta \cos\varphi, -\sin\varphi).$$

The factors 1, $\rho \sin\varphi$, and ρ between $\partial\vec{r}/\partial x^i$ and its normalized vector are sometimes called the *scaling factors* for each coordinate. Again, it is important to note that unlike the usual bases in linear algebra, the basis $\{\vec{e}_\rho, \vec{e}_\theta, \vec{e}_\varphi\}$ depends on the coordinates $(\rho, \theta\,\varphi)$ of a point in $\mathbb{R}^n$.

We are now in a position to discuss how components of various quantities defined locally, namely in a neighborhood U of a point $p \in \mathbb{R}^3$, change under a coordinate transformation on U. As mentioned before, our definitions do not possess the usual mathematical flavor, and the supporting discussion might feel like a game of symbols. However, it is important to understand the transformational properties of tensor components even before becoming familiar with the machinery of linear algebra of tensors. We begin with the simplest situation.

Definition A.1.2 Let $p \in \mathbb{R}^n$, and let U be a neighborhood of p. Suppose that $(x^1, x^2, \ldots, x^n)$ and $(\bar{x}^1, \bar{x}^2, \ldots, \bar{x}^n)$ are two systems of coordinates on U. A function $f(x^1, \ldots, x^n)$ given in the $(x^1, x^2, \ldots, x^n)$-coordinates is said to be a *scalar* if its expression $\bar{f}(\bar{x}^1, \ldots, \bar{x}^n)$ in the $(\bar{x}^1, \ldots, \bar{x}^n)$-coordinates has the same numerical value. In other words, if

$$\bar{f}(\bar{x}^1, \ldots, \bar{x}^n) = f(x^1, \ldots, x^n).$$

In Definition A.1.2, when we evaluate f and $\bar{f}$ the coordinates $(x^1, x^2, \ldots, x^n)$ and $(\bar{x}^1, \bar{x}^2, \ldots, \bar{x}^n)$ refer to the same point p in $\mathbb{R}^n$.

This definition might appear at first glance not to hold much content in that every function defined in reference to some coordinate system should possess this property, but that is not true. Suppose that f is a scalar. The quantity that gives the derivative of f in the first coordinate is not a scalar, for though $\bar{f}(\bar{x}^1, \ldots, \bar{x}^n) = f(x^1, \ldots, x^n)$, we have

$$\frac{\partial \bar{f}}{\partial \bar{x}^1} = \frac{\partial f}{\partial x^1}\frac{\partial x^1}{\partial \bar{x}^1} + \frac{\partial f}{\partial x^2}\frac{\partial x^2}{\partial \bar{x}^1} + \cdots + \frac{\partial f}{\partial x^n}\frac{\partial x^n}{\partial \bar{x}^1}.$$

As a second example of how quantities change under coordinate transformations, we consider the gradient $\vec{\nabla}f$ of a differentiable scalar function f. Recall that

$$\vec{\nabla}f = \left(\frac{\partial f}{\partial x^1}, \frac{\partial f}{\partial x^2}, \ldots, \frac{\partial f}{\partial x^n}\right)$$

in usual Cartesian coordinates. The gradient is a vector field, or we may simply consider the gradient of f at P, namely $\vec{\nabla}f_P$, which is a vector. We highlight the transformational properties of the gradient. The chain rule gives

$$\frac{\partial \bar{f}}{\partial \bar{x}^j} = \sum_{i=1}^{n} \frac{\partial x^i}{\partial \bar{x}^j} \frac{\partial f}{\partial x^i}. \tag{A.6}$$

However, it turns out that this is not the only way components of what we usually call a "vector" can change under a coordinate transformation.

Again, let $(x^1, x^2, \ldots, x^n)$ and $(\bar{x}^1, \bar{x}^2, \ldots, \bar{x}^n)$ be two coordinate systems on an open set of $\mathbb{R}^n$. Let $\vec{A}$ be a vector in $\mathbb{R}^n$, which we consider based at P. The components of the vector $\vec{A}$ in the respective coordinate systems is $(A^1, \ldots, A^n)$ and $(\bar{A}^1, \ldots, \bar{A}^n)$, where

$$\vec{A} = \sum_{i=1}^{n} A^i \frac{\partial \vec{r}}{\partial x^i} = \sum_{j=1}^{n} \bar{A}^j \frac{\partial \vec{r}}{\partial \bar{x}^j}.$$

Since $\dfrac{\partial \vec{r}}{\partial x^i} = \sum_{j} \dfrac{\partial \vec{r}}{\partial \bar{x}^j} \dfrac{\partial \bar{x}^j}{\partial x^i}$, we find that the components of $\vec{A}$ in the two systems of coordinates are related by

$$\bar{A}^j = \sum_{i=1}^{n} \frac{\partial \bar{x}^j}{\partial x^i} A^i. \tag{A.7}$$

Example A.1.3 (Velocity in Spherical Coordinates)   Let $\vec{r}(t)$ for $t \in I$ parametrize a curve in $\mathbb{R}^3$. The chain rule allows us to write the velocity vector in spherical or Cartesian coordinates as

$$\vec{r}'(t) = \rho'(t)\frac{\partial \vec{r}}{\partial \rho} + \theta'(t)\frac{\partial \vec{r}}{\partial \theta} + \varphi'(t)\frac{\partial \vec{r}}{\partial \varphi}$$
$$= x'(t)\vec{i} + y'(t)\vec{j} + z'(t)\vec{k}.$$

However, differentiating Equation (A.5), assuming ρ, θ, and φ are functions of t, allows us to identify the Cartesian coordinates of the velocity vector as:

$$\begin{pmatrix} x' \\ y' \\ z' \end{pmatrix} = \begin{pmatrix} \cos\theta\sin\varphi & -\rho\sin\theta\sin\varphi & \rho\cos\theta\cos\varphi \\ \sin\theta\sin\varphi & -\rho\cos\theta\sin\varphi & \rho\sin\theta\cos\varphi \\ \cos\varphi & 0 & -\rho\sin\varphi \end{pmatrix} \begin{pmatrix} \rho' \\ \theta' \\ \varphi' \end{pmatrix}.$$

If we label the spherical coordinates as $(\bar{x}^1, \bar{x}^2, \bar{x}^3)$ and the Cartesian coordinates as (x^1, x^2, x^3), then the transition matrix between components of the velocity vector from spherical coordinates to Cartesian coordinates is precisely $\left(\dfrac{\partial x^i}{\partial \bar{x}^j}\right)$. Thus, the velocity vector of any curve $\vec{\gamma}(t)$ through a point $P = \vec{\gamma}(t_0)$ does not change according to Equation (A.6) but according to Equation (A.7).

Though we have presented the notion of curvilinear coordinates in general, one should keep in mind the linear coordinate changes

$$
\begin{pmatrix} \bar{x}^1 \\ \bar{x}^2 \\ \vdots \\ \bar{x}^n \end{pmatrix} = M \begin{pmatrix} x^1 \\ x^2 \\ \vdots \\ x^n \end{pmatrix}
$$

where M is an $n \times n$ matrix. If either of these coordinate systems is given as coordinates in a basis of $\mathbb{R}^n$, then the other system simply corresponds to a change of basis in $\mathbb{R}^n$. It is easy to show that the transition matrix is then

$$
\left(\dfrac{\partial \bar{x}^i}{\partial x^j}\right) = M,
$$

the usual basis transition matrix. Furthermore, $\left(\dfrac{\partial \bar{x}^i}{\partial x^j}\right)$ is constant over all $\mathbb{R}^n$.

A.1.2 Tensors: Definitions and Notation

The relations established in Equations (A.6) and (A.7) show that *there are two different kinds of vectors*, each following different transformational properties under a coordinate change. This distinction is not emphasized in most linear algebra courses but essentially corresponds to the difference between a column vector and a row vector, which in turn corresponds to vectors in $\mathbb{R}^n$ and its dual $(\mathbb{R}^n)^*$. (The * notation denotes the dual of a vector space. See Appendix C.3 in [24] for background.) We summarize this dichotomy in the following two definitions.

Definition A.1.4 Let $(x^1, \ldots, x^n)$ and $(\bar{x}^1, \ldots, \bar{x}^n)$ be two coordinate systems in a neighborhood of a point $p \in \mathbb{R}^n$. An n-tuple of real numbers $(A^1, A^2, \ldots, A^n)$ is said to constitute the components of a *contravariant vector* at a point p if these components transform according to the relation

$$
\bar{A}^j = \sum_{i=1}^n \dfrac{\partial \bar{x}^j}{\partial x^i} A^i,
$$

where we assume the partial derivatives are evaluated at p.

Definition A.1.5 Under the same conditions as above, an n-tuple $(B_1, B_2, \ldots, B_n)$ is said to constitute the components of a *covariant vector* at a point p if these components transform according to the relation

$$\overline{B}_j = \sum_{i=1}^{n} \frac{\partial x^i}{\partial \bar{x}^j} B_i,$$

where we assume the partial derivatives are evaluated at p.

A few comments are in order at this point. Though the above two definitions are unsatisfactory from the modern perspective of set theory, these are precisely what one is likely to find in a classical mathematics text or a physics text presenting differential geometry. Nonetheless, we will content ourselves with these definitions and with the more general Definition A.1.6. We defer until Chapter 4 in [24] what is considered the proper modern definition of a tensor on a manifold.

Next, we point out that the quantities $(A^1, A^2, \ldots, A^n)$ in Definition A.1.4 or $(B_1, B_2, \ldots, B_n)$ in Definition A.1.5 can either be constant or be functions of the coordinates in a neighborhood of p. If the quantities are constant, one says they form the components of an affine vector. If the quantities are functions in the coordinates, then the components define a different vector for every point in an open set, and thus, one views these quantities as the components of a vector field over a neighborhood of p.

Finally, in terms of notation, we distinguish between the two types of vectors by using subscripts for covariant vectors and superscripts for contravariant vectors. This convention of notation is consistent throughout the literature and forms a central part of tensor calculus. This convention also explains the use of superscripts for the coordinates since (x^i) represents the components of a contravariant vector, namely, the position vector of a point.

As a further example to motivate the definition of a tensor, recall the transformational properties of the components of the first fundamental form given in Equation (6.7):

$$\bar{g}_{ij} = \sum_{k=1}^{2} \sum_{l=1}^{2} \frac{\partial x_k}{\partial \bar{x}_i} \frac{\partial x_l}{\partial \bar{x}_j} g_{kl},$$

where g_{kl} (resp. $\bar{g}_{ij}$) represents the coefficients of the first fundamental form in the (x^1, x^2)- (resp. $(\bar{x}^1, \bar{x}^2)$-) coordinates. This formula mimics but generalizes the transformational properties in Definition A.1.4 and Definition A.1.5. Many objects of interest that arise in differential geometry possess similar properties and lead to the following definition of a tensor.

Definition A.1.6 Let $(x^1, \ldots, x^n)$ and $(\bar{x}^1, \ldots, \bar{x}^n)$ be two coordinate systems in a neighborhood of a point $p \in \mathbb{R}^n$. A set of n^{r+s} quantities $T^{i_1 i_2 \cdots i_r}_{j_1 j_2 \cdots j_s}$ is said to constitute the components of a *tensor* of type (r, s) if under a coordinate transformation these quantities transform according to

$$\bar{T}^{k_1 k_2 \cdots k_r}_{l_1 l_2 \cdots l_s} = \sum_{i_1=1}^{n} \cdots \sum_{i_r=1}^{n} \sum_{j_1=1}^{n} \cdots \sum_{j_s=1}^{n} \frac{\partial \bar{x}^{k_1}}{\partial x^{i_1}} \cdots \frac{\partial \bar{x}^{k_r}}{\partial x^{i_r}} \frac{\partial x^{j_1}}{\partial \bar{x}^{l_1}} \cdots \frac{\partial x^{j_s}}{\partial \bar{x}^{l_s}} T^{i_1 i_2 \cdots i_r}_{j_1 j_2 \cdots j_s},$$

$$(A.8)$$

where we assume the partial derivatives are evaluated at p. The *rank* of the tensor is the integer $r + s$.

From the above definition, one could rightly surmise that basic calculations with tensors involve numerous repeated summations. In order to alleviate this notational burden, mathematicians and physicists who use the tensor notation as presented above utilize the *Einstein summation convention*. In this convention of notation, one assumes that one takes a sum from 1 to n (the dimension of $\mathbb{R}^n$ or the number of coordinates) over any index that appears both in a superscript and a subscript of a product. Furthermore, for this convention, in a partial derivative $\dfrac{\partial \bar{x}^i}{\partial x^j}$, the index i is considered a superscript, and the index j is considered a subscript.

For example, if A_{ij} form the components of a $(0, 2)$-tensor and B^k constitutes the components of a contravariant vector, with the Einstein summation convention, the expression $A_{ij} B^j$ means

$$\sum_{j=1}^{n} A_{ij} B^j.$$

As another example, with the Einstein summation convention, the transformational property in Equation (A.8) of a tensor is written as

$$\bar{T}^{k_1 k_2 \cdots k_r}_{l_1 l_2 \cdots l_s} = \frac{\partial \bar{x}^{k_1}}{\partial x^{i_1}} \frac{\partial \bar{x}^{k_2}}{\partial x^{i_2}} \cdots \frac{\partial \bar{x}^{k_s}}{\partial x^{i_s}} \frac{\partial x^{j_1}}{\partial \bar{x}^{l_1}} \frac{\partial x^{j_2}}{\partial \bar{x}^{l_2}} \cdots \frac{\partial x^{j_r}}{\partial \bar{x}^{l_r}} T^{i_1 i_2 \cdots i_r}_{j_1 j_2 \cdots j_s},$$

where the summations from 1 to n over the indices $i_1, i_2, \ldots i_r, j_1, j_2, \ldots j_s$ are understood. As a third example, if C^{ij}_{kl} is a tensor of type $(2, 2)$, then C^{ij}_{kj} means

$$\sum_{j=1}^{n} C^{ij}_{kj}.$$

On the other hand, with this convention, we do not sum over the index i in the expression $A^i + B^i$ or even in $A^i + B_i$. In fact, as we shall see, though the former expression has an interpretation, the latter does not.

In the rest of this book, we will use the Einstein summation convention when working with components of tensors.

A.1.3   Operations on Tensors

It is possible to construct new tensors from old ones. (Again, the reader is encouraged to consult Appendix C.4 in [24] to see the underlying algebraic meaning of the following operations.)

First of all, if $S^{i_1 i_2 \cdots i_r}_{j_1 j_2 \cdots j_s}$ and $T^{i_1 i_2 \cdots i_r}_{j_1 j_2 \cdots j_s}$ are both components of tensors of type (r, s), then the quantities

$$W^{i_1 i_2 \cdots i_r}_{j_1 j_2 \cdots j_s} = S^{i_1 i_2 \cdots i_r}_{j_1 j_2 \cdots j_s} + T^{i_1 i_2 \cdots i_r}_{j_1 j_2 \cdots j_s}$$

form the components of another (r, s)-tensor. In other words, tensors of the same type can be added to obtain another tensor of the same type. The proof is very easy and follows immediately from the transformational properties and distributivity.

Secondly, if $S^{i_1 i_2 \cdots i_r}_{j_1 j_2 \cdots j_s}$ and $T^{k_1 k_2 \cdots k_t}_{l_1 l_2 \cdots l_u}$ are components of tensors of type (r, s) and (t, u), respectively, then the quantities obtained by multiplying these components as in,

$$W^{i_1 i_2 \cdots i_r k_1 k_2 \cdots k_t}_{j_1 j_2 \cdots j_s l_1 l_2 \cdots l_u} = S^{i_1 i_2 \cdots i_r}_{j_1 j_2 \cdots j_s} T^{k_1 k_2 \cdots k_t}_{l_1 l_2 \cdots l_u},$$

form the components of another tensor but of type $(r+t, s+u)$. Again, the proof is very easy, but one must be careful with the plethora of indices. One should note that this operation of tensor product works also for multiplying a tensor by a scalar since a scalar is a tensor of rank 0.

Finally, another common operation on tensors is the *contraction* between two indices. We illustrate the contraction with an example. Let A^{ijk}_{rs} be the components of a $(3, 2)$-tensor, and define the quantities

$$B^{ij}_r = A^{ijk}_{rk} = \sum_{k=1}^{n} A^{ijk}_{rk} \qquad \text{by Einstein summation convention.}$$

It is not hard to show (left as an exercise for the reader) that B^{ij}_r constitute the components of a tensor of type $(2, 1)$. More generally, starting with a tensor of type (r, s), if one sums over an index that appears both in the superscript and in the subscript, one obtains the components of a $(r - 1, s - 1)$-tensor. This is the contraction of a tensor over the stated indices.

A.1.4   Examples

Example A.1.7 Following the terminology of Definition A.1.6, a covariant vector is often called a $(0,1)$-tensor, and similarly, a contravariant vector is called a $(1,0)$-tensor.

Example A.1.8 In Problem 6.4.7, one showed that the coefficients L_{ij} of the second fundamental form constitute the components of a $(0,2)$-tensor, just as the metric tensor does.

Example A.1.9 (Inverse of a $(0,2)$-tensor) As a more involved example, consider the components A_{ij} of a $(0,2)$-tensor in $\mathbb{R}^n$. Denote by A^{ij} the quantities given as the coefficients of the inverse matrix of (A_{ij}). We prove that A^{ij} form the components of a $(2,0)$-tensor.

Suppose that the coefficients A^{ij} given in a coordinate system with variables $(x^1, \ldots, x^n)$ and $\bar{A}^{rs}$ are given in the $(\bar{x}^1, \ldots, \bar{x}^n)$-coordinate system. That they are the inverse to the matrices (A_{ij}) and $(\bar{A}_{rs})$ means that A^{ij} and $\bar{A}^{rs}$ are the unique quantities such that

$$A^{ij} A_{jk} = \delta^i_k, \qquad \text{and}$$
$$\bar{A}^{rs} \bar{A}_{st} = \delta^r_t, \tag{A.9}$$

where the reader must remember that we are using the Einstein summation convention. Combining Equation (A.9) and the transformational properties of A_{jk}, we get

$$\bar{A}^{rs} \frac{\partial x^i}{\partial \bar{x}^s} \frac{\partial x^j}{\partial \bar{x}^t} A_{ij} = \delta^r_t.$$

Multiplying both sides by $\dfrac{\partial \bar{x}^t}{\partial x^\alpha}$ and summing over t, we obtain

$$\bar{A}^{rs} \frac{\partial x^i}{\partial \bar{x}^s} \frac{\partial x^j}{\partial \bar{x}^t} A_{ij} \frac{\partial \bar{x}^t}{\partial x^\alpha} = \delta^r_t \frac{\partial \bar{x}^t}{\partial x^\alpha}$$
$$\Longleftrightarrow \bar{A}^{rs} \frac{\partial x^i}{\partial \bar{x}^s} \delta^j_\alpha A_{ij} = \frac{\partial \bar{x}^r}{\partial x^\alpha}$$
$$\Longleftrightarrow \bar{A}^{rs} \frac{\partial x^i}{\partial \bar{x}^s} A_{i\alpha} = \frac{\partial \bar{x}^r}{\partial x^\alpha}.$$

Multiplying both sides by $A^{\alpha\beta}$ and then summing over α, we get

$$\bar{A}^{rs} \frac{\partial x^i}{\partial \bar{x}^s} \delta^\beta_i = \bar{A}^{rs} \frac{\partial x^\beta}{\partial \bar{x}^s} = \frac{\partial \bar{x}^r}{\partial x^\alpha} A^{\alpha\beta}.$$

Finally, multiplying the rightmost equality by $\dfrac{\partial \bar{x}^s}{\partial x^\beta}$ and summing over β, one concludes that

$$\bar{A}^{rs} = \frac{\partial \bar{x}^r}{\partial x^\alpha} \frac{\partial \bar{x}^s}{\partial x^\beta} A^{\alpha\beta}.$$

This shows that the quantities A^{ij} satisfy Definition A.1.6 and form the components of a $(2,0)$-tensor.

By a similar manipulation, one can show that if B^{ij} are the components of a $(2,0)$-tensor, then the quantities B_{ij} corresponding to the inverse of the matrix of B^{ij} form the components of a $(0,2)$-tensor.

Example A.1.10 (Gauss Map Coefficients) Recall that we denote by g^{ij} the coefficients of the inverse of the matrix associated to the first fundamental form. Example A.1.9 shows that g^{ij} is a $(2,0)$-tensor. Furthermore, recall that the Weingarten equations in Equation (6.27) give the components (associated to the standard basis on $T_p(S)$ given by a particular parametrization) of the Gauss map as

$$a^i_j = -g^{ik} L_{kj}.$$

By tensor product and contraction, we see that the functions a^i_j form the components of a $(1,1)$-tensor.

Example A.1.11 (Metric Tensors) It is important to understand some standard operations of vectors in the context of tensor notation. Consider two vectors in a vector space V of dimension n. Using tensor notation, one refers to these vectors as affine contravariant vectors with components A^i and B^j, with $i, j = 1, 2, \ldots, n$. We have seen that addition of the vectors or scalar multiplication are the usual operations from linear algebra. Another operation between vectors in V is the dot product, which was originally defined as

$$\sum_{i=1}^{n} A^i B^i,$$

but this is not the correct way to understand the dot product in the context of tensor algebra. The very fact that one cannot use the Einstein summation convention is a hint that we must adjust our notation. The use of the usual dot product for its intended geometric purpose makes an assumption of the given basis of V, namely, that the basis is orthonormal. When using tensor algebra, one makes no such assumption. Instead, one associates a $(0,2)$-tensor g_{ij}, called the metric tensor, to the basis of V with respect to which coordinates are defined. Then the first fundamental form (or *scalar product*) between A^i and B^j is

$$g_{ij} A^i B^j \qquad \text{(Einstein summation)}.$$

One immediately notices that because of tensor multiplication and contraction, the result is a scalar quantity, and hence, will remain unchanged under a coordinate transformation. In this formulation, the assumption that a basis is orthonormal is equivalent to having

$$
g_{ij} = \begin{cases} 1, & \text{if } i = j, \\ 0, & \text{if } i \neq j. \end{cases}
$$

A.1.5   Symmetries

The usual operations of tensor addition and scalar multiplication were explained above. We should point out that, using distributivity and associativity, one notices that the set of affine tensors of type (r, s) in $\mathbb{R}^n$ form a vector space. The $(r + s)$-tuple of all the indices can take on n^{r+s} values, so this vector space has dimension n^{r+s}.

However, it is not uncommon that there exist symmetries within the components of a tensor. For example, as we saw for the metric tensor, we always have $g_{ij} = g_{ji}$. In the context of matrices, we said that the matrix (g_{ij}) is a symmetric matrix, but in the context of tensor notation, we say that the components g_{ij} are symmetric in the indices i and j. More generally, if $T^{i_1 i_2 \cdots i_r}_{j_1 j_2 \cdots j_s}$ are the components of a tensor of type (r, s), we say that the components are *symmetric* in a set $\mathcal{S}$ of indices if the components remain equal when we interchange any two indices from among the indices in $\mathcal{S}$.

For example, let A^{ijk}_{rs} be the components of a $(3, 2)$-tensor. To say that the components are symmetric in $\{i, j, k\}$ affirms the equalities

$$
A^{ijk}_{rs} = A^{ikj}_{rs} = A^{jik}_{rs} = A^{jki}_{rs} = A^{kij}_{rs} = A^{kji}_{rs}
$$

for all $i, j, k \in \{1, 2, \ldots, n\}$. Note that, because of the additional conditions, the dimension of the space of all $(3, 2)$-tensors that are symmetric in their contravariant indices is smaller than n^5 but not simply $n^5/6$ either. We can find the dimension of this vector space by determining the cardinality of

$$
\mathcal{I} = \big\{ (i, j, k) \in \{1, 2, \ldots, n\}^3 \,\big|\, 1 \leq i \leq j \leq k \leq n \big\}.
$$

We will see shortly that $|\mathcal{I}| = \binom{n+2}{3}$, and therefore, the dimension of the vector space of $(3, 2)$-tensors that are symmetric in their contravariant indices is $\binom{n+2}{3} n^2$. We provide the following proposition for completeness.

Proposition A.1.12 *Let $A^{j_1 \cdots j_r}_{k_1 \cdots k_s}$ be the components of a tensor over $\mathbb{R}^n$ that is symmetric in a set $\mathcal{S}$ of its indices. Assuming that all the indices are fixed except for the indices of $\mathcal{S}$, the number of independent components of the tensor is equal to the cardinality of*

$$
\mathcal{I} = \big\{ (i_1, \ldots, i_m) \in \{1, 2, \ldots, n\}^m \,\big|\, 1 \leq i_1 \leq i_2 \leq \cdots \leq i_m \leq n \big\}.
$$

This cardinality is

$$\binom{n-1+m}{m} = \frac{(n-1+m)!}{(n-1)!m!}.$$

Proof: Since the components are symmetric in the set S of indices, one gets a unique representative of equivalent components by imposing that the indices in question be listed in nondecreasing order. This remark proves the first part of the proposition. To prove the second part, consider the set of integers $\{1, 2, \ldots, n+m\}$ and pick m distinct integers $\{l_1, \ldots, l_m\}$ that are greater than 1 from among this set. We know from the definition of combinations that there are $\binom{n-1+m}{m}$ ways to do this. Assuming that $l_1 < l_2 < \cdots < l_m$, define $i_t = l_t - t$. It is easy to see that the resulting m-tuple $(i_1, \ldots, i_m)$ is in the set $\mathcal{I}$. Furthermore, since one can reverse the process by defining $l_t = i_t + t$ for $1 \leq t \leq m$, there exists a bijection between $\mathcal{I}$ and the m-tuples $(l_1, \ldots, l_m)$ described above. This establishes that

$$|\mathcal{I}| = \binom{n-1+m}{m}.$$

$\square$

Another common situation with relationships between the components of a tensor is when components are antisymmetric in a set of indices. We say that the components are *antisymmetric* in a set S of indices if the components are negated when we interchange any two indices from among the indices in S. This condition imposes a number of immediate consequences. Consider, for example, the components of a $(0, 3)$-tensor A_{ijk} that are antisymmetric in all its indices. If k is any value but $i = j$, then

$$A_{ijk} = A_{iik} = A_{jik} = -A_{ijk},$$

and so $A_{iik} = 0$. Given any triple (i, j, k) in which at least two of the indices are equal, the corresponding component is equal to 0. As another consequence of the antisymmetric condition, consider the component A_{231}. One obtains the triple $(2, 3, 1)$ from $(1, 2, 3)$ by first interchanging 1 and 2 to get $(2, 1, 3)$ and then interchanging the last two to get $(2, 3, 1)$. Therefore, we see that

$$A_{123} = -A_{213} = A_{231}.$$

In modern algebra, a permutation (a bijection on a finite set) that interchanges two inputs and leaves the rest fixed is called a *transposition*. We say that we used two transpositions to go from $(1, 2, 3)$ to $(2, 3, 1)$.

The above example illustrates that the value of the component of a tensor indexed by a particular m-tuple $(i_1, \ldots, i_m)$ of distinct indices determines the value of any component involving a permutation $(j_1, \ldots, j_m)$ of $(i_1, \ldots, i_m)$, as

$$A_{j_1 \ldots j_m} = \pm A_{i_1 \ldots i_m},$$

where the sign $\pm$ is $+$ (resp. $-$) if it takes an even (resp. odd) number of interchanges to get from $(i_1, \ldots, i_m)$ to $(j_1, \ldots, j_m)$. A priori, if one could get from $(i_1, \ldots, i_m)$ to $(j_1, \ldots, j_m)$ with both an odd and an even number of transpositions, then $A_{i_1 \ldots i_m}$ and all components indexed by a permutation of $(i_1, \ldots, i_m)$ would be 0. However, a fundamental fact in modern algebra (see Theorem 5.5 in [14]) states that given a permutation σ on $\{1, 2, \ldots, m\}$, if we have two ways to write σ as a composition of transpositions,

$$\sigma = \tau_1 \circ \tau_2 \circ \cdots \circ \tau_a = \tau_1' \circ \tau_2' \circ \cdots \circ \tau_b',$$

then a and b have the same parity.

Definition A.1.13 We call a permutation *even* (resp. *odd*) if this common parity is even (resp. odd) and the *sign* of σ is

$$\mathrm{sign}(\sigma) = \begin{cases} 1, & \text{if } \sigma \text{ is even,} \\ -1, & \text{if } \sigma \text{ is odd.} \end{cases}$$

The above discussion leads to the following proposition about the components of an antisymmetric tensor.

Proposition A.1.14 Let $A_{k_1 \ldots k_s}^{j_1 \ldots j_r}$ be the components of a tensor over $\mathbb{R}^n$ that is antisymmetric in a set $\mathcal{S}$ of its indices. If any of the indices in $\mathcal{S}$ are equal, then

$$A_{k_1 \ldots k_s}^{j_1 \ldots j_r} = 0.$$

If $|\mathcal{S}| = m$, then fixing all but the indices in $\mathcal{S}$, the number of independent components of the tensor is equal to

$$\binom{n}{m} = \frac{n!}{m!(n-m)!}.$$

Finally, if the indices of $A_{j_1 \ldots j_s}^{i_1 \ldots i_r}$ differ from $A_{l_1 \ldots l_s}^{k_1 \ldots k_r}$ only by a permutation σ on the indices in $\mathcal{S}$, then

$$A_{l_1 \ldots l_s}^{k_1 \ldots k_r} = \mathrm{sign}(\sigma)\, A_{j_1 \ldots j_s}^{i_1 \ldots i_r}.$$

A.1.6 Numerical Tensors

As a motivating example of what are called numerical tensors, note that the quantities δ^i_j form the components of a $(1,1)$-tensor. To see this, suppose that the quantities δ^i_j are expressed in a system of coordinates $(x^1, \ldots, x^n)$ and suppose that $\bar{\delta}^k_l$ are its transformed coefficients in another system of coordinates $(\bar{x}^1, \ldots, \bar{x}^n)$. Obviously, for all fixed i and j, the values of δ^i_j are constant, and therefore

$$\bar{\delta}^k_l = \begin{cases} 1, & \text{if } k = l, \\ 0, & \text{if } k \neq l. \end{cases}$$

But using the properties of the δ^i_j coefficients and the chain rule,

$$\frac{\partial \bar{x}^k}{\partial x^i} \frac{\partial x^j}{\partial \bar{x}^l} \delta^i_j = \frac{\partial \bar{x}^k}{\partial x^i} \frac{\partial x^i}{\partial \bar{x}^l} = \frac{\partial \bar{x}^k}{\partial \bar{x}^l} = \bar{\delta}^k_l.$$

Therefore, δ^i_j is a $(1,1)$-tensor in a tautological way.

A *numerical tensor* is a tensor of rank greater than 0 whose components are constant in the variables $(x^1, \ldots, x^n)$ and hence also $(\bar{x}^1, \ldots, \bar{x}^n)$. The Kronecker delta is just one example of a numerical tensor and we have already seen that it plays an important role in many complicated calculations. The Kronecker delta is the simplest case of the most important numerical tensor, the generalized Kronecker delta. The *generalized Kronecker delta* of order r is a tensor of type (r, r), with components denoted by $\delta^{i_1 \cdots i_r}_{j_1 \cdots j_r}$ defined as the following determinant:

$$\delta^{i_1 \cdots i_r}_{j_1 \cdots j_r} = \begin{vmatrix} \delta^{i_1}_{j_1} & \delta^{i_1}_{j_2} & \cdots & \delta^{i_1}_{j_r} \\ \delta^{i_2}_{j_1} & \delta^{i_2}_{j_2} & \cdots & \delta^{i_2}_{j_r} \\ \vdots & \vdots & \ddots & \vdots \\ \delta^{i_r}_{j_1} & \delta^{i_r}_{j_2} & \cdots & \delta^{i_r}_{j_r} \end{vmatrix}. \tag{A.10}$$

It is not obvious from Equation (A.10) that the quantities $\delta^{i_1 \cdots i_r}_{j_1 \cdots j_r}$ form the components of a tensor. However, one can write the components of the generalized Kronecker delta of order 2 as

$$\delta^{ij}_{kl} = \delta^i_k \delta^j_l - \delta^i_l \delta^j_k,$$

which presents δ^{ij}_{kl} as the difference of two $(2,2)$-tensors, which shows that the coefficients δ^{ij}_{kl} indeed constitute a tensor. More generally, expanding out Equation (A.10) gives the generalized Kronecker delta of order r as a sum of $r!$ components of tensors of type (r, r), proving that $\delta^{i_1 \cdots i_r}_{j_1 \cdots j_r}$ are the components of an (r, r)-tensor.

Properties of the determinant imply that $\delta^{i_1 \cdots i_r}_{j_1 \cdots j_r}$ is antisymmetric in both the superscript indices and the subscript indices. That is to say, $\delta^{i_1\cdots i_r}_{j_1\cdots j_r} = 0$ if any of the superscript indices are equal or if any of the subscript indices are equal. Hence, the value of a component is negated if any two superscript indices are interchanged and similarly for subscript indices. We also note that if $r > n$ where we assume $\delta^{i_1\cdots i_r}_{j_1\cdots j_r}$ is a tensor in $\mathbb{R}^n$, then $\delta^{i_1\cdots i_r}_{j_1\cdots j_r} = 0$ for all choices of indices since at least two superscript (and at least two subscript) indices would be equal.

We introduce one more symbol related to the generalized Kronecker delta, namely the permutation symbol. Define

$$\begin{aligned}
\varepsilon^{i_1\cdots i_n} &= \delta^{i_1\cdots i_n}_{1\cdots n}, \\
\varepsilon_{j_1\cdots j_n} &= \delta^{1\cdots n}_{j_1\cdots j_n}.
\end{aligned} \tag{A.11}$$

Note that the use of the maximal index n in Equation (A.11) as opposed to r is intentional. Because of the properties of the determinant, it is not hard to see that $\varepsilon^{i_1\cdots i_n} = \varepsilon_{i_1\cdots i_n}$ is equal to 1 (resp. -1) if $(i_1,\ldots,i_n)$ is an even (resp. odd) permutation of $(1,2,\ldots,n)$ and is equal to 0 if $(i_1,\ldots,i_n)$ is not a permutation of $(1,2,\ldots,n)$.

We are careful, despite the notation, not to call the permutation symbols the components of a tensor, for they are not. Instead, we have the following proposition.

Proposition A.1.15 *Let $(x^1,\ldots,x^n)$ and $(\bar{x}^1,\ldots,\bar{x}^n)$ be two coordinate systems. The permutation symbols transform according to*

$$\begin{aligned}
\bar{\varepsilon}^{j_1\cdots j_n} &= J\frac{\partial \bar{x}^{j_1}}{\partial x^{i_1}} \cdots \frac{\partial \bar{x}^{j_n}}{\partial x^{i_n}}\varepsilon^{i_1\cdots i_n}, \\
\bar{\varepsilon}_{k_1\cdots k_n} &= J^{-1}\frac{\partial x^{h_1}}{\partial \bar{x}^{k_1}} \cdots \frac{\partial x^{h_n}}{\partial \bar{x}^{k_n}}\varepsilon_{h_1\cdots h_n},
\end{aligned}$$

where $J = \det\left(\dfrac{\partial \bar{x}^i}{\partial x^j}\right)$ is the Jacobian of the transformation of coordinates function.

Proof: (Left as an exercise for the reader.) □

Example A.1.16 (Cross Product) Let A^i and B^j be the components of two contravariant vectors in $\mathbb{R}^3$. If we define $C_k = \varepsilon_{ijk}A^iB^j$, we easily find that

$$C_1 = A^2B^3 - A^3B^2, \quad C_2 = A^3B^1 - A^1B^3, \quad C_3 = A^1B^2 - A^2B^1.$$

The values C_k are precisely the terms of the cross product of the vectors A^i and B^j. However, a quick check shows that the quantities C_k do *not* form the components of a covariant tensor.

One explanation in relation to standard linear algebra for the fact that C_k does not give a contravariant vector is that if $\vec{a}$ and $\vec{b}$ are vectors in $\mathbb{R}^3$ given with coordinates in a certain basis and if M is a coordinate-change matrix, then

$$(M\vec{a}) \times (M\vec{b}) \neq M(\vec{a} \times \vec{b}).$$

In many physics textbooks, when one assumes that we use the usual metric, (g_{ij}) being the identity matrix, one is not always careful with the superscript and subscript indices. This is because one can obtain a contravariant vector B^j from a covariant vector A_i simply by defining $B^j = g^{ij} A_i$, and the components (B^1, B^2, B^3) are numerically equal to (A_1, A_2, A_3). Therefore, in this context, one can define the cross product as the vector with components

$$C^l = g^{kl} \varepsilon_{ijk} A^i B^j. \tag{A.12}$$

However, one must remember that this is not a contravariant vector since it does not satisfy the transformational properties of a tensor.

The generalized Kronecker delta has a close connection to determinants, which we will elucidate here. Note that if the superscript indices are exactly equal to the subscript indices, then $\delta^{i_1 \cdots i_r}_{j_1 \cdots j_r}$ is the determinant of the identity matrix. Thus, the contraction over all indices $\delta^{j_1 \cdots j_r}_{j_1 \cdots j_r}$ counts the number of permutations of r indices taken from the set $\{1, 2, \ldots, n\}$. Thus,

$$\delta^{j_1 \cdots j_r}_{j_1 \cdots j_r} = \frac{n!}{(n-r)!}. \tag{A.13}$$

Another property of the generalized Kronecker delta is that

$$\varepsilon^{j_1 \cdots j_n} \varepsilon_{i_1 \cdots i_n} = \delta^{j_1 \cdots j_n}_{i_1 \cdots i_n},$$

the proof of which is left as an exercise for the reader (Problem A.1.10). Now let a^i_j be the components of a $(1, 1)$-tensor, which we can view as the matrix of a linear transformation from $\mathbb{R}^n$ to $\mathbb{R}^n$. By definition of the determinant,

$$\det(a^i_j) = \varepsilon^{j_1 \cdots j_n} a^1_{j_1} \cdots a^n_{j_n}.$$

Then, by properties of the determinant related to rearranging rows or columns, we have

$$\varepsilon^{i_1 \cdots i_n} \det(a^i_j) = \varepsilon^{j_1 \cdots j_n} a^{i_1}_{j_1} \cdots a^{i_n}_{j_n}.$$

Multiplying by $\varepsilon_{i_1 \cdots i_n}$ and summing over all the indices $i_1, \ldots, i_n$, we have

$$\varepsilon_{i_1 \cdots i_n} \varepsilon^{i_1 \cdots i_n} \det(a^i_j) = \delta^{j_1 \cdots j_n}_{i_1 \cdots i_n} a^{i_1}_{j_1} \cdots a^{i_n}_{j_n}.$$

Since $\varepsilon_{i_1\cdots i_n}\varepsilon^{i_1\cdots i_n}$ counts the number of permutations of $\{1,\ldots,n\}$, we have

$$n!\det(a_j^i) = \delta_{i_1\cdots i_n}^{j_1\cdots j_n}a_{j_1}^{i_1}\cdots a_{j_n}^{i_n}. \tag{A.14}$$

Problems

1. Prove that
   (a) $\delta_j^i\delta_k^j\delta_l^k = \delta_l^i$,
   (b) $\delta_j^i\delta_k^j\delta_i^k = n$.

2. Let B_i be the components of a covariant vector. Prove that the quantities
   $$C_{jk} = \frac{\partial B_j}{\partial x^k} - \frac{\partial B_k}{\partial x^j}$$
   form the components of a $(0,2)$-tensor.

3. Let $T_{j_1 j_2\cdots j_s}^{i_1 i_2\cdots i_r}$ be the components of a tensor of type (r,s). Prove that the quantities $T_{i j_2\cdots j_s}^{i i_2\cdots i_r}$, obtained by contracting over the first two indices, form the components of a tensor of type $(r-1, s-1)$. Explain why one still obtains a tensor when one contracts over any superscript and subscript index.

4. Let S_{ijk} be the components of a tensor, and suppose they are antisymmetric in $\{i,j\}$. Find a tensor with components T_{ijk} that is antisymmetric in $\{j,k\}$ satisfying
   $$-T_{ijk} + T_{jik} = S_{ijk}.$$

5. If A^{jk} is antisymmetric in its indices and B_{jk} is symmetric in its indices, show that the scalar $A^{jk}B_{jk}$ is 0.

6. Consider A^i, B^j, C^k the components of three contravariant vectors. Prove that $\varepsilon_{ijk}A^iB^jC^k$ is the value triple product $(\vec{A}\vec{B}\vec{C}) = \vec{A}\cdot(\vec{B}\times\vec{C})$, which is the volume of the parallelopiped spanned by these three vectors.

7. Prove that $\varepsilon_{ijk}\varepsilon^{rsk} = \delta_{ij}^{rs}$. Assume that we use a metric g_{ij} that is the identity matrix, and define the cross product $\vec{C}$ of two contravariant vectors $\vec{A} = (A^i)$ and $\vec{B} = (B^j)$ as $C^k = g^{kl}\varepsilon_{ijl}A^iB^j$. Use what you just proved to show that
   $$\vec{A}\times(\vec{B}\times\vec{C}) = (\vec{A}\cdot\vec{C})\vec{B} - (\vec{A}\cdot\vec{B})\vec{C}.$$

8. Let $\vec{A}$, $\vec{B}$, $\vec{C}$, and $\vec{D}$ be vectors in $\mathbb{R}^3$. Use the ε_{ijk} symbols to prove that
   $$(\vec{A}\times\vec{B})\times(\vec{C}\times\vec{D}) = (\vec{A}\vec{B}\vec{D})\vec{C} - (\vec{A}\vec{B}\vec{C})\vec{D}.$$

9. Prove Proposition A.1.15.

10. Prove that $\varepsilon^{i_1\cdots i_n}\varepsilon_{j_1\cdots j_n} = \delta_{j_1\cdots j_n}^{i_1\cdots i_n}$.

11. Let A_{ij} be the components of an antisymmetric tensor of type $(0,2)$, and define the quantities
    $$B_{rst} = \frac{\partial A_{st}}{\partial x^r} + \frac{\partial A_{tr}}{\partial x^s} + \frac{\partial A_{rs}}{\partial x^t}.$$

(a) Prove that B_{rst} are the components of a tensor of type $(0,3)$.

(b) Prove that the components B_{rst} are antisymmetric in all the indices.

(c) Determine the number of independent components of antisymmetric tensors of type $(0,3)$ over $\mathbb{R}^n$.

(d) Would the quantities B_{rst} still be the components of a tensor if A_{ij} were symmetric?

12. Let A be an $n \times n$ matrix with coefficients $A = (A_i^j)$, and consider the coordinate transformation

$$\bar{x}^j = \sum_{i=1}^{n} A_i^j x^i.$$

Recall that this transformation is called orthogonal if $AA^T = I$, where A^T is the transpose of A and I is the identity matrix. The orthogonality condition implies that $\det(A) = \pm 1$. An orthogonal transformation is called *special* or *proper* if, in addition, $\det(A) = 1$. A set of quantities $T_{j_1 \cdots j_s}^{i_1 \cdots i_r}$ is called a *proper tensor* of type (r,s) if it satisfies the tensor transformation property from Equation (A.8) for all proper orthogonal transformations.

(a) Prove that the orthogonality condition is equivalent to requiring that

$$\eta_{ij} = \eta_{hk} A_i^h A_j^k,$$

where

$$\eta_{ij} = \begin{cases} 1, & \text{if } i = j, \\ 0, & \text{if } i \neq j. \end{cases}$$

(b) Prove that the orthogonality condition is also equivalent to saying that orthogonal transformations are the invertible linear transformations that preserve the quantity $(x^1)^2 + (x^2)^2 + \cdots + (x^n)^2$.

(c) Prove that (1) the space of proper tensors of type (r,s) form a vector space over $\mathbb{R}$, (2) the product of a proper tensor of type (r_1, s_1) and a proper tensor of type (r_2, s_2) is a proper tensor of type $(r_1 + r_2, s_1 + s_2)$, and (3) contraction over two indices of a proper tensor of type (r,s) produces a proper tensor of type $(r-1, s-1)$.

(d) Prove that the permutation symbols are proper tensors of type $(n,0)$ or $(0,n)$, as appropriate.

(e) Use this to prove that the cross product of two contravariant vectors in $\mathbb{R}^3$ as defined in Equation (A.12) is a proper tensor of type $(1,0)$. [Hint: This explains that the cross product of two vectors transforms correctly only if we restrict ourselves to proper orthogonal transformations on $\mathbb{R}^3$.]

(f) Suppose that we are in $\mathbb{R}^3$. Prove that the rotation with matrix

$$A = \begin{pmatrix} \cos\alpha & -\sin\alpha & 0 \\ \sin\alpha & \cos\alpha & 0 \\ 0 & 0 & 1 \end{pmatrix}$$

is a proper orthogonal transformation.

(g) Again, suppose that we are in $\mathbb{R}^3$. Prove that the linear transformation with matrix given with respect to the standard basis

$$B = \begin{pmatrix} \cos\beta & \sin\beta & 0 \\ \sin\beta & -\cos\beta & 0 \\ 0 & 0 & 1 \end{pmatrix}$$

is an orthogonal transformation that is not proper.

13. Consider the vector space $\mathbb{R}^{n+1}$ with coordinates $(x^0, x^1, \ldots, x^n)$, where $(x^1, \ldots, x^n)$ are called the space coordinates and x^0 is the time coordinate. The usual connection between x^0 and time t is $x^0 = ct$, where c is the speed of light. We equip this space with the metric $\eta_{\mu\nu}$ where $\eta_{00} = -1$, $\eta_{ii} = 1$ for $1 \leq i \leq n$, and $\eta_{ij} = 0$ if $i \neq j$. (The quantities $\eta_{\mu\nu}$ do not give a metric in the sense we have presented in this text so far because it is not positive definite. Though we do not provide the details here, this unusual metric gives a mathematical justification for why it is impossible to travel faster than the speed of light c.) This vector space equipped with the metric $\eta_{\mu\nu}$ is called the n-dimensional *Minkowski spacetime*, and $\eta_{\mu\nu}$ is called the *Minkowski metric*.

Let L be an $(n+1) \times (n+1)$ matrix with coefficients L_β^α, and consider the linear transformation

$$\bar{x}^j = \sum_{i=0}^{n} L_i^j x^i.$$

A *Lorentz transformation* is an invertible linear transformation on Minkowski spacetime with matrix L such that

$$\eta_{\alpha\beta} = \eta_{\mu\nu} L_\alpha^\mu L_\beta^\nu.$$

Finally, a set of quantities $T_{j_1 \cdots j_s}^{i_1 \cdots i_r}$, with indices ranging in $\{0, \ldots, n\}$, is called a *Lorentz tensor* of type (r, s) if it satisfies the tensor transformation property from Equation (A.8) for all Lorentz transformations.

(a) Show that a transformation of Minkowski spacetime is a Lorentz transformation if and only if it preserves the quantity

$$-(x^0)^2 + (x^1)^2 + (x^2)^2 + \cdots + (x^n)^2.$$

(b) Suppose we are working in three-dimensional Minkowski spacetime. Prove that the rotation matrix

$$A = \begin{pmatrix} 1 & 0 & 0 & 0 \\ 0 & \cos\alpha & -\sin\alpha & 0 \\ 0 & \sin\alpha & \cos\alpha & 0 \\ 0 & 0 & 0 & 1 \end{pmatrix}$$

represents a Lorentz transformation.

(c) Again, suppose we are working in three-dimensional Minkowski spacetime. Consider the matrix

$$L = \begin{pmatrix} \gamma & -\beta\gamma & 0 & 0 \\ -\beta\gamma & \gamma & 0 & 0 \\ 0 & 0 & 1 & 0 \\ 0 & 0 & 0 & 1 \end{pmatrix},$$

where β is a positive real number satisfying $-1 < \beta < 1$ and $\gamma = \dfrac{1}{\sqrt{1 - \beta^2}}$. Prove that L represents a Lorentz transformation.

(d) Prove that (1) the space of Lorentz tensors of type (r, s) form a vector space over $\mathbb{R}$, (2) the product of a Lorentz tensor of type (r_1, s_1) and a Lorentz tensor of type (r_2, s_2) is a Lorentz tensor of type $(r_1 + r_2, s_1 + s_2)$, and (3) contraction over two indices of a Lorentz tensor of type (r, s) produces a Lorentz tensor of type $(r - 1, s - 1)$.

14. Let a^i_j be the components of a $(1, 1)$-tensor, or in other words, the matrix of a linear transformation from $\mathbb{R}^n$ to $\mathbb{R}^n$ given with respect to some basis. Recall that the characteristic equation for the matrix is

$$\det(a^i_j - \lambda\delta^i_j) = 0. \qquad (A.15)$$

[Hint: The solutions to this equation are the eigenvalues of the matrix.] Prove that Equation (A.15) is equivalent to

$$\lambda^n + \sum_{r=1}^{n} (-1)^r a_{(r)} \lambda^{n-r} = 0$$

where

$$a_{(r)} = \frac{1}{r!} \delta^{i_1 \cdots i_r}_{j_1 \cdots j_r} a^{i_1}_{j_1} \cdots a^{i_r}_{j_r}.$$

15. *Moment of Inertia Tensor.* Suppose that $\mathbb{R}^3$ is given a basis that is not necessarily orthonormal. Let g_{ij} be the metric tensor corresponding to this basis, which means that the scalar product between two (contravariant) vectors A^i and B^j is given by

$$\langle \vec{A}, \vec{B} \rangle = g_{ij} A^i B^j.$$

In the rest of the problem, call (x^1, x^2, x^3) the coordinates of the position vector $\vec{r}$.

Let S be a solid in space with a density function $\rho(\vec{r})$, and suppose that it rotates about an axis ℓ through the origin. The angular velocity vector $\vec{\omega}$ is defined as the vector along the axis ℓ, pointing in the direction that makes the rotation a right-hand corkscrew motion, and with magnitude ω that is equal to the radians per second swept out by the motion of rotation. Let $(\omega^1, \omega^2, \omega^3)$ be the components of $\vec{\omega}$ in the given basis. The *moment of inertia* of the solid S about the direction $\vec{\omega}$ is defined as the quantity

$$I_\ell = \iiint_S \rho(\vec{r}) r^2_\perp \, dV,$$

where $r_\perp$ is the distance from a point $\vec{r}$ with coordinate (x^1, x^2, x^3) to the axis ℓ.

The *moment of inertia tensor* of a solid is often presented using cross products, but we define it here using a characterization that is equivalent to the usual definition but avoids cross products. We define the moment of inertia tensor as the unique $(0,2)$-tensor I_{ij} such that

$$(I_{ij}\omega^i)\frac{\omega^j}{\omega} = I_\ell\omega, \tag{A.16}$$

where $\omega = \|\vec{\omega}\| = \sqrt{\langle \vec{\omega}, \vec{\omega} \rangle}$.

(a) Prove that

$$r_\perp^2 = g_{ij}x^i x^j - \frac{(g_{kl}\omega^k x^l)^2}{g_{rs}\omega^r \omega^s}.$$

(b) Prove that, using the metric g_{ij}, the moment of inertia tensor is given by

$$I_{ij} = \iiint_S \rho(x^1, x^2, x^3)(g_{ij}g_{kl} - g_{ik}g_{jl})x^k x^l \, dV.$$

(c) Prove that I_{ij} is symmetric in its indices.

(d) Prove that if the basis of $\mathbb{R}^3$ is orthonormal (which means that (g_{ij}) is the identity matrix), one recovers the following usual formulas one finds in physics texts:

$$I_{11} = \iiint_S \rho((x^2)^2 + (x^3)^2) \, dV, \quad I_{12} = -\iiint_S \rho x^1 x^2 \, dV,$$

$$I_{22} = \iiint_S \rho((x^1)^2 + (x^3)^2) \, dV, \quad I_{13} = -\iiint_S \rho x^1 x^3 \, dV,$$

$$I_{33} = \iiint_S \rho((x^1)^2 + (x^2)^2) \, dV, \quad I_{23} = -\iiint_S \rho x^2 x^3 \, dV.$$

(We took the relation in Equation (A.16) as the defining property of the moment of inertia tensor because of the theorem that $I_\ell\omega$ is the component of the angular moment vector along the axis of rotation that is given by $(I_{ij}\omega^i)\frac{\omega^j}{\omega}$. See [13, p. 221–222], and in particular, Equation (9.7) for an explanation.

The interesting point about this approach is that it avoids the use of an orthonormal basis and provides a formula for the moment of inertia tensor when one has an affine metric tensor that is not the identity. Furthermore, since it avoids the cross product, the above definitions for the moment of inertia tensor of a solid about an axis are generalizable to solids in $\mathbb{R}^n$.)

16. This problem considers formulas for curvatures of curves or surfaces defined implicitly by one equation.

(a) In Problem 1.3.22, the reader was asked to prove a formula
    for the geodesic curvature κ_g at a point p on a curve given
    implicitly by the equation $F(x, y) = 0$. Show that the formula
    found there can be written as

$$\kappa_g = -\frac{1}{(F_x^2 + F_y^2)^{3/2}} \varepsilon^{i_1 i_2} \varepsilon^{j_1 j_2} F_{i_1} F_{j_1} F_{i_2 j_2}, \qquad (A.17)$$

where, in the Einstein summation convention, F_1 means $\frac{\partial F}{\partial x^1} = \frac{\partial F}{\partial x}$ and F_2 means $\frac{\partial F}{\partial x^2} = \frac{\partial F}{\partial y}$, and where all the functions are evaluated at the point p.

(b) Problem 6.6.19 asked the reader to do the same exercise but
    to find a formula for the Gaussian curvature at a point on a
    surface given implicitly by the equation $F(x, y, z) = 0$. Show
    that the Gaussian curvature K at a point p on a curve given
    implicitly by the equation $F(x, y, z) = 0$ can be written as

$$K = -\frac{1}{(F_x^2 + F_y^2 + F_z^2)^2} \varepsilon^{i_1 i_2 i_3} \varepsilon^{j_1 j_2 j_3} F_{i_1} F_{j_1} F_{i_2 j_2} F_{i_3 j_3} \quad (A.18)$$

with the same conventions as used for a curve defined implicitly.

(c) Prove that Equation (A.18) can be written as

$$K = \frac{1}{(F_x^2 + F_y^2 + F_z^2)^2} \begin{vmatrix} F_{xx} & F_{xy} & F_{xz} & F_x \\ F_{yx} & F_{yy} & F_{yz} & F_y \\ F_{zx} & F_{zy} & F_{zz} & F_z \\ F_x & F_y & F_z & 0 \end{vmatrix}. \qquad (A.19)$$

Bibliography

[1] M. A. Armstrong. *Basic Topology. Undergraduate Texts in Mathematics.* Springer-Verlag, New York, 1983.

[2] V. I. Arnold. *Ordinary Differential Equations.* MIT Press, Cambridge, MA, 1973.

[3] J. Arroyo, O. J. Garay, and J. J. Mencía. When is a periodic function the curvature of a closed plane curve? *Am. Math. Mont.,* 115(5):405–414, 2008.

[4] P. O. Bonnet. La théorie générale des surfaces. *J. Ecolé Polytechnique,* 19:1–146, 1848.

[5] P. O. Bonnet. Sur quelques propriétés des lignes géodésiques. *C. R. Acad. Sci. Paris,* 40:1311–1313, 1855.

[6] R. Bonola. *Non-Euclidean Geometry – A Critical and Historical Study of its Developments.* Dover Publications, New York, 1955.

[7] A. Browder. *Mathematical Analysis: An Introduction.* Undergraduate Texts in Mathematics. Springer-Verlag, New York, 1996.

[8] J. N. Cederberg. *A Course in Modern Geometries. Undergraduate Texts in Mathematics.* Springer-Verlag, New York, 2nd edition, 2001.

[9] R. Courant, H. Robbins, and I. Stewart. *What Is Mathematics?* Oxford University Press, Oxford, 2nd edition, 1996.

[10] H. S. M. Coxeter. *Non-Euclidean Geometry.* Mathematical Association of America, Washington, DC, 6th edition, 1998.

[11] M. P. do Carmo. *Differential Geometry of Curves and Surfaces.* Prentice Hall Inc, Upper Saddle River, NJ, 1976.

[12] W. Dunham. *The Mathematical Universe.* John Wiley and Sons, Inc., New York, 1994.

[13] G. R. Fowles. *Analytical Mechanics*. Saunders College Publishing, Philadelphia, PA, 4th edition, 1986.

[14] J. A. Gallian. *Contemporary Abstract Algebra*. Houghton Mifflin, New York, 6th edition, 2006.

[15] C. F. Gauss. Disquisitiones generales circa superficies curvas. *Comm. Soc. Göttingen*, 6:99–146, 1823.

[16] M. C. Gemignani. *Elementary Topology*. Dover Publications, New York, 2nd edition, 1972.

[17] H. Gluck. Higher curvatures of curves in Euclidean space. *The American Mathematical Monthly*, 73(7):699–704, 1966.

[18] R. Hersh and P.J. Davis. *The Mathematical Experience*. Birkhäuser, Boston, MA, 1981.

[19] H. Hopf. Uber die drehung der tangenten und sehnen ebeneïkurven. *Compositio Mathematica*, 2:50–62, 1935.

[20] R. A. Horn. On Fenchel's Theorem. *American Mathematical Monthly*, 78(4):381–382, 1971.

[21] Serge Lang. *Algebra*. Addison-Wesley, Reading, MA, 3rd edition, 1993.

[22] H. Blaine Lawson. *Lectures on Minimal Submanifolds*. Monografias de Matemática, IMPA, Rio de Janeiro, 1973.

[23] W. B. Raymond Lickorish. *An Introduction to Knots, volume 175 of Graduate Texts in Mathematics*. Springer-Verlag, New York, 1997.

[24] S. T. Lovett. *Differential Geometry of Manifolds*. A K Peters, Ltd., Wellesley, MA, 2010.

[25] S. T. Lovett. *Differential Geometry of Manifolds*. CRC Press, Boca Raton, FL, 2nd. edition, 2010.

[26] Ib. H. Masden and J. Tornehave. *From Calculus to Cohomology*. Cambridge University Press, Cambridge, 1997.

[27] J. Munkres. *Topology*. Prentice Hall, Englewood Cliffs, NJ, 1975.

[28] J. C. Nitsche. *Nitsche. Lectures on Minimal Surfaces*, Volume 1. Cambridge University Press, Cambridge, 1989.

[29] R. Osserman. *A Survey of Minimal Surfaces*. Dover Publications, New York, 1969.

[30] P. J. Ryan. *Euclidean and Non-Euclidean Geometry*. Cambridge University Press, Cambridge, 1986.

[31] J. Stewart. *Calculus*. Thomson Brooks / Cole, Belmont, CA, 6th edition, 2003.

[32] J. J. Stoker. *Differential Geometry*. Wiley-Interscience, New York, 1969.

[33] D. J. Struik. *Lectures on Classical Differential Geometry*. Dover Publications, New York, 2nd edition, 1961.

[34] E. R. van Kampen. The theorems of Gauss-Bonnet and Stokes. *Am. Jour. Math.*, 60(1):129–138, 1938.

Index

negatively oriented, 15, 145
negatively-oriented, 69
positively oriented, 15, 145
positively-oriented, 69
regular, 15, 69
Riemann sum, 14
rotation index, 48
ruled surface, 207
ruling, 208

scalar, 339
scaling factors, 339
secant surface, 117, 204
second derivative test, 183, 188, 201
second fundamental form, 178–186
self-adjoint operator, 179
self-intersection, 21
sensed-parallel, 304
simply connected, 252
singular point of a surface, 125
slices, 116
smoothness condition, 13
space cardioid, 68, 78, 95
special orthogonal, 354
Spectral Theorem, 190
speed, 15, 70, 313
sphere, 112, 151, 180, 200, 277
spherical geometry, 297
spiral
    exponential, 22
    linear, 13, 22
stereographic projection, 137
Stokes' Theorem, 92
surface
    parametrized, 114
    regular, 118, 128, 322
surface of revolution, 113, 127, 160, 196, 205, 228, 278

tangent line, 70
tangent plane, 119, 121
tangent space, 119
tangent vector, 72, 119
tangential indicatrix, 48
tangential surface, 127, 160, 204, 217
Tchebysheff net, 237
tensor, 343
    addition, 344
    antisymmetric, 348
    metric, 157
    moment of inertia, 357
    multiplication, 344
    numerical, 350
    proper, 354
    Riemann curvature, 231
    symmetric, 347
Theorem of Beltrami-Enneper, 206
Theorem of Turning Tangents, 254, 258
Theorema Egregium, 230
topology, vii
torsion, 75
torus, 80, 115, 159, 182, 228, 250, 283
    flat, 326
torus knot, 80, 250
total angle function, 50
trace, 199
tractrix, 202
trajectory, 25
translation surface, 198, 228
transposition, 348
transversal intersection, 32
triangulation, 261
triple-vector product, 211, 245
trochoid, 7
tube, 138, 160, 205, 228
twisted cubic, 67

ultraparallel, 304
umbilical point, 234

Printed in the United States
by Baker & Taylor Publisher Services